THE IMPACT OF EU LAW ON THE REGULATION OF INTERNATIONAL AIR TRANSPORTATION

To my mother

The Impact of EU Law on the Regulation of International Air Transportation

MARTIN BARTLIK

ASHGATE

Published by
Ashgate Publishing Limited
Gower House
Croft Road
Aldershot
Hampshire GU11 3HR
England

Ashgate Publishing Company
Suite 420
101 Cherry Street
Burlington, VT 05401-4405
USA

Ashgate website: http://www.ashgate.com

British Library Cataloguing in Publication Data
Bartlik, Martin
The impact of EU law on the regulation of international air transportation
1. Aeronautics, Commercial – Deregulation – European Union countries 2. Aeronautics, Commercial – Law and legislation
3. Aeronautics, Commercial – Law and legislation – European Union Countries 4. Airspace (International law)
I. Title
387.7

Library of Congress Cataloging-in-Publication Data
Bartlik, Martin.
The impact of EU law on the regulation of international air transportation / by Martin Bartlik.
p. cm.
Includes bibliographic references and index.
ISBN-13: 978-0-7546-4951-9
1. Aeronautics, Commercial--Law and legislation--European Union countries. 2. Air traffic rules-- European Union countries. 3. Aeronautics, Commerical--Government policy--European Union countries. 4. Aeronautics, Commerical--Deregulation--European Union countries. 5. Air traffic rules, International. I. Title.

KJE6920.B37 2007
343.2409'7--dc22

2006033555

Printed and bound in Great Britain by MPG Books Ltd, Bodmin, Cornwall.

Contents

Preface

When aviation experts talk about the present situation of the aviation industry, most of them will refer to the challenges that this industry faces nowadays as a result of the events of the recent past. Thereby 9/11 and SARS are mentioned as the most significant crisis. Only little attention is given to the legal and political challenges that the air transportation sector will face as a result of the 'open-skies' judgement delivered by the European Court of Justice in November 2002. This verdict may have an impact on the regulation of international air transportation as significant as the conclusion of the Bermuda I agreement in the 1950s or the introduction of the 'open-skies' policy in the 1970s. As one can witness from the efforts by the European Commission to establish a Transatlantic Common Aviation Area, the European model of an integrated supranational market for air transportation services could one day extend to other countries outside the European Union.

Taking into account the developments in the European Union I will show in the following why and how the rising of this new player on the stage of international air transportation regulation will influence the international air transportation sector. Irrespective whether the participation of the European Union in this field is hailed or disliked, it is a matter of fact that in the long run, the EU will become a major player in the regulation of international air transportation. Therefore, it is recommendable to study how aviation is regulated in the EU, how the powers are distributed between the European Commission and the Member States and which legal and political difficulties presently exist and may arise in the future with respect to aviation.

Martin Bartlik

Preface

[illegible]

Acknowledgements

This work would not have been accomplished without the help of other people, to whom I am very grateful.

First I have to thank Prof. Dr Dr h.c. Manfred Zuleeg for his academic guidance. His vast knowledge of European Community law and his experience as a Professor at the Johann Wolfgang Goethe University in Frankfurt, Germany and as a judge at the European Court of Justice have been an incredible help in writing this book. In this respect I also need to mention Prof. Dr Michael Milde from McGill University in Montréal, Canada to whom I am very grateful for his guidance during my studies at the Institute of Air and Space Law. His idealism and enthusiasm for aviation law have been always a great source of inspiration. I would also like to place on record my gratitude towards Prof. Dr Ronald Schmid, who has not only given me the idea on writing this book, but has continuously encouraged and motivated me throughout the entire period. I feel very honoured for having had the support and guidance of these outstanding academics, each of them a master in its field.

At this point I also would like to thank my former classmates at the Institute of Air and Space Law at McGill University, and in particular Gautam, Philippe, Ranjana, Sanat, Stephan, Urban and Yaw, for the unforgettable year I was able to enjoy with them.

I shall also express my gratitude to the Flughafen Frankfurt Main Stiftung [Frankfurt Main Airport Foundation] for their financial support and to Ashgate for publishing my book.

Last but not least, I need to thank my mother, without whose support most of my achievements, including this book, would never have been accomplished.

Abbreviations

CAA	Civil Aviation Authority (Great-Britain)
CAB	Civil Aeronautic Board (USA)
C.M.L.R.	Common Market Law Report
CTA	Canada Transport Act
DOT	Department of Transport (USA)
EEC	European Economic Community
EC	European Community *or* reference to EC Treaty
ECJ	European Court of Justice
E.C.R.	European Court Report
ERTA	European Road Transport Agreement
EU	European Union *or* reference to EU Treaty
IASC	International Air Service Commission (Australia)
ICAO	International Civil Aviation Organization
GATS	General Agreement on Trade in Services
GATT	General Agreement on Tariffs and Trades 1947
NAFTA	North American Free Trade Area
OECD	Organization for Economic Cooperation and Development
O.J.	Official Journal of the European Communities
TCAA	Transatlantic Common Aviation Area
U.N.T.S.	United Nations Treaty Series
WTO	World Trade Organization

Abbreviations

Introduction

In this book, two aspects of air transportation are examined, whereby the focus is set on the situation in the European Community and how it may change after the European Court of Justice, the European Union's highest court, delivered its landmark 'open-skies' judgement on 5 November 2002. In Part One, which institution has the treaty-making power to conclude air service agreements with non-EC countries is scrutinized. So far, air service agreements have been concluded by the Member States. But since the European Community is becoming increasingly involved in regulating air transportation and today most regulations with respect to intra-Community air transportation are enacted by the EC, it is necessary to ask whether the Community may also have the treaty-making power to conclude air service agreements with third countries. In this respect the 'open-skies' judgement, as will be shown below, may be of highest importance for the future of European aviation and could even have an impact on the worldwide regulation of air transportation.

Part Two deals with the question of the distribution of air traffic rights. Few articles have been written and published on this topic so far. So far this matter was, from a legal and practical perspective, only of secondary importance. Most countries had only few air carriers that were able to offer international air transportation service and thus air traffic rights, which are necessary for the conduct of international air transportation services, were mostly granted to those airlines by their respective countries. Under these circumstances the allocation of air traffic rights did not cause any considerable problems. Again due to the 'open-skies' judgement this may change in the future. In the EC, the number of air carriers eligible for obtaining air traffic rights for international operations, may significantly increase in the future and the regulators responsible for this task will have to develop a scheme, how these rights shall be allocated. For this purpose several possibilities for the distribution of air traffic rights will be discussed and their advantages and disadvantages shown. Finally an allocation procedure will be presented that shall serve the needs and interests of air carriers and consumers as well as comply with EC and international laws.

In the exercise of its sovereignty every nation has the right as well as the duty to itself to develop its air power, as represented in part by its air transport, to the extent needed by its domestic and foreign commerce and other legitimate objectives. World organizations may well require sufficient international control so that air transport does not become an instrument of unfair nationalistic economic competition or political aggression and thus the source of serious international misunderstanding and dangerous ill feelings.

John Cobb Cooper[1], *The Right to Fly*, 1947

1 (1887–1967); Aviation advisor to US President F. D. Roosevelt, part of the US delegation in the 1944 Chicago Convention on International Civil Aviation, Vice-President of Pan American World Airways, legal advisor to the International Air Transport Association and founder and first director (1951–1955) of the Institute of Air and Space Law at McGill University, Montréal.

PART ONE
The Treaty-making Power to Conclude Air Service Agreements

Chapter 1

Introduction to International Air Law

Before the question is discussed, whether the European Community or the Member States have the treaty-making power to conclude air service agreement, for a better understanding a short overview of the relevant international public air law shall be given first. International public air law is dominated by the principle of sovereignty over the airspace above a State's territory. This principle is not only explicitly laid down in Articles 1 and 2 of the Chicago Convention,[1] but is also part of international customary law. It is the basis of all other regulations in international public air law.[2] As a result of the application of the principle of sovereignty, every state decides autonomously to which extent, by whom and when its airspace is used. The permission to use a state's airspace is formally called 'air traffic right,' but commonly also called 'freedom'.[3] Depending on the extent of the possibilities to use a state's airspace, one can distinguish between eight freedoms in international air law:[4]

- The *first freedom* gives the right to overfly the territory of a state without any landings.[5]

- The *second freedom* grants the privilege to land in a foreign country for technical reasons.[6]

---

1 *Convention on International Civil Aviation*, 7 December 1944, 15 U.N.T.S. 102 [Chicago Convention].

2 Alfred Verdross and Bruno Simma (1984), *Universelles Völkerrecht (Universal International Law)*, 3rd edn., (Berlin: Duncker and Humblot), 664 [Verdross/Simma].

3 I. H. Philepina Diederiks-Verschoor (1993), *An Introduction to Air Law*, 5th edn., (Deventer, The Netherlands: Kluwer), 12f. [Diederiks-Verschoor]; Max Hofmann, and Edwin Grabherr, in: Max Hofmann and Edwin Grabherr, eds. (1994), *Luftverkehrsgesetz (Aeronautics Act)*, looseleaf (Munich: Verlag C.H. Beck, 2nd supplement) § 21 at para 1 [Hofmann/Grabherr]; the term 'freedom' will be used also in the following.

4 Hofmann/Grabherr, *ibid.* at para 1; Elmar Giemulla, in: Elmar Giemulla and Ronald Schmid, eds. (2005), *Frankfurter Kommentar zum Luftverkehrsrecht – Luftverkehrsgesetz (Frankfurt Commentary to Aeronautics Law – Aeronautics Act)*, vol. 1.1, looseleaf (Neuwied, Kriftel, Germany: Luchterhand Verlag) Pref. to §§ 20-24 LuftVG at para 29 [Giemulla, in: Giemulla/Schmid, Pref. to §§ 20-24].

5 Article 5 Chicago Convention, *International Air Service Transit Agreement* (7 December 1944), ICAO Doc. 7500, Article I, Section 1, No 1 [Transit Agreement].

6 Chicago Convention, *ibid.*; Transit Agreement, *ibid.*

- The *third freedom* gives the right to fly passengers, mail and cargo from the registration state of the air carrier to a foreign state.[7]

- The *fourth freedom* correlates with the third freedom and grants the right to fly passengers, mail and cargo from a foreign state to the registration state of the air carrier.[8]

- The *fifth freedom* extends the third and fourth freedom and grants the right to fly passengers, mail and cargo between two foreign states, as long as the origin or the final destination of such a chain of air transportation services is in the registration state of the air carrier.[9]

- The *sixth freedom* is not an explicitly granted right, but describes a combination of the third and fourth freedom, independently granted to a state by two other states. The sixth freedom enables an air carrier to offer air transportation services between two foreign countries by making a stop-over in the home state of that air carrier. Thereby an airline makes use of the fourth freedom by country A and flies passengers, mail or cargo from country A into its home state, where it makes a stop-over, and continues the flight with the passengers, mail or cargo from country A to country B, thereby making use of the third freedom granted by country B.[10] The sixth freedom is usually used to establish a hub-and-spoke system, but can be also applied, when necessary fifth freedom rights are not granted by two foreign states.

- The *seventh freedom* grants a right to operate between two foreign countries, which are both not the registration state of the operating air carrier. Unlike fifth freedom traffic, there is no link between the flights of the air carrier and its home country. The aircraft used to offer seventh freedom traffic is permanently stationed in one of the two foreign countries concerned.[11]

- The *eighth freedom*, also called cabotage, grants a right to provide air transportation services within one single country.[12]

The first and second freedom are also called 'technical freedoms', while the remaining commercially more important air traffic rights are referred to as 'economic freedoms'. In the recent past, in particular the fifth freedom has become quite important as it allows to combine several flight services and thus to use the capacities

7 See definition in the *International Air Transport Agreement*, 7 December 1944, 171 U.N.T.S. 502, Article I, Section 1, No 3 [Transport Agreement].

8 See definition in Article I, Section 1, No. 4 Transport Agreement.

9 See definition in Article I, Section 1, No. 5 Transport Agreement.

10 Diederiks-Verschoor, *supra* note 2 at 13, n. 9; Nicolas Mateesco Matte (1981), *Treatise on Air-Aeronautical Law*, (Montréal: McGill University and Carswell) at 143f. [Matte].

11 Diederiks-Verschoor, *ibid.*; Matte, *ibid.* at 144.

12 Article 7 Chicago Convention.

of aircrafts more efficiently. From a commercial perspective, this freedom may be even the most important freedom. If, for example, on flights between Australia and Europe an air carrier can make only use of the third and fourth freedom, it is limited to offer air transportation services between two destinations only, for example Sydney/Paris or Sydney/Frankfurt. If the same airline is also granted the fifth freedom, it can combine these flights to one single flight, for example Sydney/Frankfurt/Paris. At the same time it will be enabled to offer purely intra-Community flights. Not only will the airline safe one aircraft that it can use for different flight connections, it will also increase its load factor on the route between Australia and Europe. The fifth freedom thus allows for a more efficient and profitable use of an airlines fleet. Taking into account the increasing competition in the air transportation sector and the reduction of state aids as a result of the privatizations of airlines, the success and survival of an airline will depend very much on the efficiency and profitability it can achieve. The problem of the fifth freedom is that it must be granted by all countries that would be part of such a chain of flights. In the above mentioned example for an Australian air carrier to operate on the route Sydney/Frankfurt/Paris, Germany and France would have to grant the fifth freedom to Australia. If this fails, an airline might be able to bypass this disadvantage and still be able to increase its efficiency and profitability by combining its available third and fourth freedom, which leads to the above mentioned sixth freedom. The idea is that an airline coming from a third state and before continuing flying to another country, makes a stop-over in its own country. As a result it can transport passengers and cargo between two foreign countries. The disadvantage of this model compared to the fifth freedom is that an airline must always make a stop-over in its own country. Whether the sixth freedom can be applicable depends therefore to a large extent on the geographical location of a country. If, for example, Germany, France and Australia do not grant each other the fifth freedom, a German air carrier can still fly passengers, mail and cargo from Paris to Sydney, if it makes a stop-over in Frankfurt. An Australian air carrier, on the other hand, would have difficulties in selling tickets for a flight Paris/Frankfurt via Sydney.

The granting of the different freedoms depends very much on the type of the air transportation services offered. In international air transportation a distinction is made between commercial and non-commercial flights. In the following, non-commercial flights, due to their limited significance for this work, will not be further taken into regard. As far as commercial aviation is concerned, another distinction is made between non-scheduled international air services and scheduled international air services. Scheduled international air services are defined pursuant to Articles 5, 6, 96(a) Chicago Convention as 'public and planned air transportation of passengers, cargo and mail'. When it comes to non-scheduled air transportation, international law does not provide for any regulation and national laws contain only a few provisions in respect of this mode of transportation.[13] A comprehensive and coherent set of regulations comparable to the one existing for schedule air transportation does not

13 Walter Schwenk and Elmar Giemulla (2005), *Handbuch des Luftverkehrsrechts (Manual of Aeronautics Law)*, 3rd edn., (Cologne: Carl Heymanns Verlag), 723 [Schwenk/Giemulla].

exist.[14] It lacks, in particular, a positive definition of what is to be understood as non-scheduled air transportation. This is because non-scheduled air transportation covers all kinds of different services, for example taxi flights, sightseeing flights, medical flights, advertisement flights or flights as part of holiday packages.[15] Therefore, non-scheduled flights are only negatively defined as not being scheduled air transportation.[16] The difference between both modes of transportation arises from the two features 'planned' and 'public' contained in the definition of scheduled air services. Air transportation services are planned, if the flights are repeated in a pre-determined order and the users can rely on the existence of a regular transportation service.[17] While this is the case with scheduled flights, non-scheduled air transportation services are undertaken only, if there is an actual demand.[18] The second feature of scheduled air transportation services distinguishing it from non-scheduled flights is that the flights must be made available to the general public, which means everybody must have the opportunity to make use of the air transportation services offered. Of course, not everybody asking for a seat on a flight must be also served, as capacities are limited. Still, a customer can be only refused for objective reasons that are equally valid for everybody, but not based on subjective criteria.[19] Unlike non-scheduled airlines, air carriers offering scheduled flights are obliged to contract. Non-scheduled flight services, on the other hand, are not available to the general public, but only to a limited group of persons characterized by specific subjective criteria.[20] Non-scheduled airlines are also not burdened with an obligation to contract.[21]

Air traffic rights, which are, as indicated above, indispensable for the carrying out of international air transportation services, are granted partly in multilateral treaties and partly in bilateral agreements. The two most important multilateral treaties in this respect were concluded in 1944 and are the Chicago Convention and the Transit Agreement. According to Article 5(1) Chicago Convention the signatory states grant the first and second freedom to all air carriers engaged in international non-scheduled air transportation. This category includes commercial non-scheduled air transportation and private, non-commercial traffic. Flights with state aircrafts do not

14 Hofmann/Grabherr, *supra* note 38 at § 22, para 3; Giemulla, in: Giemulla/Schmid, Pref. to §§ 20-24, *supra* note 4 at para 15.

15 Hofmann/Grabherr, *ibid.*; Giemulla, *ibid.* at para 7.

16 See also § 22 German Aeronautics Act or Article 5(1) Chicago Convention where reference is made to 'aircrafts not engaged in international scheduled air transportation;' Giemulla, *ibid.* § 22, para 4; Schwenk/Giemulla, *supra* note 13 at 724; Alfred Rudolf, 'Die sogenannte Pauschalreise-(IT)Charter im Spannungsfeld zwischen Fluglinien – und Gelegenheitsverkehr' (The so-called holiday package charter flight in the tension between scheduled and non-scheduled air transportation) (1970) *Zeitschrift für Luft – und Weltraumrecht* (*Air and Space L. Rev.*) 110 at 113 [Rudolf].

17 Schwenk/Giemulla, *ibid.* at 613.

18 *Ibid.* at 614.

19 Hofmann/Grabherr, *supra* note 3 at § 21, paras 2, 18; Giemulla, in: Giemulla/Schmid, *supra* note 4 at § 21, para 6.

20 Giemulla, *ibid.*; Schwenk/Giemulla, *supra* note 13 at 726f.

21 Hofmann/Grabherr, *supra* note 3 at § 22, para 6; Giemulla, in: Giemulla/Schmid, *supra* note 4, Pref. to §§ 20-24, at para 7; Schwenk/Giemulla, *ibid.* at 615.

fall under this definition as pursuant to Article 3a they are completely excluded from the scope of the Chicago Convention. The more important economic freedoms are also granted to non-scheduled air carriers pursuant to Article 5(2) Chicago Convention, but additional conditions up to total restrictions can be imposed by every single state on the exercise of these freedoms. Air carriers offering schedules international air transportation services require pursuant to Article 6 Chicago Convention a particular permission for any kind of use of a state's airspace. This regulation contained in Article 6 Chicago Convention is considered as the basis for the conclusion of numerous bilateral agreements, in which States regulated the details of international scheduled air transportation that is to be conducted between their territories.[22] These agreements are the factual as well as the legal basis for international air transportation. Apart from the exchange of air traffic rights, these agreements contain various other clauses that are important for international aviation. It can include provisions on how many air carriers each country is allowed to designate to make use of the air traffic rights that are granted, which aircrafts shall be used and thus how much capacity is offered in a market, how often and when exactly flights have to take place (frequency) or which fares airlines are allowed to charge.[23] The content of air service agreements can vary in every single case and depends very much on the States involved.[24] Air service agreements can be very restrictive and contain detailed regulations, leaving the airlines with only few possibilities to structure their operations. Such an agreement will allow the designation of only one or very few air carriers, the frequency and the type of aircrafts to be used will be pre-determined and the fares to be charged will be also fixed. The reason for such an approach is the desire of a state to protect its national airlines from too much competition.[25] However, since the end of the 1970s a global trend to more liberalization in commercial aviation has set in, when the United States of America decided to deregulate not only their domestic air transportation sector, but also to push for more liberalization of international air transportation.[26] According to this approach, economic questions should not be decided and pre-determined by politicians, but only by the entities involved in the business. Furthermore, it was believed that greater benefits could be achieved, if airlines were placed in a more competitive surrounding. For that purpose, the US introduced a new and more liberal model air service agreement, commonly known

22 Schwenk/Giemulla, *ibid.* at 630; Giemulla, in: Giemulla/Schmid, *ibid.* at para 26 (Germany, for example, has concluded about 170 bilateral air service agreements, while the worldwide number is assumed to be at about 4000 bilaterals); Virginia Rodriguez Serrano, 'Trade in Air Transport Services: Liberalizing Hard Rights' (1999) 24 *Ann. Air and Sp. L.* 199 [Serrano].

23 Matte, *supra* note 10 at 144f.

24 See e.g. Diederiks-Verschoor, *supra* note 3 (description of Bermuda I as a compromise between the liberal approach of the USA and the protective position of Great Britain, at 49); Robert Burkhardt (1974), *The Civil Aeronautics Board* (Virgina: Green Hills Company) at 136 [Burkhardt].

25 See e.g. Diederiks-Verschoor, *supra* note 3 (denunciation of the Bermuda I Agreement by Great Britain in order to conclude a more restrictive treaty (Bermuda II) to protect its airlines, at 51); Schwenk/Giemulla, *supra* note 13 at 634.

26 Diederiks-Verschoor, *ibid.* at 51.

as the 'open-skies' standard. This kind of an agreement is less restrictive as far as designation, capacity, frequency and especially fares are concerned and leaves it to the airlines to take most of the commercial decisions.[27]

As far as the granting of air traffic rights is concerned, a particularity exists in respect of the first and second freedom. According to Article 1 Transit Agreement at least the first and second freedom are granted for scheduled international air transportation services in a multilateral agreement. However, some of the geographically big states like Canada, Brazil, Russia and China are not parties to this agreement. This makes it necessary to include the technical freedoms in the bilateral agreements these states conclude with other countries. Due to their size and geographical location the first ones have a strong bargaining power with the first and second freedom in all negotiations on air service agreements. As far as the third, fourth and fifth freedom are concerned, a similar multilateral approach was envisaged in the Transport Agreement. Unlike the Transit Agreement the Transport Agreement was not very popular and was signed only by 11 states. It was and is therefore of no significance and most of the current state parties to the Transport Agreement are, from the perspective of international aviation, still only of secondary importance.[28]

As far as the air traffic rights, agreed upon in bilateral air service agreements, are concerned, it is important to know that these rights are not granted directly to the airlines, but are exchanged among the contracting states, who, in a second step, pass these rights to the airline(s) registered within their jurisdictions.[29] This transfer of air traffic rights to an air carrier goes along with a notification to the other state about which air carrier was granted the air traffic rights. This whole procedure is called designation. The designation is a crucial step before air transportation services between two countries can be offered. Designation is regularly understood as the permission to enter a market and to operate between two or more destinations located in two or more countries.[30] Being designated enables airlines to operate from a point within the designating country and it can apply to the other country for permission to fly into and out of its airspace, thereby making use of the rights agreed in the corresponding air service agreement. Although the designated airline must also obtain a separate permission of the other country to enter and leave its airspace, this

27 *Ibid.* at 52f.

28 Knut Ipsen (2004), *Völkerrecht (International Law)*, 5th edn., (Munich: Verlag C.H. Beck) at 902, para 6 [Ipsen, Intl. Law]; Verdross/Simma, *supra* note 2 at 664, note 12; it is presently signed by 11 countries of which only The Netherlands are of substantial significance in respect of air transportation services.

29 Matte, *supra* note 10 at 143; Bin Cheng (1962), *The Law of International Air Transport*, (London: Stevens and Sons) at 359 [Cheng].

30 See e.g. Eugene Sochor, 'From the DC-3 to Hypersonic Flight: ICAO in a Changing Environment' (1989–1990) 55 *J. Air L. and Com.*, 407 (author refers to designation as part of market entry, at 431, n. 78) [Sochor]; Jennifer Yang (1995), Air Transport Regulation in a new Era, (LL.M. Thesis, Institute of Air and Space Law, McGill University), (air traffic right is defined as 'a market access right which is expressed as an agreed physical or geographical specification, or combination of specifications, of who or what may be transported over an authorized route to parts thereof in the aircraft authorized', at 101) [Yang].

is usually not very difficult as the only reasons, for which an airline can be withhold this permission, are set out in the air service agreement. Thus, each country has only a limited discretion in refusing an air carrier designated by the other contracting state.[31] This also means that whether an air carrier can make use of air traffic rights depends mostly on whether it is designated by its registration state. One of the few reasons for which a permission to enter and leave a State's airspace could be withhold is, if the majority of the ownership of an air carrier does not belong to the designating state, its citizens or both combined. A clause, commonly referred to as 'nationality clause' or 'substantial ownership clause', which is usually part of every air service agreement, stipulates that both state parties to an air service agreement can refuse to grant to an air carrier, designated by the other party, the necessary permissions to operate within its airspace, if the majority of the ownership of such an air carrier is not held by the designating state, its nationals and/or both combined.[32] This provision guarantees the bilateral nature of an air service agreement and precludes that air carriers from other states can benefit from the granted privileges.[33] At the same time the designating state can be held responsible for any incorrect conduct of the air carriers it has designated as it has the authority flowing from its territorial sovereignty to impose duties and obligations on its airlines.

31 See also Matte, *supra* note 10 at 147; Cheng, *supra* note 29 at 359, 363; Burkhardt, *supra* note 24 (sometimes it is even possible that a foreign air carrier is granted the necessary permissions solely for political reasons, although it could be rejected under the applicable bilateral, at 138f.).

32 Schwenk/Giemulla, *supra* note 13 at 639.

33 *Ibid.*

Chapter 2

Regulation of Air Transportation in the European Community

As seen above, international air transportation has always been a matter that was dealt with by States through bilateral air service agreements. With the creation of the European Community and its continuous expansion to various policies, the question has arisen, whether the Community could or even should be responsible for the regulation of international aviation. With respect to the regulatory powers of the European Community, one needs to distinguish, whether the EC adopts acts that are binding on the Member States and/or the citizens of the Community or whether the Community concludes an international agreement with a non-EC country. For both kinds of activities the EC is granted competences in the EC Treaty. Competences that allow the EC to adopt acts that are binding upon the Member States and/or the citizens will be referred to in the following as internal competences or internal powers. Competences that enable the Community to conclude an international agreement are called hereafter treaty-making powers.

THE ESTABLISHMENT OF A COMMON MARKET FOR AIR TRANSPORTATION SERVICES

The main internal competence for adopting regulations relating to air transportation is laid down in Article 80 EC.[34] This provision says:

> (1) The provisions of this title shall apply to transport by rail, road and inland waterways.
>
> (2) The Council may, acting by a qualified majority, decide whether, to what extent and by what procedure appropriate provisions may be laid down for sea and air transport.

---

34 *Treaty Establishing the European Community*, 25 March 1957, 298 U.N.T.S. 11, as amended by the *Treaty Establishing the European Community*, 7 February 1992, [1992] 1 C.M.L.R. 573 as amended by the *Treaty on the European Union*, 7 February 1992, [1992] 1 C.M.L.R. 719, as amended by the *Treaty of Amsterdam*, 2 October 1997, [1997] O.J. C 340/01, and the *Treaty of Nice*, 24 December 2002, [2002] O.J. C 325/33. References to the EC Treaty and the EU Treaty are abbreviated with EC and EU and it is referred to the consolidated version of the Treaty of Nice.

Since Article 80(2) EC leaves it to the discretion of the EC Council, whether it wants to adopt measures in the area of air transportation, this field was left out of the agenda of the Community for a long time.[35] Mainly because of some judgements of the European Court of Justice, the Council was forced to become involved with air transportation matters. In the 'French Seamen' judgement,[36] the ECJ ruled that Article 80(2) EC did not excluded the applicability of the EC Treaty to sea and air transport. Still, even then the Council did not held it as necessary to adopt any substantial regulations and, after three years of deliberations, enacted only a Directive on inter-regional scheduled air services,[37] which was of limited significance. Although this regulation was of little significance, it, at least, set the liberalization process in motion.[38] This process was further accelerated when in 1986, in the 'Nouvelles Frontières'[39] judgement, the Court of Justice concluded that the EC competition laws were applicable on air transportation. From that moment on, the Council became more active in the area of air transportation. In 1987, and thus shortly after the 'Nouvelles Frontières' judgement the Council approved a first package of rules regarding air transportation.[40] The most important rule of this package was possibly the inclusion of the air transportation sector under the EC competition laws, from which the air transportation industry had been excluded until that moment. However, a transitional period of three years was also adopted until EC competition laws would become fully applicable. During this time, block exemptions were granted, providing immunity for most of the airlines' activities from the competition rules.[41] Other parts of this first liberalization package included the following areas:[42]

---

35 Elmar Giemulla, Ronald Schmid and Walter Mölls (February 1993), *European Air Law*, looseleaf, (Deventer, The Netherlands: Kluwer Law and Taxation Publishers), Introduction at para 15 [Giemulla/Schmid/Mölls, Europ. Air Law]; Kenneth Button and Dennis Swann, 'Aviation Policy in Europe' in: Kenneth Button, ed. (1991), *Airline Deregulation – International Experiences*, (Washington Square, NY: New York University Press) 85 at 113 [Button and Swann].

36 ECJ, *Commission v French Republic*, C-167/73, [1974] E.C.R. 359 [French Seamen].

37 EEC, Council Directive 83/416 of 25 July 1983 concerning the approval of inter-regional scheduled air service for the transport of passengers, mail and cargo between EEC countries, [1983] O.J. L 237/19.

38 Button/Swann, *supra* note 35 at 114.

39 ECJ, *Ministère Public v Asjes*, C-209-213/84, [1986] E.C.R. 1425 [Nouvelles Frontières].

40 George Williams (1993), *The Airline Industry and the Impact of Deregulation*, (Brookfield, VA: Ashgate), 79f. [Williams].

41 *Ibid.*; Button/Swann, *supra* note 35 at 115f.; Andreas Loewenstein (1991), *European Air Law – Towards a new System of International Air Transportation Regulation* (Baden-Baden, Germany: Nomos Verlagsgesellschaft) (exemptions were granted for pooling and revenue sharing agreements, tariff proposals and consultations, slot allocations and Computer Reservation System agreements, at 72) [Loewenstein].

42 Williams, *supra* note 40 at 80f.; Button/Swann, *ibid.* at 116; Elmar Giemulla, Ronald Schmid and Walter Mölls, (June 2000) *Europäisches Luftverkehrsrecht* (*European Air Law*), looseleaf (Neuwied, Kriftel, Germany: Luchterhand Verlag, 32nd supplement),

- capacity, regulated in the bilaterals between the Member States, could have been adjusted on a country-pair basis to a ratio of first 45:55 and then 60:40, that means it was not mandatory any more that the airlines of two states would share traffic in an equal manner;

- multiple designation was made mandatory on a country-pair basis for routes, which generated a certain minimum number of traffic. The threshold was set initially at 250,000 passengers per year in 1988 and reduced to 180,000 passengers per year in 1990;

- fares had to be filed within 60 days before becoming effective and the single-disapproval-rule[43] was made applicable. Member States had to approve fares, if they were reasonably related to the long term, fully allocated costs of the airline and could not refuse them solely for the reason that they were lower than fares, offered by other air carriers.[44] Zones of flexibility were introduced that provided for an automatic approval of the applicable fares, if certain conditions with respect to the offered air transportation services were met. These conditions covered aspects like length of stay of the travellers, booking of a round trip, certain time of advance reservation, age of travellers, and so on. The following zones were established:

  – discount fares of 90–66 per cent of the reference fare (pursuant to Article 2(c) of Directive 87/601 reference fare is the normal economy air fare charged by a third- or fourth-Freedom Air carrier on the routes in question or a corresponding average, if there are more than one-third/fourth-Freedom Air carrier operating on this route; if no normal economy fare existed, the lowest fully flexible fare was taken as a reference fare);

  – deep discount fares of 65–45 per cent of the reference fare;

  – super discount fares of not less then 35 per cent of the reference fare;

- automatic market access was granted to all EC air carriers for routes between their countries and other Member States, if the destination in the other Member State was a regional airport. Additionally, EC air carriers were granted fifth

Introduction, at paras 31f. [Giemulla/Schmid/Mölls]; EEC, Council Regulation 3975/87 of 14 December 1987 laying down the procedure for the application of the rules on competition to undertakings in the air transport sector, [1987] O.J. L 374/01; EEC, Council Regulation 3976/87 of 14 December 1987 on the application of Article 85 (3) of the Treaty to certain categories of agreements and concerted practices in the air transport sector, [1987] O.J. L. 374/09; EEC, Council Directive 87/601 of 14 December 1987 on fares for scheduled air services between Member States, [1987] O.J. L 374/12; EEC, Council Decision 87/602 of 14 December 1987 on the sharing of passenger capacity between air carriers on scheduled air services between Member States and on access for air carriers to scheduled air-service routes between Member States, [1987] O.J. L 374/19.

43 See below at 90 for definition.

44 Loewenstein, *supra* note 41 at 74.

freedom rights for flights between two (foreign) Member States, if such a route included the landing on a regional airport and the respective air carrier did not use more than 30 per cent of its annual capacity on such a route for fifth freedom passengers.

The first package's impact on the promotion of competition in the air transportation sector was limited and the Commission was eager to overcome the shortcomings in a second liberalization package.[45] Meanwhile, the European Court of Justice decided another case that created further need to amend the existing regulations.[46] This finally led to the adoption of a new bundle of rules in 1990. This second package further liberalized the regulations of the first package:[47]

- fares were subjected to the double-disapproval-rule,[48] if they were higher than 105 per cent of the reference fare. Fares, lower than 40 per cent of the reference fare, were subjected to the single-disapproval-rule. The flexibility zones, with the automatic approval mechanism, were extended and included:
  - a normal economy fare of 105–95 per cent of the reference fare;
  - discount fares of 94–80 per cent of the reference fare;
  - deep discount fares of 79–30 per cent of the reference fare;

- capacity regulations were amended to that extent that airlines from one Member State could obtain 75 per cent of all available capacity on one route, hence possibly leaving only 25 per cent to air carriers from the other Member State;

- the multiple designation limit was reduced and made applicable already to such routes that generated 140,000 passengers or 800 return flights per year. These numbers were further reduced to 100,000 passengers or 600 return flights from 1992 on;

- with respect to market access all EC air carriers were granted complete third and fourth freedom rights. Fifth freedom rights remained limited to the extent that its use must not exceed 50 per cent of the annual capacity of a carrier on a route.

---

45 Williams, *supra* note 40 at 81; Giemulla/Schmid/Mölls, *supra* note 42 at para 32f.

46 ECJ, *Ahmed Saeed v Office for Combating Unfair Competition*, C-66/86, [1989] E.C.R. 229.

47 Williams, *supra* note 40 at 82f.; Giemulla/Schmid/Mölls, *supra* note 42 at paras 44–59; EEC, Council Regulation 2342/90 of 24 July 1990 on fares for scheduled air services, [1990] O.J. L 217/01; EEC, Council Regulation 2343/90 of 24 July 1990 on access for air carriers to scheduled intra-Community air service routes and on the sharing of passenger capacity between air carriers on scheduled air services between Member States, [1990] O.J. L 217/08; EEC, Council Regulation 2344/90 of 24 July 1990 amending Regulation (EEC) No. 3976/87 on the application of Article 85 (3) of the Treaty to certain categories of agreements and concerted practices in the air transport sector, [1990] O.J. L 217/15.

48 See below at 90 for definition.

Finally, in 1993, the third liberalization package was adopted, bringing the EC deregulation process to an end. The third package contains the following rules:[49]

- available exemptions from EC competition laws are reduced;
- capacity constraints are completely removed, except for routes between regional airports, where the traffic volume does not exceed 30,000 seats/year;
- full fifth freedom traffic rights are granted to all EC air carriers;
- cabotage is gradually introduced and fully allowed since 1997;[50]
- common rules regarding the licensing of air carriers are introduced;[51]
- no fare control exists since 1996. Still, Member States can require the filing of fares 24 hours before they become effective. Air fares can be suspended to prevent anti-competitive behaviour through overpricing or predatory pricing based on the double-disapproval rule. Furthermore, only Community air carriers can be price leaders on routes within the EC. Thus, no non-EC air carrier can offer lower fares for air transportation services between points in the EC, when making use of fifth freedom rights.[52]

The third liberalization package is a compromise between those Member States that had already liberalized their air transportation sector, mainly the Netherlands and Great Britain, and who were eager to see their airlines entering into full competition and those countries that had not yet deregulated their air transportation industries, in particular France, Germany, Greece, Italy and Spain, and who intended to grant their formerly protected airlines sufficient time to gradually adapt to the new system.[53] However, with the adoption of the third package, the common market was also established in the air transportation sector. Since this moment every air carrier is entitled to offer air transport services between any Member State and even with a single Member State (cabotage). As a result of the development in the EC, a particularity exists with respect to European air carriers and European air transportation. Although intra-Community air transportation should be classified as

49 Williams, *supra* note 40 at 84f.; Button/Swann, *supra* note 35 at 117; Giemulla/Schmid/Mölls, *supra* note 42 at paras 61–69; EC, Council Regulation 2407/92 of 23 July 1992 on Licensing of Air Carriers, [1992] O.J. L 240/01 [Regulation 2407/92]; EC, Council Regulation 2408/92 of 23 July 1992 on Access for Community Air Carriers to Intra-Community Air Routes, [1992] O.J. L 240/08 [Regulation 2408/92]; EC, Council Regulation 2409/92 of 23 July 1992 on Fares and Rates for Air Services, [1992] O.J. L 240/15 [Regulation 2409/92].

50 Giemulla/Schmid/Mölls, *ibid.* (the granting of all air traffic rights within the European Union does not exempt the air carriers from obtaining the necessary licences by the concerned Member States, at 95f.).

51 Regulation 2408/92, *supra* note 49, Article 2 (a) (all air carriers which are licensed by one of the Member States according to Regulation 2407/92 are called community air carriers).

52 As will be seen below, this rule was one of the provisions that were crucial in the 'open-skies' judgement.

53 Giemulla/Schmid/Mölls, Europ. Air Law, *supra* note 35, Introduction at para 70.

international air transportation from the perspective of international public air law, a different approach is necessary as a result of the establishment of the common market for air transportation services in the EC. While flights to, from and within the Community by non-Community air carriers continue to be classified as international air transportation, this is not the case for Community air carriers engaged in intra-Community air transportation. Air transport services offered by Community air carriers are rather comparable to purely national flights than to international air transportation. Therefore, if reference is made to international air transportation here, it is understood as traffic between two countries offered by non-Community air carriers or, if Community air carriers undertake such flights, these two countries must not be both Member States of the European Community. Purely intra-Community flights of Community air carriers shall not fall under the definition of international air transportation any more.

THE TREATY-MAKING POWER OF THE EC WITH RESPECT TO AIR SERVICE AGREEMENTS

AIR SERVICE AGREEMENTS CONCLUDED BY THE EC

Already in 1990, and thus even before a common market for air transportation services was established, the Commission intended to replace the Member States in the negotiation and conclusion of air service agreements and requested a mandate from the Council for the conclusion of air service agreements with third countries.[54] The Council denied to grant the Commission a comprehensive mandate. However, the Commission was allowed to enter into negotiations on an air service agreement with Norway and Sweden.[55] This has been necessary as Norway, Sweden and Denmark were involved in the airline consortium Scandinavian Airline System (SAS) and since, at that time, only Denmark has been a member of the European Union, this airline consortium was already partially subject to EC laws.[56] Two years later the European Commission filed in another proposal with the Council for a mandate to conclude air service agreements. This proposal was again refused by the Council, arguing that air service agreements were part of the transport policies of each Member State and since these policies would differ quite significantly from each other, every Member State must remain responsible for its external air transportation relations.[57] The EC was given another single mandate by the Council to conclude an air service

---

54 Giemulla/Schmid/Mölls, *supra* note 42, Introduction at para 150.

55 *Ibid.*

56 Albrecht Frohnmeyer, in: Eberhard Grabitz and Meinhard Hilf, eds. (2000), *Das Recht der Europäischen Union* (*The Law of the European Union*), looseleaf (Munich: Verlag C.H. Beck, 16th supplement), Article 80 at para 25 [Frohnmeyer in Grabitz/Hilf, Article 80]; see also EEC, Council Decision 92/384 of 22 June 1992 concerning the conclusion of an Agreement between the European Economic Community, the Kingdom of Norway and the Kingdom of Sweden on civil aviation [1992] O.J. L 200/20 [Agreements with Norway and Sweden].

57 Giemulla/Schmid/Mölls, *supra* note 42, Introduction at para 151.

agreement with Switzerland that was part of a bundle of agreements concluded with Switzerland in 1999.[58] Still, the Commission's main goal, the conclusion of an air service agreement with the United States of America, has not been achieved so far. Although a mandate was granted to the Commission already once in 1996 to enter into negotiations with the US on air transportation matters, these negotiations failed and an agreement was not finalized mainly because the Commission's mandate was limited in its scope and excluded the most interesting aspects of air transportation, like market access including air traffic rights, capacity, designation and fares.[59] After the 'open-skies' judgement, the Commission was mandated to conclude a particular kind of 'horizontal agreement' with third countries, which are limited on the substitution of a specific clause, the nationality clause, included in the existing air service agreements of the Member States.[60] Apparently the European Community can conclude air service agreements, but does not do it on a regular basis. In the following it shall be scrutinized, to which extent the European Community has a treaty-making power with regard to air service agreements to be concluded with third countries. Thereby attention will not be put on the relation between the European Community and a third country with whom an air service agreement shall be concluded. The here discussed problem regarding a treaty-making power of the Community for the conclusion of the aforementioned kind of agreements affects the relation of the European Community and the Member States. The sole question is, whether the Community can substitute the Member States in the conclusion of air service agreements that the latter ones have been concluding in the past 50 years.

TREATY-MAKING POWERS IN THE EC TREATY

Explicit treaty-making powers of the Community can be found in Articles 111, 133, 181, 182, 301, 302, 303, 304, 308, 310 EC. It shall be examined, which of these competences could enable the Community to conclude air service agreements.

Article 133(3) EC

Article 133(3) EC is the most important explicit treaty-making power in EC law and was even, together with Article 310 EC, the only provision empowering the

---

58 Jürgen Erdmenger, in: Hans von der Groeben and Jürgen Schwarze, eds. (2003), *Vertrag über die Europäische Union und Vertrag zur Gründung der Europäischen Gemeinschaft (Treaty on the European Union and Treaty establishing the European Community),* vol. 1, 6th edn. (Baden-Baden, Germany: Nomos Verlagsgesellschaft) Article 80 EC at para 83 [Erdmenger, in: Groeben/Schwarze]; see also Agreement between the European Community and the Swiss Confederation on Air Transport [2002] O.J. L 114/73 [Agreement with Swiss Confederation].

59 Erdmenger, in: Groeben/Schwarze, *ibid.* para 82; Ulrich M. Gassner and Alexander Deichstetter, 'Luftverkehrsabkommen als Gemeinschaftsaufgabe' (Air Service Agreements as a Task of the Community) (2003) *Europäisches Wirtschafts- und Steuerrecht* (*Europ. J. of Commercial and Tax L.*) 265 at 267 [Gassner/Deichstetter].

60 See below for the corresponding passage of the 'open-skies' judgement at 92f.

European Community to conclude international agreements until the development of implied powers.[61] The scope of this provision was disputed for a long time. First, the European Court of Justice decided in Opinion 1/78[62] that Article 133 EC must not be construed in a way that would limit the provision solely to the traditional forms of foreign trade. Although the term 'commercial policy' contained in Article 133(3) EC referred to international commercial law as it was defined by the General Agreement on Tariffs and Trades (1947) (GATT, 1947) and the Organization for Economic Cooperation and Development (OECD), when it was included in the EC Treaty in 1957, it should not be limited to the objectives of GATT (1947) and the OECD, that is the liberalization of foreign trade. The scope of the term 'commercial policy' should be applied extensively and also cover new situations that have emerged since. This dynamic interpretation was also justified by the non-exhaustive enumeration of the principles set out in Article 133(1) EC that should be observed within the scope of the common commercial policy. Furthermore, the judges conceded that, taking into account further developments in international commerce, a too restrictive definition of the term 'commercial policy' could render Article 133 EC insignificant one day.[63] Such a degrading of Article 133 EC would be incompatible with the *effet-utile* principle. Taking into account the increasing significance of services in international commerce and referring to Opinion 1/78, one could conclude that an application of Article 133(3) EC on international agreements on services would be possible.[64] This is even more true, if one considers that the scope of international commerce law has changed with the establishment of the World Trade Organization (WTO) and the conclusion of the General Agreement on Trade in Services (GATS) now also includes more than the classical trade with goods.[65] However, in Opinion 1/94 on the treaty-making powers of the Community to sign the international agreements establishing the WTO and in particular with respect to the GATS, the Court surprisingly turned away from its dynamic interpretation of the term 'commercial policy' to a very restrictive one.[66] The judges examined the four kinds of services covered by the GATS and concluded that only one of the services, that is the provision of cross-border services, where neither the service provider nor the recipient of the service need to cross the border, fell into the responsibility of the Community. This kind of service was very similar to foreign trade with goods and could be therefore covered

---

61 Christoph Vedder, in: Eberhard Grabitz and Meinhard Hilf, eds. (2001), *Das Recht der Europäischen Union* (*The Law of the European Union*), vol. 2, (Munich: Verlag C.H.Beck, 17th supplement) Article 133 EC at para 4 [Vedder, in: Grabitz/Hilf, Article 133].

62 ECJ, *Opinion 1/78*, [1979] E.C.R. 2871, 2917f. [Opinion 1/78].

63 *Ibid.* at 2913.

64 See also Commission in ECJ, *Opinion 1/94* [1994] E.C.R. I-5267 at I-5400 [Opinion 1/94].

65 Till Müller-Ibold, in: Carl-Otto Lenz and Klaus Borchardt, eds. (2003), *EU- und EG-Vertrag* (*EU and EC Treaty*), 3rd edn. (Cologne: Bundesanzeiger) Preface to Article 131–134 EC at para 2 [Müller-Ibold, in: Lenz/Borchardt, Pref. Article 131–134]; Michael Hahn, in: Christian Calliess and Matthias Ruffert, eds. (2002), *Kommentar zum EU-Vertrag und EG-Vertrag* (*Commentary to the EU Treaty and EC Treaty*), 2nd edn. (Neuwied, Kriftel, Germany: Luchterhand Verlag), Article 133 EC at para 29 [Hahn, in: Calliess/Ruffert, Article 133].

66 Müller-Ibold, *ibid.* at para 7.

by the common commercial policy pursuant to Article 133 EC.[67] Contrary to this kind of service, the remaining three possible services included in the GATS, required that either the recipient or the service provider proceeded to another state. From the ECJ's perspective, citizens and entities from non-EC countries crossing the border to the EC to provide services were not covered by the common commercial policy.[68] Firstly, Article 3 EC distinguished between the 'common commercial policy' (Article 3(1)(b) EC) and 'measures concerning the entry and movement of persons' (Article 3(1)(d) EC). Secondly, regulations regarding the free movement of natural and legal persons were laid down in particular provisions in the EC Treaty. That is why these subject matters could not be part of the common commercial policy and hence a corresponding treaty-making power could not be based on Article 133(3) EC.[69] Similarly, with respect to the common transport policy the ECJ also argued that it was regulated in an own title in the EC Treaty and therefore could not be part of the common commercial policy any more.[70] Thereby no distinction was made between agreements containing safety rules in respect of transportation matters or agreements of a commercial nature.[71] Both kinds of agreements had to be based on the provisions on the common transport policy. A different conclusion could be only reached, if aspects included in a commercial agreement relating to transport were only of a marginal significance.[72] In such a case an agreement could be nevertheless completely based on Article 133(3) EC. However, such a subordinated importance of transport related aspects cannot be affirmed in respect of air service agreements. Here, transportation is the sole subject of the treaty and the main reason, why it is concluded. To safeguard the result achieved in Opinion 1/94, the Member States included a new paragraph in Article 133 EC, when the EC Treaty was amended at Nice. Article 133(6) EC now prescribes expressis verbis that international agreements on transport cannot be based on Article 133 EC, but only on the provisions of Title V on the common transport policy.

Article 310 EC

Article 310 EC grants the Community the power to conclude association agreements. The objective of these agreements is to establish close relations with non-EC states and to allow for limited participation in the integration process, without granting

67 Opinion 1/94, *supra* note 64 at I-5401.

68 *Ibid.* at I-5402.

69 *Ibid.*; see also Christoph Wilhelm Vedder (1980), *Die auswärtige Gewalt des Europa der Neun* (*The Treaty-making power of the Europe of the Nine*), (Göttingen, Germany: Schwartz) at 23 [Vedder, Treaty-making power].

70 Opinion 1/94, *supra* note 64 at I-5402, in: Karl Stefan Eisermann ed., 'Die Luftfahrtaußenkompetenz der Gemeinschaft', (The External Power of the Community in Air Transportation) (1995) *Europäische Zeitschrift für Wirtschaftsrecht* (*Europ. J of Commercial L.*) 331 at 332 [Eisermann]; Vedder, in: Grabitz/Hilf, Article 133, *supra* note 51 at para 39.

71 Opinion 1/94, *ibid.* at I-5402f.

72 Opinion 1/94, *ibid.* at I-5403; Eisermann, *supra* note 70 at 332.

them a membership in the Community.[73] Association agreements are intended to establish close economic bounds and a privileged relation between the European Community and third countries. They go beyond mere trade agreements.[74] Due to the reference to 'common action' and 'special procedure,' an indispensable feature of any association agreement is the establishment of common institutions with equal representation. These bodies are entitled to adopt legally binding law within their responsibility.[75] Several kinds of association agreements have emerged in the past years.[76] The most common ones are accession association agreements that are concluded in preparation of a later accession to the European Union. There are also free trade association agreements that establish close economic relations that can lead to the establishment of a free trade area or a customs union. Finally also development association agreements can be concluded, which foresee free access and one-sided exemptions from customs to underdeveloped countries and which are thus a tool of a development aid policy. The conclusion of air service agreements in the form of association agreements based on Article 310 EC does not seem very likely, because of the lack of sufficiently close economic relations. It is of course not impossible that the Community concludes an agreement with a third country that also extends to air transportation matters. The air service agreement would be insofar part of a greater association agreement. In most cases, however, air service agreements are concluded separately without establishing a particularly close relationship with the Community. Under these circumstances, Article 310 EC cannot serve as the basis for the conclusion of air service agreements.

Article 308 EC

Another provision that might empower the Community to conclude international agreements could be Article 308 EC. Article 308 EC is applicable when (a) an action by the Community is necessary to (b) attain, in the course of the operation of the common market, an objective (c) without that the EC Treaty provides the necessary powers for this purpose. The objective of this provision is to ensure that the Community posses the necessary powers to adopt necessary measures, when

---

73 Kirsten Schmalenbach in: Christian Calliess and Matthias Ruffert, eds. (2002), *Kommentar zu EU-Vertrag und EG-Vertrag* (*Commentary to EU Treaty and EC Treaty*), 2nd edn. (Neuwied, Kriftel, Germany: Luchterhand Verlag) Article 310 EC at para 6 [Schmalenbach in Calliess and Ruffert, Article 310]; Rudolf Streinz (2001), *Europarecht* (*European Law*), 5th edn. (Heidelberg, Germany: C.F. Müller Verlag) at para 613 [Streinz, European Law].

74 Dierk Booß, in: Carl-Otto Lenz and Klaus Borchardt eds. (2003), *EU- und EG-Vertrag* (*EU and EC Treaty*), 3rd edn. (Cologne: Bundesanzeiger) Article 310 EC at para 1 [Booß, in: Lenz/Borchardt, Article 310].

75 Schmalenbach, in: Calliess/Ruffert, Article 310, *supra* note 73 at paras 3, 11; Albert Bleckmann, in: Albert Bleckmann, ed. (1997), *Europarecht* (*European Law*), 6th edn. (Cologne, Berlin, Bonn, Munich: Carl Heymanns Verlag) at paras 1369 [Bleckmann, European Law].

76 See overview Matthias Herdegen (2003), *Europarecht* (*European Law*), 5th edn. (Munich: C.H. Beck Verlag) at paras 452ff. [Herdegen, European Law]; Schmalenbach, *ibid.* at 35ff.

situations that fall into the scope of the EC Treaty arise, but which were not foreseen at the time when the EC Treaty was signed and for which the Community was not granted the necessary powers. Article 308 EC closes the gap between objectives of the Community and the lack of the corresponding necessary powers, which by mistake were not included in the EC Treaty.[77] It is a supplementary provision and only applicable, if other, more appropriate and special powers are not available.[78] The scope of Article 308 EC is limited due to the reference to the objectives of the EC Treaty. It cannot be applied to amend existing or to introduce new objectives in the EC Treaty.[79] Article 308 EC does not grant a power to change the EC Treaty and therefore also does not question the principle of limited attribution.[80] Recourse to Article 308 EC can be made, to adopt internal regulations or to conclude international agreements.[81] The general references in Article 308 EC to 'action of the Community' and the undertaking of 'appropriate measures' show that all kinds of activities are possible under this provision. The application of Article 308 EC as the basis for the conclusion of air service agreements could be precluded, if Article 80(2) EC is a more appropriate competence that has precedence over Article 308 EC. At first glance, Article 80(2) EC does not contain an explicit treaty-making power. On the other hand, since the wording of Article 80(2) EC is quite broad and since it neither prescribes any particular objectives nor means that must be followed with respect to the regulation of air and sea transportation one could equally held the opinion that Article 80(2) EC can provide for the necessary treaty-making power to conclude air service agreements.[82] A first limitation of the scope of Article 80(2) EC could be brought forward due to the fact that the Council may decide whether, to what extent and by what procedure 'appropriate provisions may be laid down for sea and air transport'. One could hold the opinion that the term 'appropriate provisions' only includes the acts mentioned in Article 249 EC. In this respect, the German version of the EC Treaty could be interesting as the phrase 'provision may be laid down' in Article 80(2) EC is translated with 'geeignete Vorschriften … zu erlassen'. The proper translation of 'erlassen' is not 'lay down' but 'enact'[83] or 'issue'[84]. The

77 Ivo E. Schwartz, in: Hans von der Groeben and Jürgen Schwarze, eds. (2004), *Vertrag über die Europäische Union und Vertrag zur Gründung der Europäischen Gemeinschaft* (*Treaty on the European Union and Treaty establishing the European Community*), vol. 4, 6th edn. (Baden-Baden, Germany: Nomos-Verlagsgesellschaft) Article 308 EC at paras 9ff. [Schwartz, in: Groeben/Schwarze, Article 308]

78 *Ibid.* at para 61ff.

79 *Ibid.* at para 14ff.

80 *Ibid.* at para 14.

81 Opinion of Advocate General Tizzano in Case No. C-476/98 of 31.1.2002, [2002] E.C.R. I-9431 at I-9452 [Opinion AG Tizzano]; Schwartz, in: Groeben/Schwarze, Article 308, *supra* note 77 at 198.

82 Eisermann, *supra* note 70 at 333.

83 Heinz Messinger (1988), *Langenscheidts Großes Schulwörterbuch Englisch–Deutsch* (*Langenscheidt's Dictionary English–German*), 2nd edn., (Berlin, New York Langenscheidt).

84 Alfred Romain and B. Sharon Byrd, *Wörterbuch der Rechts – und Wirtschaftssprache (Dictionary of Legal and Economical Terms)*, 4th edn., (Munich: Verlag C.H. Beck).

translation of the German version of Article 80(2) EC must be therefore read as 'the Council ... enacts proper provisions'. The term 'enacting' is, in the German version, also used in Article 249 EC, where it refers to regulations, directives, decisions, recommendations and opinions. Hence, from the similarity of the wordings of Articles 80(2) and 249 EC in the German translation of the EC Treaty, one could draw the conclusion that Article 80(2) EC only empowers to the enactment of the various acts listed in Article 249 EC, but not to the conclusion of international treaties. On the other hand, it can be argued that international agreements are adopted by decisions of the Council pursuant to Article 300 EC. Therefore, if one would limit Article 80(2) EC to the adoption of measures set out in Article 249 EC, it would also include the power to adopt decisions of the conclusion of international agreements. In this respect additional difficulties arise, when it comes to the term 'decision' as it is used in the German version of the EC Treaty. Firstly, decisions on the conclusion of international agreements are not are not enacted, but taken. Article 249 EC in the German translation speaks of enacting also with respect to decisions. Secondly, when it comes to 'decisions' two different terms are used in the German version of the EC Treaty. When reference is made to the decision of the Council on the conclusion of an international agreement, the term 'Beschluss' is used. 'Decision' as it is used in, for example, Article 249(1) EC is translated with 'Entscheidung'. Hence, in the German version of the EC Treaty the decision the Council takes on the conclusion of an international treaty (Beschluss) is a different one than the decision that is taken pursuant to Article 249(1) EC (Entscheidung). Since Article 249 EC, in the German translation, does not include the 'Beschluss' on the conclusion of international treaties, one could be of the opinion that this power is also not encompassed by Article 80(2) EC due to the similarity of the terms used in this provision and in Article 249 EC. Another difficulty with respect to the wording of Article 80(2) EC results from the fact that a distinction is made in the EC Treaty between provisions and agreements. All provisions in the EC Treaty granting a treaty-making power to the EC use *expressis verbis* the term 'agreement'.[85] Such a reference to 'agreements' is missing in Article 80(2) EC.

As far as the grammatical difficulties detected in the German translation of the EC Treaty are concerned, it is sufficient to say that no similar problems exist in the other translations of the EC Treaty. The German differentiation between 'Entscheidung' and 'Beschluss' is unknown in other languages. The English, French, Spanish, Italian and Portuguese versions use a uniform term (decision, decision, decisione).[86] Additionally, in the English version, according to Article 249(1) EC decisions are 'taken' and not 'enacted' like in the German translation. Hence, solely the wording of Article 80(2) EC does not preclude to derive a treaty-making power from this provision. However, difficulties may still exist with respect to the structure of the EC Treaty and the reference of Article 80(2) EC to Article 71 EC, which itself further refers to Article 251 EC. The problem is that only the acts set out in Article 249 EC are enacted pursuant to Article 251 EC. International agreements, on the one hand, are adopted according to the procedure set out in Article 300 EC. For this reason,

85 Compare Articles 111, 133, 181, 310 EC.

86 Schwartz, in: Groeben/Schwarze, Article 308, *supra* note 77 at para 207.

one could hold the opinion to limit the scope of Article 80(2) EC to the adoption of secondary law and to exclude a treaty-making power. On the other hand, in the ERTA judgement the European Court of Justice decided that Article 71(1)(a) EC itself provided for a treaty-making power.[87] Thus, as far as the question is concerned which measures can be undertaken under Article 80(2) EC, it must be irrelevant that a reference to Article 251 EC takes place. With respect to the systematic of the EC Treaty, another interesting situation emerges as a result from the regulation included in Article 14(1) EC. This provision concerns the establishment of the common market and for this purpose refers, among others, to Article 80 EC. Thus, it could be concluded that Article 80(2) EC can be only used to adopt measures that serve the establishment of the common market. According to such an approach, treaty-making powers would have to be excluded from Article 80(2) EC as it is hardly possible that the conclusion of international agreements with non-Member States can have an effect on the establishment of the common market. This objective can be achieved without any international agreement. However, although Article 14(1) EC refers to Article 80 EC for the establishment of the common market, it does not have to mean that Article 80 EC would have to be automatically limited to the attainment of this goal only and could not be applied in a different context as well. This is in particular true when one takes into account the international nature of sea and air transportation. It is a long established custom that civil air transportation is partially governed by bilateral and multilateral agreements. This had been already the case even before the European Community was established in 1957 and thus well known to the Member States. Under these circumstances it must be assumed that, by leaving the wording of Article 80(2) EC quite open, the Member States intended to grant the necessary powers to the Community that the latter one would require to regulate this field in an effective manner. This would have to encompass the possibility to conclude international agreements.[88] Otherwise and considering in particular the common approach in international aviation, the Member States could have included a corresponding regulation in the EC Treaty, if they would have wanted that no treaty-making powers of the EC should emerge in the areas of sea and air transportation. This possibility still existed later on during the various amendments of the EC Treaty that took place, in particular, during the 1990s. Still, the Member States refused to undertake any amendments hereto, even in view of the increasing power of the Community in the area of air transportation as a result of the ongoing liberalization process. Therefore, the conclusion can be drawn that it corresponds with the intention of the Member States, if Article 80(2) EC is not limited solely to intra-Community regulations, but also contains an implicit treaty-making power. Such an understanding is further fostered, when Article 133(6) EC is taken into regard. This regulation refers for the conclusion of international transportation agreements to the provisions of Title V and Article 300 EC. So far, treaty-making

---

87 See below at 25.

88 Ernst-Werner Fuß, 'Die Befugnis der Europäischen Wirtschaftsgemeinschaft zum Abschluß völkerrechtlicher Verträge mit Drittstaaten', (The Competence of the European Economic Community to Conclude International Agreements with Third Countries), (1972) *Deutsches Verwaltungsblatt* (*German J. of Admin. L.*) 237 at 240 [Fuß].

powers in the area of transportation have been derived only from Articles 70, 71 EC. Since Article 133(6) EC refers to the whole Title V and not only to Articles 70, 71 EC, it is an indication that further treaty-making powers exist in this Title. Another indication for the existence of a treaty-making power included in Article 80(2) EC is the fact that the Community has already concluded several air service agreements with third countries and thereby had recourse to Article 80(2) EC.[89] Concluding, it is therefore submitted that Article 80(2) EC grants the power to conclude air service agreements.[90] However, the Community can make use of this power only, if the Council holds it as necessary that an air service agreement is concluded and grants a corresponding mandate.[91] Since Article 80(2) EC provides for a treaty-making power of the EC, no recourse to Article 308 EC can be taken regarding the conclusion of air service agreements with third countries.

Other treaty-making powers

Other treaty-making powers are laid down in Articles 111, 181, 182, 170, 174, 301–304 EC. These provisions are, however, only applicable in very particular cases, none of them relating to air transportation. Hence, they are not relevant for the questions dealt here and any further examination of these provisions is therefore not necessary. It shall be mentioned that Article 300 EC is not a treaty-making power itself, but only contains procedural regulations on the conclusion of international agreements by the Community.[92]

Result

The previous analysis has shown that Article 80(2) EC contains a treaty-making power of the European Community to conclude air service agreements. The use of this power, however, depends on a positive assessment of the Council on the necessity of the conclusion of an air service agreement and on a corresponding mandate.

---

89 Agreement with Sweden and Norway, *supra* note 56; Agreement with the Swiss Confederation, *supra* note 58.

90 See also Open-Skies, *infra* note 105 at I-9895; Christian Pitschas,'Anmerkungen zum "open-skies" Urteil des EuGH' (Remarks on the 'Opens-Skies' Judgement of the ECJ), (2003) *Europäische Zeitschrift für Wirtschaftsrecht* (*Europ. J. of Commercial L.*) 92 (according to Pitschas, Article 80(2) EC provides for implied and exclusive powers of the EC, at 92) [Pitschas]; contra Daniel Thym, 'Der Binnenmarkt und die "Freiheit der Lüfte"' (The Internal Market and the Freedoms of the Air) (2003) *Europarecht* (*Europ. L. Rev.*) 277 (Thym considers Article 80(2) EC as a concurrent treaty-making power of the European Community, at 287) [Thym].

91 See also Open-Skies, *infra* note 105 at I-9894f., I-9900.

92 Till Müller-Ibold, in: Carl-Otto Lenz and Klaus Borchardt, eds. (2003), *EU- und EG-Vertrag* (*EU and EC Treaty*), 3rd edn. (Cologne: Bundesanzeiger) Article 300 at para 1; Wolfgang Müller-Huschke in: Jürgen Schwarze, ed. (2000), *EU-Kommentar* (*EU Commentary*), (Baden-Baden, Germany: Nomos Verlagsgesellschaft), Preface to Article 131–134 EC at para 8; Vedder, in: Grabitz/Hilf, Article 133, *supra* note 61 at para 4.

IMPLIED POWERS

With respect to explicit treaty-making powers, the EC Treaty initially provided for only two such competences: Article 133(3) and Article 310 EC. Since these provisions were insufficient for the European Community to be able to attain its objectives, the European Court of Justice has expanded the EC's treaty-making powers in the ERTA judgement and in Opinion 1/76. Both rulings have transformed into customary law in the area of the Community's external relations.

The ERTA judgement

The ERTA judgement[93] dealt with the following case: five of the then existing six Member States signed within the framework of the Economic Commission of the United Nations for Europe, the European Road Transport Agreement (ERTA) in 1962, which contained provisions on the working conditions of persons engaged in international road transportation services. Due to a low number of ratifications this agreement originally did not enter into force. In 1967, the Economic Commission resumed negotiations hoping that the ERTA could be still ratified by the necessary number of states. At the same time, the European Community was drafting an act to harmonize the driving and rest periods of drivers engaged in road transportation services. This finally resulted in the adoption of Regulation No. 543/69/EEC on 25 March 1969. On 20 March 1970 the Council discussed a common position to be adopted by the Member States in respect of the envisaged ERTA at the next meeting of the Economic Commission. The Council reached the conclusion that the negotiations at the Economic Commission should be finalized and an amended ERTA should be drawn up for signing. The Commission filed a suit at the European Court of Justice against the Council and requested the cancellation of the Council's decision. The judges first conceded that the Community not only had explicit powers to conclude international agreements as they were in particular laid down in Articles 133(3) and 310 EC, but that other provisions of the EC Treaty could also provide for an implicit treaty-making power, if the effectiveness of these provisions would be impaired without such a corresponding treaty-making power. The ECJ further added that treaty-making powers could also result from measures adopted within the framework of the provisions of the EC Treaty.[94] With the term 'measures' the ECJ referred to 'secondary law,' which includes the acts listed in Article 249 EC, that is directives, regulations, recommendations and opinions by the European Commission, the European Council and/or the European Parliament according to the competences laid down in the EC Treaty. In contrast, the term 'primary law' covers the provisions laid down in the treaties establishing the European Union and its Communities, that is the EU Treaty, the EC Treaty and the Euratom Treaty. While primary law can be only amended unanimously by all Member States, secondary law can be adopted, amended or revoked by the governing bodies of the EU and, depending on the subject matter, a simple majority of votes in the competent body

93 ECJ, *Commission v Council*, 22/70, [1971] E.C.R. 263 [ERTA judgement].
94 *Ibid.* at para 16.

might be sufficient for this purpose. According to the European Court of Justice, secondary law could be the basis for a treaty-making power, if the Community had enacted a regulation, whose effective application could be impaired in case the Member States undertook obligations with respect to the same subject matter towards other, non-EC states, because, to fulfil their obligations, Member States could be compelled to ignore the adopted regulation and thus infringe their duties laid down in the EC Treaty. Hence, with an increasing integration the European Community would gain more and more treaty-making powers. Furthermore, under these circumstances, it was also not always possible to clearly distinguish between internal competences and treaty-making powers. The ECJ concluded that, according to Article 71(1)(a) EC the Council could adopt common rules for international transport to and from a Member State. This should also encompass situations, in which transportation occurred between a Member State and a non-EC state.[95] The judges further argued that in order to ensure the *effet-utile* of Article 71(1)(a) EC and due to the international character of the transportation sector, the Community should be entitled to conclude international agreements with non-Member States in this area and thus required a corresponding treaty-making power. The *effet-utile* principle is a very important tool in construing EC law and stipulates that the interpretation of a provision of EC law shall ensure the widest possible application of a provision pursuant to its wording and its objective, thereby helping to realize and pursue effectively the objectives of the EC Treaty as a whole.[96] Under these circumstances, the judges also conceded that a remaining concurrent treaty-making powers of the Member States next to the one of the European Community could not exists as this was held incompatible with the unity of the common market and the uniform application of EC law. In this respect the ECJ did not follow the proposal of its Advocate General, who rejected to attribute the necessary treaty-making power to the EC.[97] In the end, however, the ECJ rejected the Commission's claim as the decision of the Council referred to the negotiations undertaken in 1962 and hence to a time, when a corresponding EC regulation had not existed yet. Additionally, the judges added that, if in the second round of talks in 1970 the Community would have substituted the Member States, the other states involved might have been irritated and the conclusion of the ERTA could have been at risk. The ECJ concluded that in this situation the Council's decision was in the interest of the Community as it obliged the Member States to take a common position and to aim at modifying the

95 *Contra* De Lamothe, Opinion of Advocate General Dutheillet De Lamothe in Case 22/70 of 10 March 1971, [1971] E.C.R. 284 (De Lamothe defined international transportation as intra-Community transportation, at 290) [Opinion AG De Lamothe].

96 See also Manfred Zuleeg, 'Die Auslegung des Europäischen Gemeinschaftsrechts' (The Interpretation of European Community Law) (1969) *Europarecht* (*Europ. L. Rev.)* 97 at 107 [Zuleeg, interpretation]; Reinhard Böhm (1985), *Kompetenzauslegung und Kompetenzlücken im Gemeinschaftsrecht* (*Interpretation of Competences and Omissions in Community Law*) (Frankfurt, Bern, New York) at 193 [Böhm].

97 Opinion AG De Lamothe, *supra* note 95 (see the argumentation of Advocate General De Lamothe, at 290ff.).

ERTA agreement pursuant to the respective EC regulation.[98] Although the judgement seems to be very pro-European and one might have expected a speeding up of the integration process in the EC, the contrary happened. The Member States were afraid of losing further treaty-making powers, if they adopted more EC regulations. As a consequence, the Council simply stopped to enact any regulation, which resulted in a total standstill in the common transport policy in the following years.[99]

Opinion 1/76

In Opinion 1/76[100] the ECJ tried to undo the halt it had partially caused through the ERTA judgement. Opinion 1/76 dealt with a question referred to by the Commission, as to whether an agreement on the establishment of a fund for the compensation of entities, engaged in transportation on inland waterways, for voluntarily and temporary laying up their ships for periods when overcapacities existed, was in accordance with EC law. The intention of such a fund was to avoid extensive and destructive competition in the internal waterways transportation sector and to keep rates at a level that allowed for profitable activities. The agreement was set up for signing by the European Community, Switzerland, Belgium, Germany, France, Luxembourg, the Netherlands and Great Britain. The additional participation of these Member States was necessary due to Article 307(2) EC and because of their involvement in two other previously concluded agreements that would have been affected by the new agreement. The Commission was of the opinion that, instead of the multilateral agreement, the EC should first enact a regulation pursuant to Article 71 EC and, since Switzerland would not be bound by this act, a subsequent bilateral agreement between the EC and Switzerland would have to be concluded. The necessary treaty-making power of the EC could be derived, in accordance with the ERTA judgement, implicitly from Articles 70, 71 EC and the newly adopted EC regulation. The European Court of Justice first concluded that Article 71 EC was the correct competence for the proposed agreement. However, the judges did not consider it as necessary that an EC regulation was adopted before an implicit treaty-making power of the Community pursuant to the ERTA judgement could emerge. According to the ECJ treaty-making powers could be also derived implicitly from primary law, if the participation of the Community in an international agreement was necessary to attain

---

98 *Ibid.* (In addition to that an amendment of the EC regulation was also necessary, in order to allow the Member State to fulfil their duties resulting from the EC regulation as well as from the ERTA agreement, at 285.)

99 Jürgen Erdmenger, in: Hans von der Groeben and Jürgen Schwarze, eds. (2003), *Vertrag über die Europäische Union und Vertrag zur Gründung der Europäischen Gemeinschaft* (*Treaty on the European Union and Treaty establishing the European Community*), vol. 1, 6th edn. (Baden-Baden, Germany: Nomos Verlagsgesellschaft), Preface to Articles 70 to 80 EC at paras 35f. [Erdmenger in: Groeben/Schwarze, Pref. Article 70–80]; Axel Werbke, 'Staatliche und gemeinschaftliche Kompetenz zum völkerrechtlichen Handeln' (Treaty-making powers of States and of the Community) (1971) *Neue Juristische Wochenschrift* (*New Weekly L. Rev.*) 2103 at 2109 [Werbke]; Opinion AG De Lamothe, *ibid.* at 292 (this possibility was already shown before the ECJ delivered its decision by Advocate General De Lamothe and Werbke).

100 ECJ, *Opinion 1/76*, [1977], E.C.R., 741 [Opinion-1/76].

the objectives of the EC, which could not be achieved merely through the adoption of internal EC law.[101] In the end, the ECJ concluded that the proposed agreement was not in accordance with EC law, although not for reasons relating to treaty-making powers of the Community. The exact reasons for its incompatibility are irrelevant for the here discussed problems and therefore will not be dealt with any further.

Opinion 1/76 is based on the idea that under certain conditions the Community can achieve its objectives only, if it also has the ability to conclude international agreements.[102]

Contrary to the ERTA judgement, in Opinion 1/76 the ECJ did not expressly consider, whether the provision of primary law that it held applicable implicitly also provided for a treaty-making power. For this reason, some scholars have drawn the conclusion that all provisions of the EC Treaty provided for a concurrent implied competence of the Community to conclude international agreements.[103] Only when the EC really needed to conclude an international agreement to attain its objectives, the respective concurrent treaty-making power became an exclusive treaty-making power of the EC.[104] Not only does this approach give rise to doubts from a methodological perspective, in particular when attention is given to Article 5(1) EC, if in all competences of the EC Treaty that allow to enact secondary law a concurrent treaty-making power is acknowledged, but this approach and interpretation of Opinion 1/76 is also from a historical perspective not convincing. The fact, that in Opinion 1/76 the ECJ refrained from explicitly examining whether Article 71 EC provided for an implied power to conclude international agreements, by itself is not sufficient to assume that all competences empowering to enact internal law implicitly also include a treaty-making power. The omission of any further examination of Article 71 EC by the ECJ may be also due to the fact that the Court has done this and already derived a treaty-making power from Article 71 EC a few years before in the ERTA judgement. For this reason, Opinion 1/76 cannot be understood that every competence laid down in the EC Treaty implicitly provides for a treaty-making power. In accordance with common principles of law, it is still necessary

101 *Ibid.* at 755f.; this description of the ECJ in Opinion 1/76 was referred to by scholars as the theory of parallelism of internal and external competences. This term was later adopted by the ECJ itself in its Opinion 1/94.

102 Rudolf Geiger, 'Vertragsschlußkompetenzen der Europäischen Gemeinschaft und auswärtige Gewalt der Mitgliedstaaten' (Treaty-making powers of the European Community and External Powers of the Member States), (1995) *Juristenzeitung* (*Lawyer's Journal*) 973 at 976 [Geiger]; Michael Schweitzer and Waldemar Hummer (1996), *Europarecht – Das Recht der Europäischen Union – Das Recht der Europäischen Gemeinschaften (EGKS, EG, EAG)* (*European Law – The Law of the European Union – The Law of the European Communities*), 5th ed. (Neuwied, Kriftel, Berlin: Luchterhand Verlag) at para 654 [Schweitzer/Hummer].

103 Fuß, *supra* note 88 at 241; Christian Tomuschat, in: Hans von der Groeben and Jürgen Schwarze, eds. (2004), *Vertrag über die Europäische Union und Vertrag zur Gründung der Europäischen Gemeinschaft* (*Treaty on the European Union and Treaty establishing the European Community*), vol. 4, 6th edn. (Baden-Baden, Germany: Nomos Verlagsgesellschaft) Article 300 EC at para 11 [Tomuschat, in: Groeben/Schwarze, Article 300].

104 Opinion 1/94, *supra* note 64 at I-5414; ECJ, *Opinion 2/92* [1995] E.C.R. I-525 at 559 [Opinion 2/92].

to determine the scope of every competence by referring to its wording, structure, historical background and objective to decide, whether it provide for a treaty-making power. The difference *vis-à-vis* the ERTA judgement is that the existence of treaty-making powers pursuant to Opinion 1/76 does not require the previous adoption of secondary law. A treaty-making power of the EC can also emerge, if the conclusion of an international agreement by the Community is necessary to attain one of its objectives set out in the EC Treaty.

The ERTA judgement and Opinion 1/76 were of particular importance for the development of external relations of the European Community and, as will be shown in the following, were decisive in the 'open-skies' judgement with respect to the question, whether the European Community could conclude air service agreements with third countries.

The 'open-skies' judgement[105]

In the cases leading to the 'open-skies' judgement the Commission accused Great Britain, Denmark, Sweden, Finland, Belgium, Luxembourg, Austria and Germany of having breached their duties under the EC Treaty by concluding 'open-skies' air service agreements with the United States of America. From the Commission's point of view the Member States did not have the necessary treaty-making power to conclude these agreements and only the European Community had this power. Therefore, according to the Commission, the Member States have infringed the treaty-making power of the EC and thus have infringed Article 10 EC.

Submissions of the parties

The Commission argued that the Community had the treaty-making power to conclude air service agreements and referred for this purpose to the ECJ's ERTA judgement and Opinion 1/76. A treaty-making power of the Community according to the ERTA judgement was justified by the Commission in the 'open-skies' cases by relying on the multitude of secondary law that had been adopted in the field of air transportation. The Commission argued that the *effet-utile* of these acts could be hampered, if the Member States were permitted to continue concluding own bilateral air service agreements with non-EC states. Thereby it should not be decisive, whether the respective EC regulation was in fact endangered, but, according to the Commission, it should be sufficient for a treaty-making power to emerge on the part of the European Community, if the EC had simply enacted a regulation. The Member States disagreed with this notion and considered a treaty-making power of

105 ECJ, *Commission v United Kingdom*, C-466/98, [2002] E.C.R. I-9427; ECJ, *Commission v Denmark*, C-467/98, [2002] E.C.R. I-9519; ECJ, *Commission v Sweden*, C-468/98, [2002] E.C.R. I-9575; ECJ, *Commission v Finland,* C-469/98, [2002] E.C.R. I-9627; ECJ, *Commission v Belgium*, C-471/98, [2002] E.C.R. I-9681; ECJ, *Commission v Luxembourg*, C-472/98, [2002] E.C.R. I-9741; ECJ, *Commission v Austria*, C-475/98, [2002] E.C.R. I-9797; ECJ, *Commission v Germany*, C-476/98, [2002] E.C.R. I-9855; any references made to the 'open-skies' judgement in the following refer to the published ruling in the *Commission v Germany* case [Open-Skies].

the Community only as justified, if existing secondary law was actually impaired or its effective application really in danger due to an international agreement of the Member States. They argued that the Commission failed to show that such a danger emerged from the existence of the bilateral air service agreements. Furthermore, they brought forward that the Council deliberately did not adopt any secondary law in the past that would have regulated the establishment of entities from non-EC countries in the Community and from EC entities in non-Member States. The regulation of these subject matters therefore must have remained in the responsibility of the Member States. Finally, the Member States argued that air transportation was not fully harmonized by EC law yet. Hence, the Commission could not request the complete treaty-making power for the Community in this field. Finally, the Member States also brought forwards that, according to the ERTA judgement the Community could claim only those external competences that were already laid down in the EC Treaty. Any treaty-making power in the area of air transportation was to be based on Article 80(2) EC. But since this provision granted a wide discretion to the Council and did not provide directly for any treaty-making powers, the EC was also precluded from an external activity in this area solely based on the ERTA judgement.[106]

Apart from the ERTA judgement, the Commission also referred to Opinion 1/76 to justify a treaty-making power of the Community in the area of air transportation and argued that such a power of the EC was necessary for the attainment of the objectives of the EC Treaty and, in particular, for the avoidance of discriminations, distortions of competition and impairments of EC law in general. Additionally, due to the international character of air transportation, the Commission held it as impossible to distinguish between internal and external markets. Under these circumstances the Commission did not consider it as useful and efficient to grant a power to the EC to regulate aviation with respect to the common market, but to deny it a corresponding treaty-making power. The Member States, on the other hand, rejected the applicability of Opinion 1/76 in the existing situation by arguing that objectives to be attained in the air transportation sector could be only determined by the Council due to its discretion granted in Article 80(2) EC and for this purpose required the adoption of respective secondary law. Hence, the existence of a treaty-making power to conclude air service agreements by the Community was also dependent on a positive assessment of its necessity by the Council, which the later one had denied so far. Insofar the Member States also had doubts on the inseparability of internal competences and corresponding treaty-making powers in the area of air transportation.

Findings of the European Court of Justice

When considering a treaty-making power pursuant to Opinion 1/76, the Advocate General concentrated in his opinion mainly on the question of the necessity of such a competence. The biggest difficulty of this assessment was, according to the Advocate General, to identify the criteria, upon which the necessity of the conclusion of an international agreement by the EC was to be determined, in particular, when the

106 Open-Skies, *ibid.* at I-9893f.

Community had not previously exercised a corresponding internal power.[107] Such a determination could be neither undertaken by the Member States, as they lacked the necessary powers, nor by the Council or the Commission, as they were involved in the execution of the respective treaty-making power. The European Court of Justice was also precluded from deciding this question, because the judges could review the legality of actions and omissions only. The question, whether the EC required a treaty-making power in a particular situation, was a discretionary decision and the ECJ could not substitute discretionary decisions of other institutions of the EC with its own discretion.[108] Advocate General Tizzano therefore suggested to determine the necessity of an international agreement of the EC pursuant to the procedure that was foreseen for the execution of the corresponding internal power. Consequently, if such an internal power was not available, recourse could be taken to the procedure set out in Article 308 EC.[109] This meant that as long as the competent institutions of the EC did not consider an agreement as necessary, an implied power of the Community pursuant to Opinion 1/76 did not exist. With respect to the 'open-skies' cases, the Commission considered an air service agreement between the EC and the United States of America as necessary, but the Council did not. Since Article 80(2) EC was the appropriate provision for any activities and regulations related to air transportation and since according to the procedure set out therein, the Council's assessment prevailed in this area, the Advocate General agreed with the Member States and refused a necessity for an air service agreement between the EC and the US according to Opinion 1/76.[110]

When it came to a possible treaty-making power of the Community pursuant to the ERTA judgement, Advocate General Tizzano first examined the degree of impairment of EC law that was necessary for a treaty-making power of the Community to emerge. In this respect he agreed with the Commission that a treaty-making power of the Community existed in all areas, where secondary law had been adopted. Member States were precluded from concluding international agreements in these fields, even if these agreements might be in accordance with existing EC law as this would contravene the unity of the common market and the uniform application of Community law.[111] This was even the case, if existing Community law was included verbatim in an international agreement. Firstly, this would bear the danger that a uniform application of EC law was not ensured and secondly, amendments of EC law might not be adopted completely and timely in the respective agreements.[112] Hence, in general the mere existence of an EC act is sufficient to acknowledge a danger for such an act, if Member States are permitted to continue concluding international agreements, and hence result in the emergence of a corresponding treaty-making power of the Community. According to the Advocate General it is not important to which extent international agreements of the Member States concluded with non-

107 Opinion of AG Tizzano, *supra* note 81 at I-9453ff.
108 *Ibid.* at I-9454.
109 *Ibid.*
110 *Ibid.* at I-9456.
111 *Ibid.* at I-9461.
112 *Ibid.*

Member States really may endanger EC law. Consequently the Advocate General allowed for an exception from this general rule only for those agreements that did not touch upon a subject matter regulated in secondary law or, if international agreements contained subject matters that, although previously included in EC regulations, were limited to issues that were left out in the respective EC act. With respect to air transportation matters the Advocate General concluded that such a situation existed with respect to the regulation of the activities of citizens and entities from third countries in the Community as well as for activities of EC citizens and entities in non-Member States included in international air service agreements. Insofar air service agreements had a different scope than similar secondary law adopted with respect to the air transportation sector, as the latter one was limited to activities of EC citizens and entities within the European Community only. Therefore, in general, EC law could not be affected by international air service agreement concluded by the Member States with third countries.[113] Still, due to the proximity of the subject matters included in air service agreements and in EC acts on air transportations, Advocate General Tizzano saw a potential danger that EC law could be affected by existing air service agreements of the Member States. This, however, needed further examination. For this purpose Tizzano analysed to what extent EC law could be restricted in its effectiveness due to the bilateral air service agreements of the Member States. An impairment of Regulation No. 2407/92/EC on the licensing of air carriers,[114] of Regulation No. 2408/92/EC on the access for Community air carriers to intra-Community air routes[115] and of Regulation No. 95/93/EC on common rules for the allocation of slots at Community airports[116] as was argued by the Commission was rejected by him. The same was true for the Commission's opinion that bilateral air service agreements led to a distortion of competition and discrimination in the European Community. The Advocate General rejected most of the Commission's arguments on the ground that they were too general and too abstract. Insofar the Commission failed to bring sufficient prove for its accusations, in particular in respect of a possible impairment of Regulation 95/93, as is its duty in a procedure pursuant to Article 226 EC.[117] Still, Tizzano affirmed an impairment of Regulation No. 2409/92/EC on fares and rates for air services[118] and of Regulation No. 2299/89/EC on a code of conduct for computerized reservation systems.[119]

The European Court of Justice first examined, whether the European Community could claim a treaty-making power for air service agreements pursuant to Opinion 1/76. As far as air transportation was concerned the judges concluded that Article 80(2) EC alone could not be the basis for justifying a treaty-making power

113 *Ibid.* at I-9466.

114 Regulation 2407/92, *supra* note 49.

115 Regulation 2408/92, *supra* note 49.

116 EC, Council regulation 95/93 of 18 January 1993 on Common Rules for the Allocation of Slots at Community Airports, [1993] O.J. L 014/01 [Regulation 95/93].

117 Opinion of AG Tizzano, *supra* note 81 of at I-9473.

118 *Ibid.* at I-9476f.; Regulation 2409/92, *supra* note 49.

119 Opinion of AG Tizzano, *supra* note 81 at I-9476f.; EC, Council Regulation 2299/89 of 24 July 1989 on a Code of Conduct for Computerized Reservation Systems [1989] O.J. L 220/01 [Regulation 2299/89].

of the EC in the area of air transportation, but that such a power could exist only, if the Council considered it as necessary that a comprehensive air service agreement was to be concluded between the EC and a third country in order to attain objectives of the EC. Although the Commission had been given a mandate by the Council in the past to negotiate an air service agreement with the US, this mandate was limited and excluded core elements usually part of air service agreements, like for example the exchange of air traffic rights, regulations on capacities and fares and questions of designations. These issues were left in the responsibility of the Member States. The fact that the Council did not grant any further powers to the EC showed that it did not consider the conclusion of a comprehensive air service agreement between the EC and the US as necessary as is a condition of the existence of treaty-making power pursuant to Opinion 1/76. Furthermore, the judges added that for the attainment of objectives of the EC Treaty in the area of air transportation no conclusion of an international agreement was necessary, but that these goals could be also attained by adopting secondary law, where the Member States could be obliged to adopt a common position in their external relations. This was held to be sufficient in order to avoid discriminations and the distortion of competition within the EC. To underline the validity of this proposal the judges pointed out that the freedom of services in the area of air transportation had been established in the Community merely with the adoption of the liberalization packages consisting of internal regulations only and without the conclusion of any international agreements. For these reasons, a treaty-making power of the Community to conclude air service agreements pursuant to Opinion 1/76 seemed unnecessary for the purpose of attaining objectives of the EC Treaty and was rejected accordingly by the European Court of Justice, who thereby followed Advocate General Tizzano's proposal. Subsequently the ECJ scrutinized, whether a treaty-making power of the EC could exist pursuant to the ERTA judgement. First the ECJ repeated and affirmed its position that treaty-making powers could be derived implicitly from provisions of the EC Treaty and also from secondary law adopted in the framework of these provisions. The judges then contradicted the Member States, who were of the opinion that no treaty-making powers could be derived from secondary law that was based on Article 80(2) EC.[120] In the next step the ECJ dealt with the Commission's argument that, due to the comprehensive system of air transportation legislation adopted by the EC in the context of the liberalization of the air transportation sector, a treaty-making power of the Community to conclude air service agreements should exist. To verify this notion, the ECJ analysed every single act regulating air transportation matters that could be affected by bilateral air service agreements that the Member States had concluded or might conclude with non-Member States. Concurring with Advocate General Tizzano the judges assumed that an impairment of EC law was possible, if the obligations of the Member States undertaken in air service agreements fell in the scope of a provision of secondary law or if they referred to an area that was covered to a large extent by EC law. Another possibility for a treaty-making power of the EC pursuant to the ERTA judgement to arise was, if secondary law contained an explicit regulation regarding the treatment of citizens from third countries in the EC or if

120 Open-Skies, *supra* note 105 at I-9901f.

it directly foresaw a competence of the Community to enter into negotiations with third countries.[121]

As far as the granting of the fifth freedom by the 'open-skies' agreement was concerned, the Commission held the opinion that it might affect the effective application of Regulations 2407/92 and 2408/92. The first one contains rules on the licensing of air carriers in the EC. The latter one automatically grants the third, fourth and fifth freedom to those air carriers that were previously given an operating licence pursuant to Regulation 2407/92. The Commission argued that the fifth freedom enabled air carriers from non-Member States to offer air transportation services within the Community comparable to the ones offered by Community air carriers, but without obliging the former ones to obey the rules set out in Regulations 2407/92 and 2408/92. The ECJ disagreed and pointed out that the 'open-skies' agreements did not provide for any particular link to air carriers of the Community or to intra-Community air transportation. Hence, they did not have the same scope as these two regulations and the Commission failed to show how EC law could be affected by the 'open-skies' agreements. Therefore, there was no reason to affirm a corresponding treaty-making power of the Community pursuant to the ERTA judgement. The Commission argued further that distortions of competition and discriminations would occur in the European Community as a result of the totality and combination of all 'open-skies' agreements concluded by the Member States with the US. Air carriers registered in those Member States that had concluded such an agreement would have competitive advantages over those air carriers, whose Member States did not conclude 'open-skies' agreements. Wassenbergh, analysing the Commission's argumentation, drew the conclusion that a possible distortion of competition in the common market was the main reason, why the Commission filed in the suit. From his perspective a negative impact on the common air transportation market was possible, if a Community air carrier established in one Member State made use of the sixth freedom and diverted passengers and cargo from another Member State, thereby diminishing the business possibilities of the air carriers that were established in the latter Member State.[122] As an example he showed how Dutch airline KLM transported passengers and cargo from Germany via Amsterdam to the US and that this was hampering the business of, for example, German Lufthansa. Since the Netherlands had concluded an 'open-skies' agreement with the US long before Germany did, KLM not being confined by any restrictions on frequencies, capacities and fares, could offer services for lower fares than Lufthansa, which, due to the lack of a comparable liberal agreement, could not act as flexible as KLM. The Commission was not only concerned about the economic effects such a situation might have for the common market, but also about the possible political consequences. The selective approach of the US could drive a wedge between the Member States and

121 *Ibid.* at I-9902.

122 Henri Wassenbergh, 'Common Market, Open Skies and Politics: A Bald Eagle's-Eye View of today's Air Transport Regulation', (2000) 25(4) *Air and Space L.*, 5, 174, 177, 180 [Wassenbergh, Open-Skies].

endanger the European integration.[123] The Court of Justice, however, did not share the concerns of the Commission and refused once again to grant a treaty-making power to the EC pursuant to the ERTA judgement for the conclusion of air service agreements. The judges held the opinion that distortions in the flow of services in the internal market due to different bilateral air service agreements separately concluded by the Member States could not affect by themselves common rules adopted in the air transportation sector.[124] Subsequently, the Court scrutinized whether Regulation 2409/92, which deals with fares applicable on routes within the Community, could be affected by bilateral air service agreements. Generally, the ECJ conceded that pursuant to Article 1(2)(a) of Regulation 2409/92, this act was not applicable to air carriers from non-Member States. Still, a peculiarity existed in Article 1(3) of this regulation. Article 1(3) says:

> (3) Only Community air carriers shall be entitled to introduce new products or lower fares than the ones existing for identical products.

This provision prescribes that air carriers from non-EC countries are precluded from offering lower fares than Community air carriers. According to the ECJ, this provision contained a regulation in respect of citizens from other, non-EC states. Therefore, an exclusive treaty-making power of the Community existed in this regard. This provision could be affected by the 'open-skies' agreements as they not only granted the fifth freedom, but also allowed for a free determination of the applicable air fares. This enabled non-EC air carriers to offer intra-Community flights at fares lower then the ones applied by Community air carriers. Hence, an impairment of Regulation 2409/92 by the 'open-skies' agreements was possible and therefore the EC was granted the treaty-making power to 'enter into commitments with non-member countries relating to that limitation on the freedom of non-Community carriers to set fares and rates'.[125] Next the European Court of Justice referred to Regulation 2299/89, which deals with computerized reservation systems, and in particular to Articles 1 and 7 of this regulation. Article 1 says:

> This Regulation shall apply to computerized reservation systems (CRSs) when offered for use and/or used in the territory of the Community for the distribution and sale of air transport products irrespective of the status or nationality of the system vendor, … ;

Articles 7(1), (2) stipulates:

> (1) The obligations of a system vendor under Articles 3 and 4 to 6 shall not apply in respect of a parent carrier of a third country to the extent that its CRS outside the territory of the Community does not offer Community air carriers equivalent treatment to that provided under this Regulation and under Commission Regulation (EEC) No 83/91.

---

123 Cited in Onno Rijsdijk, 'EC Aviation Scene', (2001) *Air and Space L.* 16: 6, 333, 338; see also Loyola De Palacio cited in Onno Rijsdijk, 'EC Aviation Scene', (2000) *Air and Space L.* 15: 4/5, 184–191f. (referring to the divisive impact of the open-skies agreement); in this respect the US policy resembles the philosophy of divide et impera.

124 Open-Skies, *supra* note 105 at I-9903.

125 *Ibid.* at I-9905.

> (2) The obligations of parent or participating carriers under Articles 3a, 4 and 8 shall not apply in respect of a CRS controlled by (an) air carrier(s) of one or more third country (countries) to the extent that outside the territory of the Community the parent or participating carrier(s) is (are) not accorded equivalent treatment to that provided under this Regulation and under Commission Regulation (EEC) No 83/91.

Due to the reference to air carriers of non-EC countries, the Court concluded that the Community must be granted the exclusive treaty-making power to contract with non-EC states on computerized reservation systems used or offered for use by non-EC air carriers in the territory of the Community.[126] Finally, the Court examined Regulation 95/93 on common rules for the allocation of slots at Community airports. The objective of this regulation is to ensure a neutral, transparent and non-discriminatory distribution of slots. With regard to the dispute, Article 12(1) was of particular interest, which says:

> (1) Whenever it appears that a third country, with respect to the allocation of slots at airports:
>
> (a) does not grant Community air carriers treatment comparable to that granted by Member States to air carriers from that country; orb) does not grant Community air carriers *de facto* national treatment; orc) grants air carriers from other third countries more favourable treatment than Community air carriers, appropriate action may be taken to remedy the situation in respect of the airport or airports concerned, including the suspension wholly or partially of the obligations of this Regulation in respect of an air carrier of that third country, in accordance with Community law.

Generally, the Court held the opinion that, since this regulation is also applicable on air carriers from non-EC states, the Community should be granted an exclusive treaty-making power to conclude international agreements with respect to this subject matter. However, the Commission failed to show that the bilateral air service agreements concluded by the Member States contained a provision that dealt with this subject matter and thus could led to an impairment of Regulation 95/93. Therefore, the judges came, at last, to the conclusion that the Member States did not infringe the treaty-making power of the Community based on Regulation 95/93. In summary and as far as the treaty-making powers of the Community are concerned, the Court reached the decision that some of the Member States infringed their obligations according to Article 10 EC and Regulations 2409/92 and 2299/89 by concluding 'open-skies' agreements with the US as these agreements contained regulations concerning the applicable fares on intra-Community routes and the use of computerized reservations systems. These subject matters fell into the exclusive treaty-making power of the European Community.

---

126 *Ibid.* at I-9907.

DISCUSSION

Comments on the 'open-skies' judgement

The European Commission was satisfied with the judgement and perceived it as a confirmation of its position. In its initial reaction, the Commission demanded from all Member States on 20 November 2002 to denounce their bilateral air service agreements with the United States of America and asked the Council for a mandate to enter into negotiations with the US for the conclusion of a new and comprehensive air service agreement between the US and the EC.[127] In a following communication the Commission presented its conclusions it derived from the judgement as well as the objectives it intended to attain in the area of air transportation by making use of the newly gained treaty-making powers.[128] The Commission underlined that for a long time it already held the opinion that it would be more beneficial for the European aviation industry, if a common approach under the leadership of the Community would be adopted in this area.[129] Since EC law regulated most aspects of aviation, it was not appropriate to leave the treaty-making powers with the Member States.[130] Moreover, the Commission assumed that every Member State had concluded between 60 and 70 bilateral air service agreements with third countries that, basically, were very similar.[131] Therefore, not only the 'open-skies' agreements dealt with by the ECJ infringed EC law, but all air service agreements containing similar provisions.[132] Consequently, after the delivery of the 'open-skies' judgement the Commission initiated more procedures for an infringement of EC law, although so far only against those Member States that have concluded 'open-skies' or similarly liberal air service agreement with the USA, that is the Netherlands, France, Italy and Portugal. Still, none of these cases has been filed with the ECJ yet.[133] The Commission further concluded that, according to the 'open-skies' judgement, neither the European Community nor the Member States had all

127 EC, Commission, *Communication from the Commission on the Relation between the Community and third Countries in the Field of Air Transport*, COM (2003), 94 at 8, online: Eur-Lex <http://europa.eu.int/eur-lex/en/com/pdf/2003/com2003_0094en01.pdf> (date accessed: 27 July 2003) [Commission, Communication].

128 EC, Commission, *Communication from the Commission on the consequences of the Court judgements of 5 November 2002 for European air transport policy*, COM (2002) 649, online: Eur-Lex <http://europa.eu.int/eur-lex/lex/LexUriServ/site/en/com/2002/com2002_0649en01.pdf> (date accessed: 2 June 2005) [Commission, Consequences].

129 *Ibid.* at 2.

130 EC, Commission, *Communication from the Commission on a Civil Aviation Policy towards the People's Republic of China – Strengthening Cooperation and Opening Markets*, COM (2005), 78 final of 11 March 2005, online: Eur-Lex <http://europa.eu.int/eur-lex/lex/LexUriServ/site/en/com/2005/com2005_0078en01.pdf> at 10 (date accessed: 8 August 2005) [Commission, Aviation Policy China].

131 Commission, Consequences, *supra* note 128 at 4.

132 *Ibid.* at 10; Commission, Communication, *supra* note 127 at 3, 8.

133 Frederick Sorensen, Wilko van Weert and Angela Cheng-Jui Lu, 'ECJ Ruling on Open Skies Agreements v. Future International Air Transport', (2003) *Air and Space L.* 28: 2 at 3, 8 [Sorensen/van Weert/Cheng-Jui Lu].

the necessary treaty-making powers to conclude air service agreements alone.[134] As indicated above, air service agreements are not limited to the granting of air traffic rights, but contain various other regulations as well, for example in respect of safety, security, competition or environmental protection. Since, according to the Commission, the competences for these areas were divided between the European Community and the Member States and would require that, when in the future air service agreements would be concluded, the Member States and the Community would have to work closely together.[135] Thereby the Commission was of the opinion that those subject matters that fell into the Community's responsibility should be also dealt with by the Commission. Besides, the Member States should be allowed to continue negotiating on issues that remained in their responsibility. However, the Commission hoped that Member States would join the Commission in its future negotiations so that all issues could be negotiated with the respective third country within one procedure and thus a comprehensive air service agreement could be concluded.[136] In a following Communication the Commission not only further specified this assessment, but even argued that air service agreements should not be concluded between Member States and third countries at all any more, but solely between the Community and the third countries.[137] Still, in another statement, the Commission set out that, 'future amendments to bilateral air service agreements concerning air traffic rights will be negotiated by each EC Member State, and not by the European Commission'.[138] Hence, it seems that the Commission itself is not fully aware of the scope of its treaty-making powers. What the Commission is definitely certain about is the fact that the participation of the Community in the negotiations and the conclusion of air service agreements would bring political and economical advantages for Community air carriers and consumers. The European Community is one of the most important participants in international aviation with one of the biggest single markets and some of the biggest and strongest air carriers. The European airline industry generates more than 30 per cent of the worldwide air transport revenues (€120 million) and out of the first 15 passenger and cargo carriers in the world, four are European passenger carriers and five European cargo carriers. Taking only international traffic into regard, Community air carriers have double the international operations of their American counterparts.[139] Due to the existing system of bilateral agreements concluded between the Member States alone and

134 Commission, Consequences, *supra* note 128 at 10.

135 *Ibid.* at 7f.

136 Commission, Communication, *supra* note 127 at 9, 10f.

137 *Ibid.*, EC, Commission, *Developing the Agenda for the Community's External Aviation Policy*, COM (2005), 79 final of 11 March 2005, online: Eur-Lex <http://europa.eu.int/eur-lex/lex/LexUriServ/site/en/com/2005/com2005_0079en01.pdf> at 2 (date accessed: 2 August 2005) [Commission, Aviation Policy Agenda].

138 EC, Commission, *Staff working document: Annex to the Communication from the Commission 'Developing the agenda for the Community's external aviation policy (COM (2005) 79 final)' of 11 March 2005*, online: European Commission <http://www.europa.eu.int/comm/transport/air/international/doc/sec_2005_336-en.pdf> at 7 (date accessed: 8 August 2005) [Commission, Staff Working Document].

139 *Ibid.* at 2f.

third countries, the EC was so far precluded to make use of its particular significance to promote the interest of Community air carriers and consumers.[140] Because of the 'open-skies' judgement the situation has changed and would enable the Community to achieve the EC's objectives in international air transportations. Representing an area that encompasses 25 Member States, the European Community would have a strong political power that it could use to enforce its goals.[141] The Commission's main interest would be to promote a safe and efficient air transportation industry that was be beneficial for consumers. To achieve this objective, the EC should therefore:

i) be given the mandate to modify the existing air service agreements of the Member State and to bring them in accordance with EC law, whereby the potential of the common market should be maximized,
ii) propose amendments on the international stage that are beneficial to the air transport industry and that would allow for more cross-border investments,
iii) foster competition,
iv) ensure the maintenance of a high safety level, and
v) take care that environmental and consumer protection requirements are followed.[142]

The Commission admitted that it was aware of the complexity, flexibility and economic significance of the present system of bilateral air service agreements. Therefore, any amendments that were necessary within this system to ensure its accordance with EC law should be as little as possible to preserve the balance of rights and obligations agreed upon in the air service agreements by the Member States and their contracting partners.[143] This however did not mean that existing air service agreements could not be substituted with a new agreement concluded by the EC and the respective third country. If such a new air service agreement with a third country might become necessary from the perspective of the Commission, but be rejected by the third country, the Commission would have to coordinate any further steps with the Member States. As *ultima ratio* all Member States could be also asked to terminate their air service agreements with the respective third country to achieve the adoption of a single and uniform air service agreement. However, the Commission assumed that most countries would not risk their relations with the EC and the Member States in respect of air transportation as, from an economic perspective, they were too important for them.[144]

This radical and progressive approach of the Commission was received with scepticism. Even before the ECJ delivered its judgement, the Commission was already criticized for having even initiated the procedure and was blamed for endangering the system of bilateral air service agreements that has proven its effectiveness in regulating international aviation for over 50 years. The Commission

140 Commission, Consequences, *supra* note 128 at 11.
141 Commission, Communication, *supra* note 127 at 9f.
142 Commission, Consequences, *supra* note 128 at 12.
143 Commission, Communication, *supra* note 127 at 7.
144 *Ibid.* at 11.

was accused of intending to open up the internal US market for Community air carriers, thereby extending the liberalized common air transportation market to the territory of the US.[145] Wassenbergh criticized the 'open-skies' judgement blaming the ECJ for having ignored leading principles of public international air law, existing international agreements, the sovereignty of the Member States and the interests of third countries.[146] From his perspective, civil aviation should remain in the exclusive competence of the Member States, at least as long as the Community was still merely an economic union without any own sovereignty over its airspace and such a sovereignty was exercised only by the Member States.[147] Still, even Wassenbergh acknowledged that after the 'open-skies' judgement the treaty-making powers in the area of air transportation did not belong exclusively to the Member States any more. Although the ECJ refused to grant a comprehensive and exclusive treaty-making power to the Community to conclude air service agreements, it has nevertheless reduced the possibilities of the Member States to conclude such agreements alone and independently. While the EC was responsible now for regulations regarding slots, tariffs on intra-Community flights and computerized reservation systems, the Member States were still responsible for the remaining issues dealt with in air service agreements, among others, important issues like capacities, granting of air traffic rights, designation of air carriers and, to a limited degree, also tariffs.[148] However, these remaining powers of the Member States would be only of a limited significance compared to the situation before the 'open-skies' judgement.[149] Concurring with the Commission, most scholars held the opinion that future regulations on international air transportation would require either the conclusion of two separate air service agreements between the Member States and the EC with third countries or that mixed agreements would have to be concluded, in which the EC and the Member States would act jointly as one party.[150] Since it can be assumed that third countries would be very reluctant to conclude two fragmented agreements instead of a single and uniform one, the latter solution seems to be the more probable alternative.[151] If it comes to modifying the existing agreements to ensure their compatibility with EC law, it was suggested that this task should be accomplished by the EC on behalf of all Member States.[152] However, emphasizing that the European Court of Justice explicitly refused to grant a comprehensive treaty-making power for the conclusion

145 John Trevett, 'EC Dark Cloud across Open Skies', (2001) *The Avmark Aviation Economist*, 10:18 [Trevett].

146 Henri Wassenbergh, 'The Decision of the ECJ of 5 November 2002 in the "Open Skies" Agreements Cases', (2003) *Air and Space L.* 28:1 at 19 [Wassenbergh, ECJ Decision].

147 *Ibid.* at 20.

148 Sorensen/van Weert/Cheng-Jui Lu, *supra* note 133 at 12f, 15f.

149 Wassenbergh, ECJ Decision, *supra* note 146 at 21.

150 Henri Wassenbergh, 'A Mandate to the European Commission to Negotiate Air Agreements with Non-EU States: International Law versus EU Law', (2003) *Air and Space L.* 28: 3 139, 141 [Wassenbergh, Mandate]; Thym, *supra* note 90 at 285; Gassner/Deichstetter, supra note 59 at 271.

151 Thym, *ibid.*; Sorensen/van Weert/Cheng-Jui Lu, *supra* note 82 at 12f., 14, 16f.

152 Wassenbergh, Mandate, *supra* note 150 at 142.

of air service agreements to the EC, the Commission would continue to require a mandate of the Member States, assembled in the Council, before it could conclude an air service agreement. But since the Member States would see their interests best fulfilled with the existing bilaterals, it was considered as quite unrealistic that such a mandate would be granted in the short future.[153] Therefore, one could expect that the present air service agreement would remain in force and would not be substituted by new 'Community air service agreements'.[154] This assessment was strengthened by the first reaction of the USA to the 'open-skies' judgement. The US administration announced that the existing bilateral air service agreements remained the basis for international air traffic between the Member States and the US.[155] Some scholars, on the other hand, expected that a new EC–US agreement could be signed until 2010.[156] In this respect, reference is often made to the establishment of a Transatlantic Common Aviation Area (TCAA), a common air transportation market of the EC and the US. Wassenbergh is quite sceptical about this project, since it would require that the internal US market is opened for air carriers from the Community. For him this seems not only impossible from a political perspective, but also from a legal one. The main obstacle is caused by a US act that reserves the domestic US air market to US airlines only.[157]

Sorensen, van Weert and Cheng-Jui Lu, on the other hand, have a more positive opinion of the 'open-skies' judgement and the future of international air transportation. The new treaty-making powers of the Community in various areas of air law and the common air transport market in the EC could have a positive impact on the conclusion of a new air service agreement between the US on one side and the EC and the Member States on the other side. Such an agreement could serve as a model for future agreements between the EC, the Member States and third countries.

---

153 Wassenbergh, ECJ Decision, *supra* note 146 at 24.

154 David Solon, 'What Next for Europe – Wide Multilateral Air Service Agreements?', (2002) *The Avmark Aviation Economist* 7:2 at 2f. [Solon].

155 US Under Secretary of Commerce Grant Aldona, Unofficial transcript of a press conference 'Interview with Secretary Aldona on trade talks with EU' of 17 December 2003, online: US Mission to the EU <http://www.useu.be/Categories/Biotech/Dec1702AldonasPress.html> (date accessed: 2 June 2005) [Aldona, Interview]; US State Department, Communication 'US calls for discussion of open skies agreement in Europe' of 20 November 2002, online: US Mission to the EU <http://www.useu.be/Categories/Transportation/Nov2002USOpenSkiesCommission.html> (date accessed: 2 June 2005) [US Communication].

156 Solon, *supra* note 154 at 3; but see Sorensen, van Weert and Cheng-Jui Lu, *supra* note 133 at 12 (the authors construe the statements of the US officials at the Department of Transport that there is a willingness to enter into negotiations with the EU on matters relating to air transport).

157 Wassenbergh, Mandate, *supra* note 150 at 142; Wassenbergh, ECJ Decision, *supra* note 146 at 30 (Wassenbergh defines the TCAA as 'an extension of the EU Common Market to include the North-Atlantic'); but see Wassenbergh, Open-Skies, *supra* note 122 at 183 (here Wassenbergh not only held a TCAA between the USA and the EU as possible, but also wanted to included Canada and Mexico into this project).

From their perspective, the 'open-skies' judgement could be the beginning of a new era in international aviation.[158]

In a reaction to the Commission's proposal for a Community agenda on an external aviation policy, the Council presented its opinion.[159] It first recognized that negotiations of new or amended air service agreements shall deliver mutually beneficial outcomes for the parties involved. Furthermore, the Council acknowledged that differences in the stages of development of the Member States' bilateral air service agreements with third countries could result in unequal business opportunities for Community air carriers. Still, these bilateral agreements would remain, for the time being, the principal basis for international relations between the Member States and third countries, to ensure continuity of services for the users as well as a stable operating environment for the industry and hence benefiting the economy as a whole. The Council also affirmed that the Member States and the Community would have to work together in a concerted manner and needed to strengthen their co-operation and coordination, to bring all air service agreements into conformity with EC law. With respect to treaty-making powers, the Council wants to allow the Member States to continue negotiating and concluding air service agreements with third countries in parallel with Community-level negotiations. This, however, is limited during a transition period until an open aviation area is established or other agreements are concluded. On the other hand, before the Community shall be given a mandate for the conclusion of an air service agreements between the EC and a third country, the Council demands that the added value of any such Community-level agreement is shown, with particular respect to the prospects of obtaining new opportunities for EC industry and users, as well as achieving greater levels of regulatory convergence with a view to ensuring a competitive level playing field.

From my point of view it is inappropriate to blame the Commission for the destruction of the system of bilateral air transportation agreements that is held to have successfully served the interests of all participants in international aviation in the past. There is no indication for such an intention of the Commission. Although international air service agreements were concluded in the past among states and although the European Community is not a state pursuant to international law yet, its participation in this system does not mean that it will be abolished. It will be rather modified according to the new circumstances that have emerged in international aviation. Air transportation in the EC is not regulated by the Member States any more, but by the Community. It is therefore only about time that this reality is also recognized on the international stage and that the Community is attributed a corresponding significance and possibility of participation. Similarly, the accusation that the national aviation market of the US will be overtaken by the EC and flooded with Community air carriers is wrong. The only conclusion that was drawn from the 'open-skies' judgement is that future air service agreements cannot be concluded by

---

158 Sorensen, van Weert and Cheng-Jui Lu, *supra* note 133 at 14.

159 *Ibid.*; EC, Council *Conclusions on Developing the Agenda for the Community's External Aviation Policy*, online: European Commission, <http://www.europa.eu.int/comm/transport/air/international/doc/2005_06_council_conclusions_en.pdf> (date accessed: 26 October 2005) [Council, Conclusions].

the Member States alone any more, but that some sort of cooperation between the Member States and the European Community will be necessary. In this respect the Council's conclusion and introduction of an added-value-criterion, when it comes to EC treaty-making powers in the aviation external relations, must be strictly refused. Such a criterion has never been part of the assessment, whether the Community could have specific treaty-making powers. This is in particular true, when treaty-making powers of the EC were based on the ERTA judgement or on Opinion 1/76. In such a situation, the Community was always exclusively competent and there was no further assessment of an added value of an activity undertaken by the EC.[160]

Activities undertaken by the EC in the aftermath of the 'open-skies' judgement

A first practical experience with the new situation arisen as a result of the 'open-skies' judgement was made only a few months after the ECJ delivered its ruling. On 5 June 2003 the Council authorized the Commission

(i) to enter into negotiations with the United States of America on the establishment of a Transatlantic Common Aviation Area, and
(ii) to negotiate with other non-EC countries the replacement of existing air service agreements with new Community air service agreements.

As far as the negotiations with the US are concerned, both sides reached an agreement, but in the end, the Council of the European Union, who needs to accept an international agreement concluded by the European Community pursuant to Article 300 EC, rejected it. The US was very disappointed about this final outcome of what it called could have been 'an historic agreement representing a major advance beyond traditional Open Skies'.[161] However, it also did not spare with negative remarks and comments on the situation that existed as a result of the 'open-skies' judgement. One of the critical remarks of the US was directed towards the fragmented powers of the European Community. For example, the mandate of the Council empowered the Commission only to grant fifth freedom rights for routes within the European Community. Here the EC was requested by the US to 'ensure that it could address the full range of rights traditionally included in air service agreements including fifth freedom rights out side the EU'.[162] Although

---

160 From the most recent Communication it seems that the Commission has accepted the added-value criterion; see EC, Commission, *Developing a Common Aviation Area with Ukraine*, COM (2005) 451 final of 27 September 2005, online: European Commission <http://www.europa.eu.int/comm/transport/air/international/doc/2005-09-28/com_2005_0451_en.pdf> at 6 (date accessed: 5 November 2005) [Commission, Common Aviation Area with Ukraine].

161 Deputy Assistant Secretary for Transportation Affairs Byerly US Department of State, 'US Determining How to Proceed on Open Skies Talks with EU' of 13 July 2004, online: US Mission to the EU <http://www.useu.be/Article.asp?ID=A1024B73-08FC-4485-B500-ACC710FDA142> (date accessed: 2 June 2005) [Byerly, Open-Skies Talks with EU].

162 *Ibid.*

the Community and the Member States, from a legal perspective, were unable to speak with one voice on all subjects, the US demanded from the Community to 'find a more workable way to deal with all relevant issues'.[163] Although a US–EC agreement finally failed due to the refusal by the Council, it would not be appropriate to blame solely the EC for the failure. The main problem was not a limited mandate of the Commission, but different convictions on both sides of the Atlantic on how air transportation between the EC and the US shall be organized. While the EC favoured the establishment of a TCAA, the US intended to extend its 'open-skies' air service agreements to all Member States. There is a significant difference between both approaches and major political and economical considerations come into play. The US has consistently denied the conclusion of an agreement that would establish a TCAA for various and changing reasons. US officials stress on every occasion that their interest is to foster competition and a free market in air transportation 'that serves consumer, helps communities and creates jobs'.[164] For these reasons they hail their 'open-skies' agreements as they result in 'no restriction on the number of airlines that can fly, no restriction on the number of flights they can fly; all cities are open, an open pricing regime and the structure of accompanying doing business rights, things like ground handling and code sharing, customs provisions that let the market, not government, decide what's going to happen in international aviation markets'.[165] In short, 'open-skies' agreements 'help expand the overall market for aviation and produce enormous benefits for millions of passengers in the form of better quality, lower-priced, and more competitive services'.[166] In this respect it is interesting to notice that a common aviation area, as favoured by the Commission, brings the same advantages. When the Commission asked the Council for a mandate to conclude a comprehensive air service agreement with Ukraine, Jacques Barrot, the Vice-President of the Commission who is also in charge of transport matters, made the following statement:

> By sharing the benefits of the single aviation market, both Ukraine and the EU will improve their relationship and create benefits for aviation industries and consumers.[167]

---

163 *Ibid.*

164 *Ibid.*

165 Chief negotiator for transportation at the US State Department Byerly, US Department of State, 'Byerly: US aims for comprehensive accord in air service talks with EU' of 29 September 2003, online: US Mission to the EU <http://www.useu.be/Categories/Transportation/Sept2903ByerlyOpenSkies.html> (date accessed: 2 June 2005) [Byerly, Comprehensive Accord].

166 US Secretary of Transportation Norman Mineta, 'Mineta urges successful conclusion of US-EU open skies talks,' online: US Mission to the EU <http://www.useu.be/Article.asp?ID=06364054-1B62-43D3-9F6A-44AB36AEEC63> (date accessed: 2 June 2005) [Mineta, US/EU Talks].

167 *Ibid.*; EC, Commission, Press Release, *The European Commission proposes to open aviation negotiations with Ukraine*, IP/05/1190 of 27 September 2005, online: European Commission, <http://www.europa.eu.int/comm/transport/air/international/doc/2005-09-28/ip_1190_en.pdf> (date accessed: 4 November 2005).

Hence, since a TCAA would be the most liberal manner to organize an air transportation market, to ensure competition and to attain better quality, lower-priced, and more competitive services, the establishment of a TCAA should be, also from the US perspective, a concept worth supporting it. Still, the US refuse to accept the establishment of a TCAA.[168] On the one hand US officials argue that a TCAA would bring cabotage to the US and that this is not possible under current domestic legislation, since Congress is not willing to amend the respective Act to allow foreign airlines operate on domestic routes. On the other hand, it is also stressed that a comprehensive 'open-skies' agreement with the European Community would be only a transitional step, before a more progressive agreement can be concluded. Firstly, it can be agreed with the US that a TCAA would result in cabotage rights for Community air carriers in the US. The chief negotiator for transportation at the US State Department Byerly describes cabotage as problematic, because it entails 'full access to the US domestic market, which is huge, by a foreign carrier that operates under foreign law, under foreign safety and security oversight, under foreign environmental law, foreign labour law, subject to foreign taxation, presumably. That raises very fundamental issues of the equality of the playing field to begin with'.[169] But the granting of fifth freedom rights to the US results in exactly the same, namely *de facto* cabotage for US airlines in the Community. The US does not regard it as problematic that US carriers can operate within the European Community and not being subject to EC law. This is hypocritical. The US rejects to equate cabotage and fifth freedom rights by referring to the fact that the Member States of the EC are sovereign states under international law each with a separate vote in the International Civil Aviation Organization (ICAO).[170] From a legal perspective this is correct. However, it remains to be explained what influence the voting power of the Member States at ICAO has to do with the granting of commercial air traffic rights. Although the EC is not formally a state, it is, from the perspective of aviation, state-like. With the adoption of the third liberalization package, one uniform European aviation market was established.[171] So, even if the Community would become one single state, this would have, from a Community perspective, no impact on aviation. The degree of integration would not be increased, if the EC would be completely united and Community air carriers would not obtain any additional rights or business possibilities. Under these circumstances, fifth freedom granted to the US results in *de facto* cabotage. Secondly, it is not convincing, if the US describes the conclusion of a comprehensive 'open-skies' agreement with the European Community as a

168 In this respect the Head of Office of the European Union at ICAO, Timothy Fenoulhet, remarked during a conference of the American Bar Association-Forum on Air and Space Law and the Institute of Air and Space Law of McGill University in Montréal on 30 September 2005 that 'it felt strange to have to justify and to defend the advantages of liberalizing international air transportation to the US'.

169 Byerly, Comprehensive Accord, *supra* note 165.

170 Byerly, Open-Skies Talks with EU, *supra* note 161.

171 US Officials are aware of this fact, see Secretary of Transport Norman Mineta, 'Bush seeks more foreign ownership of US airline stock' of 2 June 2003, online: US Mission to the EU <http://www.useu.be/Categories/Transportation/June0203AirlineOwnership.html> (date accessed: 2 June 2005) [Mineta, Foreign Ownership].

transitional step and justifies the postponing of the establishment of a TCAA on domestic legislation adopted by Congress. If the necessary amendments to US laws presently cannot be obtained from Congress, it is not certain that this will succeed in the future. There is no guarantee that, what the US describes as a transitional step, might not turn into a permanent situation. Additionally, US officials argue that cabotage rights are from a commercial perspective absolutely insignificant and no European air carrier really intends to enter into competition with US airlines in their domestic markets.[172] If this is the case, then one may wonder why the US is reluctant to grant these rights to the EC. Finally, the US also brings forward safety and security concerns.[173] As far as security and safety is concerned, it is hardly imaginable that European carriers shall be less safe and secure than US airlines. Safety and security standards are in Europe at least at the same level as in the US. So far the US reference to safety and security is therefore unfounded. Hence, what the US is strongly denying to the Community by refusing a TCAA, it is requesting for itself, by suggesting to extend its 'open-skies' standard to all Member States. The same is true with respect to the new Community nationality clause that was included in the EC–US air service agreement. The US consider this shift to an EC carrier model as 'badly overdue' and underline that they are willing to agree to a new kind of nationality clause that is in accordance with EC law as it is requested by the ECJ in the 'open-skies' judgement and would allow for every Community air carrier to operate to the US from any and all points in the Community.[174] While the US is explicitly underlining this aspect, it does not mention that by signing a comprehensive 'open-skies' agreement with the EC, US airlines would obtain the same rights, namely to operate from any point in the US to any point in the EC and even more to operate between any points within the Community. The true reason why the US denies the establishment of a TCAA is that US airlines would be exposed to full competition, even in their own national markets. There would be no possibility for the government to protect US airlines. Taking into account the current conditions of US and European air carriers, it could happen that US airlines might be simply driven out of the market by their European counterparts. Although US officials stress that they intend to introduce more competition into the air transportation business, the 'open-skies' policy is, in fact, designed to protect and support US airlines. Since the US is the only country that is consequently trying to put through its 'opens-skies' standard throughout the world,[175] they will have the most benefits from it. If the US concluded 'open-skies' agreements

172 Byerly, Open-Skies Talks with EU, *supra* note 161; Deputy Assistant Secretary of State John Byerly, 'US-EU air service negotiations' of 8 February 2005, online: US Mission to the EU <http://www.useu.be/ Categories/Transportation/Feb0805ByerlyAirTalksSpeech.html> (date accessed: 3 June 2005) [Byerly, Negotiations].

173 Byerly, Comprehensive Accord, *supra* note 165.

174 Byerly, Open-Skies Talks with EU, *supra* note 161; Mineta, US/EU Talks, *supra* note 166; Under Secretary Jeffrey Shane, 'US urges more liberalization in air services with China, EU' of 3 December 2004, online: US Mission to the EU <http://www.useu.be/Articles.asp?ID=2C502D09-9A64-4C2B-8109-E9020A29CA5F> (date accessed: 15 June 2005) [Shane, Liberalization].

175 Byerly, Comprehensive Accord, *supra* note 165; Secretary Norman Y. Mineta, US Department of Transportation, 'US, France sign open skies aviation agreement' of 22

with every single nation, its air carriers would be enabled to offer air transportation services between every single nation in the world. As no other nation is pursuing the goal of having 'open-skies' agreements to the same degree as the US, no other nation would be in a similar position and could offer the same business opportunities to its airlines. It is true as is underlined by US officials that they are intending to strengthen competition. On a market-by-market basis competition would indeed be strengthened, but on a worldwide basis, US carriers would have considerable advantages. Under these circumstances the refusal of the EC–US agreement by the Council must be approved as this agreement provided for an unbalanced distribution of commercial possibilities and would have granted significantly more rights to US carriers than Community air carriers would have obtained in return. After the negotiations with the US failed, the Commission apparently gave up its strategy to first conclude a comprehensive air service agreement with the US and instead decided to also approach other countries with respect to this subject matter. For this purpose it requested from the Council a mandate to enter into negotiations with Russia and China on the conclusion of a comprehensive air service agreement between each of the countries and the European Community. Additionally, Ukraine asked for a closer cooperation with the EC in the area of air transport and the opening of negotiations on a comprehensive aviation agreement. Therefore, the Commission asked the Council for a corresponding mandate on the establishment of a common aviation area with Ukraine.[176] Although the requests were neither denied nor granted yet, the Council made already clear that it will grant such a mandate only, if the Commission can prove that the conclusion of a comprehensive air service agreement between the Community and a third country would bring an added-value in particular with regard to the prospectus of obtaining new business opportunities for the EC industry and its users as well as achieving greater levels of regulatory convergence as far as the establishment of a competitive level playing field was concerned.[177] Apart from the mandate granted to the Commission to negotiate with the US, the Council also mandated the Commission to negotiate with countries from the Western Balkans on the establishment of a European common aviation area and with Morocco on a comprehensive air service agreement.[178]

Apart from the mandate to enter into negotiations with the US and other third countries, the Community also enacted Regulation No. 847/2004/EC on the negotiation and implementation of air service agreements between Member States and third

January 2002, online: US Mission to the EU <http://www.useu.be/Categories/Transportation/Jan2202USFranceOpenSkies.html> (date accessed. 2 June 2005).

176 See the Commission's request in: Commission, Aviation Policy Agenda, *supra* note 137 at 7ff., Commission, Aviation Policy China, *supra* note 130 at 11f.; EC, Commission, *Communication to the European Parliament and the Council on a framework for developing relations with the Russian Federation in the field of air transport*, COM (2005) 77 final of 14 March 2005, online: Eur-Lex <http://europa.eu.int/eur-lex/lex/LexUriServ/site/en/com/2005/com2005_0077en01.pdf> (date accessed: 8 August 2005) [Commission, Aviation Relation with Russia]; Commission, Common Aviation Area with Ukraine, *supra* note 176 at 3, 6.

177 Council, Conclusions, *supra* Note 159 at 5f.

178 Commission, Common Aviation Area with Ukraine, *supra* note 176 at 3.

countries on 29 April 2004.[179] Although it is mentioned in the preamble of this Act that the aspects usually regulated in air service agreements belong to the exclusive treaty-making power of either the EC or the Member States, it is unfortunately not determined how these powers are divided and how future agreements on air transportation shall be concluded. It is only set out vaguely that, due to the necessity of a uniform representation of the Community on the international stage, a close coordination and cooperation between the Member States and the EC will be necessary and it is the objective of Regulation 847/2004 to structure this cooperation.[180] The goal of this cooperation is the amendment of all air service agreements that infringe EC law or to conclude new agreements, if this should be necessary to ensure their compatibility with EC law.[181] Further it is stressed that the regulation shall have no determining effect on the distribution of competences between the Community and the Member States.[182] Although the Commission initially intended to intensively use its newly gained treaty-making powers, it seems that it has realized that the re-negotiation of all existing air service agreements of the Member States is a very extensive task and maybe too extensive to be achieved by the Community alone. For this purpose, the Member States were empowered in Regulation 847/2004 to modify their air service agreements on behalf of the Community, until the Commission can conclude a new Community air service agreement with the respective non-EC country.[183] When undertaking these amendments, the Member States shall take into account existing EC law, the interests of the Community and current negotiations that the Commission is undertaking.[184] Therefore, the regulation contains various procedural provisions that require the Member States to report to the Commission and to other Member States, if and how they intend to act, when negotiating air transportation matters with other countries. Additionally, the Member States shall use standardized clauses developed by the EC and the Member States in their air service agreements.[185] Until mid-2005 and thus three years after the 'open-skies' judgement, out of about 1500, only 50 bilateral air service agreements of the Member States have been amended by the Member States.[186] In summary, these are the most important provisions of Regulation 847/2004:

179 EC, European Parliament and Council regulation No 847/2004 of 29 April 2004 on the Negotiation and Implementation of Air Service Agreements between Member States and Third Countries, [2004] O.J. L 157/7 [Regulation 847/2004].

180 *Ibid.*, Article 1(1), recitals 2 and 4.

181 *Ibid.*, recital 6.

182 *Ibid.*, recital 5.

183 *Ibid.*, recital 7; see also EC, Commission–Directorate-General for Energy and Transport, *Towards a Coherent External Policy for Aviation: Recent Developments – Bringing Air Service Agreements in Line with Community Law*, online: European Commission <http://europa.eu.int/comm/transport/air/international/dev-en.htm#conformity> (date accessed: 27 July 2004); EC, Commission, *Communication from the Commission to the European Parliament*, COM (2004) 08 at 1, online: European Commission <http://europa.eu.int/eur-lex/en/com/pdf/2004/com2004_0008en01.pdf> (date accessed : 27 July 2004).

184 Regulation 847/2004, *ibid.*, recital 8.

185 *Ibid.*, Article 1(1).

186 Commission, Aviation Policy Agenda, *supra* note 137 at 6.

- Article 1(1) allows the Member States to conclude new or to amend existing air service agreements, emphasizing that this does not influence the distribution of external competences between the Member States and the European Community. According to Article 1(2) Member States, wishing to conclude such an agreement, need to file in information concerning the objectives of its negotiations with the third country. Pursuant to Article 1(3) the Commission and the other Member States are allowed to comment on the new agreement and their views should be considered by the Member State concluding the new agreement.
- Article 2 stipulates that, if air carriers wish to participate in the negotiations on the new agreement, Member States are obliged to treat all interested air carriers, having an establishment within their jurisdiction, equally.
- Article 3 prescribes that Member States must not agree to a lower number of Community air carriers that will be eligible for designation than the one provided for in the previous agreement.
- Article 5 sets out that, if air traffic rights, granted by a third country, are limited, Member States are obliged to ensure that these traffic rights are distributed in a non-discriminatory and transparent manner among all interested and eligible air carriers.

Own proposal

Implied powers of the European Community to conclude air service agreements

If all objections of the European Court of Justice in respect of the infringement of the Community's treaty-making powers by the Member States are taken into regard, one may reach the conclusion that the Member States are precluded from:

(i) entering into any obligations on applicable fares on intra-Community flights;
(ii) concluding agreements with non-EC states on computerized reservation systems, and
(iii) contracting on the allocation of slots on Community airports.

Since the judges did not express any other objections, one could hold the opinion that the Member States can continue to conclude air service agreements with non-Member States in the future, if they refrain from including such regulations in their respective agreements. Thereby it is interesting to notice that the ECJ refrained from granting the EC the treaty-making power regarding the granting and exchange of the fifth freedom, which can be exercised on intra-Community routes. On the other hand, the European Community is held competent to contract on tariffs that are applied on flights within the Community. The difficulty arising thereof is that, whenever air traffic rights are granted in an air service agreement, a regulation is also included regarding the fares applicable on the routes, where these air traffic rights can be exercised. If the Member States remain competent for the granting of the freedoms, including the fifth freedom that can be exercised within the EC, and if the Community becomes now responsible for contracting on fares applicable on intra-Community routes, a very complex situation arises. Many authors therefore hold the opinion that future air service agreements must be concluded as mixed agreements. Taking this into account and having also regard to the 'open-skies' judgement, a deeper examinations of the EC's treaty-making powers in the area of air transportation seems necessary and will be undertaken in the following.

As set out above, no explicit treaty-making powers exist in the EC Treaty that would establish an exclusive competence of the Community to conclude air service agreements with third countries. Although the ECJ concluded that Article 80(2) EC provides for a treaty-making power, the exercise of this competence requires a positive assessment by and a corresponding mandate of the Council. Under these circumstances the Community can conclude air service agreements with third countries only based on implied powers pursuant to Opinion 1/76 or the ERTA judgement.

IMPLIED POWERS PURSUANT TO OPINION 1/76

As explained above, an implied power pursuant to Opinion 1/76 is acknowledged, if the conclusion of an international agreement is necessary to attain an objective set out in the EC Treaty that cannot be achieved merely by enacting internal EC law, whose applicability is limited to the Member States and their citizens, but does not extend to subjects outside the European Community.[187] However, since Opinion 1/94 and the 'open-skies' judgement, doubts may exist, to which extent the Community can still have recourse to Opinion 1/76 when establishing a treaty-making power. These doubts are based on a statement of the ECJ made in Opinion 1/94 and repeated in the 'open-skies' judgement. On both occasions the Court refused to acknowledge a treaty-making power of the Community on the basis of Opinion 1/76 arguing that 'there is nothing in the Treaty to prevent the institutions [of the Community] arranging, in the common rules laid down by them, concerted action in relation to the United States of America, or to prevent them prescribing the approach to be taken by the Member States in their external dealings'.[188] This reasoning is quite surprising as it could be invoked against all agreements that were, or still are, to be concluded based on a treaty-making power pursuant to Opinion 1/76. It could be always argued that the Community does not require a treaty-making power to conclude an international agreement, since it could simply impose an obligation on the Member States in (internal) secondary law, to adopt a common approach or to undertake a concerted action in a specific matter in their foreign relations. Even in the case leading to Opinion 1/76 it could have been brought forward that the Community could enact a regulation regarding the transport on inland waterways on the Rhine and include a clause in this act that would impose an obligation on the Member States, to contract with Switzerland on particular subjects in a specific manner. With the above-mentioned argument of the ECJ, all treaty-making powers

187 See above at 26ff.

188 Opinion 1/94, *supra* note 64 at I-5267, 5411; Open-Skies, *supra* note 105 at I-9895f.

based on Opinion 1/76 could be rejected. Therefore, one has to ask, whether the European Court of Justice has finally given up its own theory on treaty-making powers based on Opinion 1/76. Formally the Court still adheres to it, as the judges did not denounce the applicability of this theory. *De facto* it seems quite impossible in the present situation to ever base any treaty-making powers of the Community on Opinion 1/76. Thym concludes that it would be therefore better, if the ECJ would take all its courage and would formally denounce its theory of implied powers based on Opinion 1/76.[189] Presently it is very difficult to assess, whether the Community can still refer to Opinion 1/76 to justify an external competence. Taking into account that the ECJ has clearly refused to base a treaty-making power on Opinion 1/76 in two cases and justified this with the above-mentioned argument, it seems that it has abolished its theory on implied powers pursuant to Opinion 1/76. The only question is, whether such an abolition and the argument brought forward is convincing. From a methodological perspective, Opinion 1/76 could be qualified as customary law of the Community. For customary law to emerge, two conditions must be fulfilled. There must be (i) a continuous application of a rule and (ii) its application must be supported by the general public, that is, there must be the conviction of the general public to be bound by the rule.[190] With respect to the ERTA judgement both conditions are fulfilled. Firstly, the ECJ applied this rule continuously since it introduced this rule in 1971. Secondly, the Member States, as the subjects being directly obliged by EC law and hence also by the ERTA judgement, explicitly acknowledged this ruling of the ECJ in a declaration annexed to the Treaty of Maastricht.[191] This justifies it to acknowledge the ERTA judgement as customary law of the European Community.[192] Although a similar declaration was never given with respect to Opinion 1/76, the Member States never raised objections to this rule. Unlike with respect to the ERTA judgement there was simply no reason to give a corresponding declaration with respect to Opinion 1/76. But like the ERTA judgement, Opinion 1/76 was also widely accepted as a basis for implicit treaty-making powers of the EC by the Member States. Therefore, Opinion 1/76 shall be also regarded as Community customary law. If the above-mentioned two conditions are the pre-requisites for customary law to come into existence, similar conditions must be applied for existing customary law to cease to exist. Although it will be more difficult to prove the continuous non-application of a rule, since it can happen that the conditions for its application are simply not fulfilled, it could be still argued that treaty-making powers based on

---

189 Thym, *supra* note 90 at 286.

190 Christian Tomuschat (1972), *Verfassungsgewohnheitsrecht*, (*Constitutional Customary Law*), (Heidelberg, Germany: Carl Winter Universitätsverlag); Reinhold Zippelius, 'Zum Problem der Rechtsfortbildung' (The Problem of Customary Law), (1964) *Neue Juristische Wochenschrift* (*New Weekly L. Rev.*), 1981 at 1986 [Zippelius]; Dietrich Ostertun (1996), *Gewohnheitsrecht in der Europäischen Union* (*Customary Law in the European Union*), (Göttingen, Germany: Peter Lang) at 36, 41, 80 [Ostertun].

191 Declaration to Articles 111, 174 and 181 annexed to the Treaty of Maastricht.

192 Compare Vedder, Treaty-making power, *supra* note 69 (Vedder describes that the majority of legal scholars has acknowledged that the EC Treaty grants implicit powers to the Community and to justify the existence of such powers, reference is simply made to the ERTA judgement of the ECJ, at 115).

Opinion 1/76 have been refused twice, once in Opinion 1/94 and the second time in the 'open-skies' judgement. Furthermore, taking into account the Court's opinion that treaty-making powers on Opinion 1/76 were not necessary, as the Community could always lay down common rules that would prescribe a common approach and concerted action of the Member States in their external dealings, it seems that the ECJ is convinced that the Community does not require treaty-making powers based on Opinion 1/76 to attain its objectives. Therefore, it is concluded that the ECJ has given up its theory of implied powers based on Opinion 1/76.

IMPLIED POWERS PURSUANT TO THE ERTA JUDGEMENT

As was already described above, according to the ERTA doctrine the degree of treaty-making powers depends on the scope of enacted secondary law. The more secondary law is adopted, the more treaty-making powers arise for the Community. The scope of the treaty-making powers is thereby equal to the scope of existing secondary law. Thereby it does not matter, whether the Member States international agreements might even be in accordance with EC law. The Member States shall even lose their treaty-making powers, if they include clauses in their international agreements that are fully in accordance with existing EC law as such agreements would have the disadvantage that they fixed existing Community law.[193] Due to the strong reference to secondary law, one could draw the conclusion that treaty-making powers pursuant to the ERTA judgement depend on the adoption of EC regulations only. Such a conclusion must cause serious objections as it would contravene the methodology of interpretation of legal rules and would be also incompatible with Article 5(1) EC.[194] It is impossible that the degree of secondary law determines the scope of treaty-making powers. Whether the Community can conclude an international agreement depends on the competences laid down in the EC Treaty. Hence, treaty-making powers are part of primary law. Therefore, one can not argue that the extent of treaty-making powers depends solely on the adoption of secondary law. Otherwise the EC institutions could determine the scope of their external competences by simply adopting internal acts. According to this author's opinion the ERTA judgement cannot be understood in this manner. Instead, for a treaty-making power pursuant to the ERTA judgement to exist, it must be first determined, whether a competence laid down in the EC Treaty (primary law) and empowering to adopt internal regulations can also include an implicit treaty-making power.[195] If

193 Open-Skies, *supra* note 105 at I-9906f.; ECJ, *Opinion 2/91* [1992] E.C.R. I-1061 at 1080, paras 25f. [Opinion 2/91].

194 Oliver Dörr, 'Die Entwicklung der ungeschriebenen Außenkompetenzen der EG', (The Development of Implicit External Powers of the EC), (1996) *Europäische Zeitschrift für Wirtschaftsrecht* (*Europ. J. of Commercial L.*) 39 at 41 [Dörr]; Vedder, Treaty-making power, *supra* note 69 at 130f.; Hans Krück (1977), *Völkerrechtliche Veträge im Recht der Europäischen Gemeinschaften* (*International Agreements in the Law of the European Communities*), (Berlin, Heidelberg, New York: Springer Verlag) at 40 [Krück, International Agreements].

195 Dörr, *ibid.* at 41.

this is affirmed, the Community is attributed a corresponding implicit, concurrent external competence. In a second step it is then examined, whether the Community has already made use of this respective internal power to adopt internal regulations (secondary law). If this is the case, the implicit, concurrent treaty-making power becomes an implicit, exclusive treaty-making power of the Community, but only within the scope of the EC regulation.[196] This understanding of the ERTA judgement was confirmed by the ECJ in its Opinions 1/91 and 1/94. Here the judges explicitly distinguished between the existence of treaty-making powers that were derived from primary law and the question of their exclusivity, which depended on secondary law.[197] In this context, two schools of thought can be mentioned, according to which the emergence of exclusive treaty-making powers pursuant to the ERTA judgement shall not depend on the enactment of secondary law. Geiger and Arnold hold the opinion that it is sufficient for the emergence of exclusive treaty-making powers of the Community, if the Community concludes once an international agreement in a particular area. From that moment on, the EC shall have the exclusive treaty-making power for that respective area.[198] To justify their approach, Geiger and Arnold argue

---

196 Albert Bleckmann, 'Die Kompetenz der Europäischen Gemeinschaft zum Abschluss völkerrechtlicher Verträge', (The Power of the European Community to Conclude International Agreements) (1977) *Europarecht* (*Europ. L. Rev.*) 109 at 115 [Bleckmann, Power]; Christian Tomuschat, 'Anmerkungen zum Gutachten 1/76' (Remarks on Opinion 1/76) (1977) *Europarecht* (*Europ. L. Rev.*) 157 at 161 [Tomuschat, Remarks on 1/76]; Thomas Oppermann (1999), *Europarecht* (*Europ. L. Rev.*), 2nd edn. (Munich: C.H. Beck Verlag) at para 1706 [Oppermann, European Law]; Streinz, European Law, *supra* note 73 at para 594; Christoph Sasse, 'Zur auswärtigen Gewalt der Europäischen Gemeinschaft' (The Treaty-making power of the European Community) (1971) *Europarecht* (*Europ. L. Rev.*) 208 at 230 [Sasse]; Geiger, *supra* note 102 at 975; Rudolf Geiger, 'Außenbeziehungen der Europäischen Wirtschaftsgemeinschaft und auswärtige Gewalt der Mitgliedstaaten' (External Relations of the European Economic Community and the Treaty-making power of the Member States) (1977) *Zeitschrift für ausländisches öffentliches Recht und Völkerrecht* (*J. of Foreign Public L. and Intl. L.*) 657 [Geiger, External Relations]; Vedder, Treaty Making Power, *supra* note 69 (Vedder describes implied powers as 'virtual powers' that are turned into 'real powers' through the adoption of EC secondary law, at 130f.); Meinhard Hilf, 'EG-Außenkompetenzen in Grenzen' (The Limits of the Treaty-making powers of the EC) (1995) *Europäische Zeitschrift für Wirtschaftsrecht* (*Europ. J. of Commercial L.*) 7 (Hilf argues that after Opinion 1/94 of the ECJ the Community can have only implied powers, if it has already made use of its explicit powers. Whether these implied powers are exclusive or concurrent powers is not answered by Hilf, at 8) [Hilf, Treaty-making powers]; Manfred Zuleeg, 'Verkehrspolitische Außenbeziehungen der Europäischen Gemeinschaft' (External Relations in the Transport Policy of the European Community), in: Jürgen Basedow, ed. (1987), *Europäischen Verkehrspolitik* (*European Transport Policy*) (Tübingen, Germany: J.C.B. Mohr (Paul Siebeck)) 59 (Zuleeg rejects to accept any concurrent powers in external relations; any external powers of the Community must be exclusive ones, at 59, 65.) [Zuleeg, External Transport Policy]

197 Dörr, *supra* note 194 at 42.

198 Geiger, External Relations, *supra* note 196 at 977; Rainer Arnold, 'Der Abschluss gemischter Verträge durch die Europäischen Gemeinschaften' (The Conclusion of Mixed Agreements by the European Community), (1980/81) *Archiv des Völkerrechts* (*Archives of Intl. L.*) 419 at 423 [Arnold].

that pursuant to Article 300(7) EC international agreements are, from the perspective of EC law, legally as valid and binding as ordinary secondary law.[199] They even hold the opinion that mixed agreements, where the Community and the Member States act jointly as one party to conclude an agreement with a third country, can result in the emergence of an exclusive treaty-making power of the Community.[200] Taking into regard that the EC has already concluded air service agreements, for example with Norway and Sweden[201] as well as with Switzerland,[202] it would have to be concluded according to this school of thought that, from this moment on, the Community have had the treaty-making power in respect of air service agreements. Objections to this approach can be raised by pointing out that it is contradictory to base the exclusivity of a treaty-making power on the conclusion of a mixed agreement, where the participation of the Member States was necessary due to a lack of sufficient powers on the part of the Community.[203] If Geiger's and Arnold's opinion would be applicable, no Member State would be willing to conclude a mixed agreement any more, due to the fear that it might lose more of its own treaty-making powers. This school of thought also neglects the nature and objective of the ERTA judgement. The idea behind the ERTA judgement was to safeguard the effectiveness of EC law. Primary and secondary law should not be affected by international treaties of the Member States. If Geiger and Arnold argue that the conclusion of an international agreement by the Community itself should already constitute an exclusive treaty-making power, they implicitly assume that the effectiveness of this agreement could be endangered by concurrent agreements concluded between the Member States and the respective third country with which the Community has previously contracted. Such a danger does not seem very reasonable. It is hardly imaginable that the third country itself will conclude an agreement with the Member States that would impair the treaty it has previously concluded with the Community. If such a treaty is nevertheless concluded, the third country could be liable for a breach of contract under international law.[204] At the same time, procedures under Article 226 EC against the respective Member States could be initiated for the infringement of Article 10 EC. For these reasons, the proposal to have an exclusive treaty-making power of the Community on a previously concluded agreement by the EC, is not convincing and therefore rejected. The fact that the Community has already concluded air service agreements do not lead to the emergence of a corresponding treaty-making power of the EC. Another school of thought holds the opinion that within a common policy all competences, the internal as well as the external ones, shall be transferred to

199 Geiger, *ibid.*

200 *Ibid.* at 978.

201 Agreements with Sweden and Norway, *supra* note 56; see above at 15.

202 Agreement with the Swiss Confederation, *supra* note 58; see above at 15.

203 See also Klaus D. Stein (1986), *Der gemischte Vertrag im Recht der Außenbeziehungen der Europäischen Gemeinschaft* (*The Mixed Agreement in External Relations of the European Community*) (Berlin: Duncker and Humblot) at 156 [Stein, Mixed Agreement]; Tomuschat, in: Groeben/Schwarze, Article 300, *supra* note 103 at para 88; Tomuschat and Stein hold the opinion that the conclusion of a mixed agreement should not lead to an increase or a reduction of powers on the part of the Member States or the Community.

204 Verdross/Simma, *supra* note 2 at 873ff.

the European Community.[205] This would affect the common commercial policy, the common agricultural policy and the common air transport policy.[206] It is argued that the Member States have lost any power in the areas covered by common policies and thus could not guarantee the implementation of the international obligations they may undertake in these respective fields.[207] It is, however, disputed even within this school, whether the transfer of treaty-making powers to the EC depends on the previous adoption of secondary law.[208] Irrespective of this dispute from a structural perspective it may be argued against this school of thought that the common commercial policy contains a particular provision in Article 133(3) EC that regulates the exercise of the treaty-making power. Such a provision would not be necessary, if the Community would have the treaty-making power in this area anyway. But even if one would like to follow this school, due to Articles 80(1), (2) EC it would be limited, within the common transport policy, to transport by rail, road and inland waterway. The here relevant transport by air carriers is excluded.[209] The common air transport policy is defined by the regulations adopted by the Community and thus, any treaty-making power in this field is defined by the scope of existing secondary law and cannot be affirmed in general for the whole area of air transportation.

When determining the scope of treaty-making powers pursuant to the ERTA judgement in the area of air transportation, the ECJ analysed which clauses, contained in the 'open-skies' agreements, were incompatible with the EC Treaty or could affect secondary law. Although the ECJ has extensively dealt with this question in its 'open-skies' judgement, it does not mean that its findings are final. It must be borne in mind that the 'open-skies' judgement was based on a procedure pursuant to Article 226 EC, in which the Commission must bring proof for its claims and accusations. From the point of view of Advocate General Tizzano the Commission partially failed to do so.[210] Therefore, it cannot be totally excluded that aspects were overseen by the Commission as well as by the ECJ that might have established a more extensive treaty-making power of the European Community in the area of international air transportation. Hence, it will be examined whether other provisions of EC law could

205 Christof von Arnim, 'Die Vertragsschließungskompetenz der Europäischen Wirtschaftsgemeinschaft in der Rechtsprechung des Europäischen Gerichtshofes' (The Treaty-making power of the European Economic Community in Rulings of the European Court of Justice) (1972) *Außenwirtschaftsdienst des Betriebs-Beraters* (*J. of Foreign Trade and Investment*) 215 at 219 [Arnim]; Vedder, Treaty-making power, *supra* note 69 at 136; Werbke, *supra* note 99 at 2105.

206 Werbke, *ibid.* (Werbke even extends the definition of a common policy to all areas, where the EC can enact a regulation pursuant to Article 249 (2) EC, at 2105).

207 Arnim, *supra* note 205 at 219.

208 Vedder, Treaty-making power, *supra* note 69 at 136; Arnim, *ibid.* (von Arnim requests for a transfer of external powers to the Community for those subject matters, where it has previously adopted secondary law, at 220); Werbke, *supra* note 99 (Werbke argues for the previous adoption of secondary law at least for such areas that require the existence of secondary laws to be legally binding, like for example high seas and air transportation pursuant to Article 80 (2) EC, at 2105).

209 Werbke, *ibid.*

210 See above at 33.

be affected, if the Member States continue to remain (partially) responsible for the conclusion of air service agreements. However, those areas that fall into the treaty-making power of the EC pursuant to the 'open-skies' judgement, will not be taken into regard any more in the following examination. The Commission argued that Regulations 2407/92 and 2408/92 could be impaired as a result of the 'open-skies' agreements, but failed to substantiate such a danger. Indeed, it is held possible that Regulation 2408/92 could be affected by 'open-skies' or similarly liberal agreements. To illustrate this hypothesis it is necessary to have regard to Article 9 of Regulation 2408/92, which says:

> (1) When serious congestion and/or environmental problems exist the Member State responsible may, subject to this Article, impose conditions on, limit or refuse the exercise of traffic rights, in particular when other modes of transport can provide satisfactory levels of service.

When construing this provision, a first problem arises with respect to the definition of the term 'serious congestion'. On the one hand it could refer to the limited capacities of Community airports, to enable the Member States to limit the exercise of air traffic rights, if an airport is seriously congested. Another possible understanding of 'serious congestion' could refer to the number of air transportation services offered on a particular route. Pursuant to this interpretation, Article 9(1) of Regulation 2408/92 would allow for a limitation of these services, if too many available flights led to serious congestion on this route. Hence, while the first interpretation refers to land-based congestions, the second one refers to air-based congestions. For the latter understanding one may invoke the subject matter of Regulation 2408/92, which deals with access for Community air carriers to intra-Community routes. Questions of limited capacities on airports are dealt with in Regulation 95/93. In addition, Article 9(1) prescribes that the exercise of air traffic rights can be imposed conditions on, limited or refused, if 'other modes of transportation can provide satisfactory levels of service'. Here, reference is made to the existing levels of transportation services on a particular route irrespective of the situation on an airport. Furthermore, according to Article 9(7) of Regulation 2408/92 reference is also made to the limitation of air traffic rights on a route. Article 9(7) says:

> (7) When a decision taken by a Member State in accordance with this Article limits the activity of a Community air carrier on an intra-Community route, the same conditions or limitations shall apply to all Community air carriers on the same route. When the decision involves the refusal of new or additional services, the same treatment shall be given to all requests by Community air carriers for new or additional services on that route.

Hence, for to grammatical and systematical reasons it is concluded that the term 'serious congestion' must refer to the existing level of air transportation services on a particular route. The Member States are given the opportunity to limit air traffic rights, if oversupply exists on a route. Thereby all Community air carriers must be treated equally. It is not possible to deny air traffic rights to one air carrier based on Article 9(1) and to grant it to another one at the same time. Difficulties can arise, if air carriers from third countries get involved. Air carriers, whose host country has concluded 'open-skies' or similarly liberal agreements with several Member States, in which the latter ones have granted the fifth freedom, can operate on intra-Community routes. This might also include such routes, to which a Member State has denied access based on Article 9(1). The question is now, whether air carriers from non-EC countries could be denied the exercise of their fifth freedom on intra-Community routes comparable to Community air carriers, due to the existence of oversupply on a particular route. To answer this question, Article 8 (2) of the 'open-skies' agreement between Germany and the US will be exemplary scrutinized. This provision says:

> (2) Each contracting party shall allow each designated airline to determine the frequency and capacity of the international air transportation it offers, based upon commercial considerations in the marketplace. Consistent with this right, neither contracting party shall unilaterally limit the volume of traffic, frequency or regularity of service, or the aircraft type or types operated by the designated airlines of the other contracting party, except as may be required for customs, technical, operational, or environmental reasons under uniform conditions consistent with Article 15 of the Convention.

This clause is the only provision in the 'open-skies' agreement that potentially could permit a limitation of the exercise of air traffic rights. The only question is, under which of the exceptions mentioned in sentence two of this provision, the existence of oversupply on a route could be subsumed. The most probable possibility seems to be the 'operational reason'. Fist of all it shall be stressed that, from a methodological perspective, Article 8 (2) s.2, being an exception to Article 8 (2) s.1, must be applied in a restrictive manner. From a grammatical perspective, operational means as being 'engaged or involved in operations'.[211] The term operations can mean:

- 'the action or process or method of working or operating',
- 'the state of being active or functioning',
- 'the scope or range of effectiveness of a thing's activity', or
- 'an active process; a discharge of a function'.[212]

Although the grammatical meaning of operational/operating is not absolutely clear, the terms seem rather to refer to the working of a machine or a facility, but not a market. Problems resulting from oversupply on a specific route, on the other hand, would have to be rather described as 'competitive reasons' or, with respect to the air transportation sector, as 'transport policy reasons' or 'structural reasons' and not 'operational reasons'. From a systematic perspective the other exceptions mentioned in Article 8 (2) s.2, 'customs' and 'technical reasons,' are also related to fixed facilities like airports or radio navigation facilities and not to difficulties resulting from competition. Therefore, 'operational reasons' shall be also understood as relating to difficulties arising in connection with a fixed facility, like for example shortages of capacities on airports. Such an interpretation of 'operational reasons'

211 R.E. Allen, (1990), *The Concise Oxford Dictionary of Current English*, 8th edn., Oxford: Clarendon Press.

212 *Ibid.*

would also comply with Article 8 (2) s.1, according to which it shall be left to the designated airlines to determine the frequency and capacity of their services based upon commercial considerations in the marketplace. Thus, it is submitted that the German–US 'open-skies' agreement does not permit Germany to withhold air traffic rights, including the fifth freedom that can be exercised on intra-Community routes by air carriers designated by the US, for reasons of oversupply on a route. Since the 'open-skies' agreements of the US usually contain the same standardized clauses, this conclusion is held to be applicable on all 'open-skies' agreements concluded by the US and the Member States. As a result of this situation a discrimination of Community air carriers for the benefit of air carriers from non-EC countries is held possible. While Community air carriers could lose their shares in the market due to a withholding of air traffic rights based on Article 9(1) of Regulation 2408/92, air carriers from third countries could be able to maintain or even to improve their positions, since a comparable limitation of their operations in not possible. One could even think of a situation, where non-EC air carriers make use of their fifth freedom in such a manner as to create oversupply on an intra-Community route, to preclude any further access to this route by Community air carriers according to Article 9(1). Under such circumstances, Article 9(1), that is intended to help achieving and maintaining sound competition, would be impaired in its effectiveness. The Commission was therefore right when it said that 'such international commitments, if not entered into by the Community, are contrary to Community law and deprive the latter of its effectiveness, because they have a discriminatory effect, cause distortions of competition and destabilise the Community market through the participation in it of airlines of non-member countries'.[213] Unfortunately, it failed to further substantiate its assessment. The notion of the Court of Justice initially mentioned in Opinion 1/94 and repeated in the 'open-skies' judgement that 'any distortions in the flow of services in the internal market which might arise from bilateral "open skies" agreements concluded by Member States with non-Member States do not in themselves affect the common rules adopted in that area and are thus not capable of establishing an external power of the Community'[214] is, in my opinion, incorrect. The possible impairment of Article 9(1), as was shown here and which was not sufficiently considered in the 'open-skies' judgement, is the typical situation that requires the granting of implied powers pursuant to the ERTA judgement to the Community. To prevent an impairment of Article 9(1) of Regulation 2408/92, the Community should be at least responsible for the granting and withholding of the fifth freedom that can be exercised on intra-Community routes.

Additionally, it shall be examine, whether Article 9(8) of Regulation 2408/92 provides for a regulation on the treatment of citizens from third countries. Pursuant to the European Court of Justice in such a case the Community would be attributed a corresponding treaty-making power.[215] Article 9(8) says:

---

213 Open-Skies, *supra* note 105 at I-9898.

214 *Ibid.* at I-9903.

215 See above at 34.

(8) Without prejudice to Article 8(1) and except with the agreement of the Member State(s) involved, a Member State shall not authorize an air carrier:

a) to establish new services, or

b) to increase the frequency of an existing service,between a specific airport in its territory and another Member State for such time as an air carrier licensed by that other Member State is not permitted, on the basis of slot-allocation rules as provided for in Article 8(2), to establish a new service or to increase frequencies on an existing service to the airport in question, pending the adoption by the Council and the coming into force of a Regulation on a code of conduct on slot allocation based on the general principle of non-discrimination on the grounds of nationality.

A reference to nationals from third countries can be derived from the use of the term 'air carrier' and the fact that, pursuant to Articles 2(a), (b) of Regulation 2408/92, it does not mean 'Community air carrier'. Hence, air carriers can be also such airlines that are designated by third countries. Although Article 9(8) therefore seems to contain a provision regarding nationals from non-Member States, this provision may not be applicable any more, if Regulation 95/93 is the Regulation referred to in this provision, upon which adoption this provision ceases to be applicable ('... pending the adoption by the Council and the coming into force of a Regulation ....'). Regulation 95/93 contains procedural regulations for the case that capacities on a Community airport are insufficient, to offer all air transportation services that are foreseen for that respective airport. In such a situation, the airport is declared as coordinated and an airport coordinator is designated, who is responsible for the allocation of slots from this airport. This allocation of slots is undertaken pursuant to Articles 4(3), (5) of Regulation 95/93 in a neutral, non-discriminatory and transparent procedure. Since the procedure laid down in Regulation 95/93 complies with the conditions set out in Article 9(8) of Regulation 2408/92, it is concluded that Regulation 95/93 is equal to the Regulation referred to in Article 9(8). Therefore, latter provision is not applicable any more and cannot serve as the basis for any treaty-making powers of the Community.

The here undertaken examination has shown that the Community must have an implied power pursuant to the ERTA judgement to conclude international agreements regarding the granting of the fifth freedom. Insofar it is disagreed with the 'open-skies' judgement, where the ECJ has denied granting such a power to the EC. Due to the conclusion reached, it must be further examined, how this additional treaty-making power of the EC could influence the distribution of treaty-making powers to conclude air service agreements between the Community and the Member States and, in particular, who is competent for the granting of the remaining air traffic rights.

Consequently, in the following it shall be further examined, whether, because of the responsibility of the Community for the granting of the fifth freedom, a treaty-making power of the Community to grant the third and fourth freedom in air service agreements could also exist pursuant to the ERTA judgement. Since an EC regulation on the taking up and putting down of passengers and cargo by Community air carriers in third countries, and of air carriers from non-EC states in the Community, which would constitute the basis for a treaty-making power of the Community in respect of the third and fourth freedom pursuant to the ERTA judgement, does not exist. The assumption is that the Member States are still responsible for granting these two air traffic rights in air service agreements. Under the traditional application of the ERTA judgement the Community can have only treaty-making powers for such subject matters that were previously dealt with in secondary laws. A different result could be obtained only if the ERTA judgement is applied in a less restrictive manner. One could think of granting the Community not only implied powers with respect to the conclusion of international agreements for those subject matters that have been previously part of an EC regulation, but to extend the treaty-making powers also to those areas, that are indispensable for the Community to make full use of the treaty-making powers directly based on the ERTA judgement. This might even extend to subject matters that were not yet regulated by secondary law. With respect to the conclusion of air service agreements this could be insofar important as the fifth freedom cannot be exercised without also having the third and fourth freedom. Any air transportation operation based on the fifth freedom requires that a third or fourth freedom operation is concluded before or after the fifth freedom service. Thus, the competence of the Community to grant the fifth freedom in air service agreements to third countries is useless, if those countries cannot also obtain the third and fourth freedom. It shall be therefore scrutinized, whether the ERTA judgement could be applied in a manner that would allow the Community to grant not only the fifth, but also the third and fourth freedom in air service agreements.

For this purpose the legal nature of the ERTA judgement and its methodological basis must be examined. According to one school of thought the ERTA judgement is based on the *effet-utile* principle.[216] It is argued that, to ensure the *effet-utile* of a competence of primary law laid down in the EC Treaty, it is necessary that any activity of the Member States, which could affect secondary law that is based on this primary law, is precluded. One possibility for the effectiveness of a competence laid down in the EC Treaty to be impaired exists, if the Member States are allowed to continue concluding international treaties. There is a risk that Member States, when implementing their international obligations in national laws, could need to adopt national regulations that might be incompatible with corresponding EC regulations. Thereby not only secondary law would be directly affected, but the authority of the respective primary law would be also neglected. To avoid the emergence of such a situation and for the safeguarding of existing EC law, the Community shall be granted implicit treaty-making powers for those subject matters that are dealt with in secondary laws, thereby precluding any further activity of the Member States in this field on the international level.[217] A second school of thought is of the opinion that the ERTA judgement shall be subsumed under the principle of primacy of EC

216 Böhm, *supra* note 96 at 79; Fuß, supra note 88 at 240; see above at 25 for the definition of the *effet-utile* principle.

217 Vedder, Treaty-making power, *supra* note 69 at 131 (it shall be noticed that although Vedder believes that the *effet-utile* principle and the implied-powers theory are closely related to each other, implied-powers in the EC cannot be derived from the *effet-utile* principle, at 120).

law.[218] This principle prescribes that, when Community law and national law clash, only Community law shall be applied.[219] The objective of this principle is to ensure a uniform and effective application of EC law throughout the whole European Community. This second school argues that, if the Community has adopted a regulation, the Member States are precluded from enacting an own national act that contravenes the respective EC law. This prohibition must also extend to the conclusion of international agreements.[220] Although, in fact, the principle of primacy of EC law does not prohibit the Member States to adopt contrary national laws, it however demands that these laws are not applied, if they clash with EC law. But, due to the fact that third countries are not bound by the principle of primacy of EC law, it is necessary to amend this principle with respect to international agreements to that extent that Member States are completely prohibited from contracting on subject matters that fall into the scope of an EC regulation. A third school finally argues that a prohibition for the Member States to adopt measures that could affect secondary law adopted in the area of the common transport policy can be based on Article 10 EC in connection with Article 3(1)(f) EC.[221] This would also include international agreements.

Apart from these three approaches, the ERTA judgement could be also based on the implied-powers theory. This theory was developed in the United States of America, when it was concluded that the powers granted to the federal state were insufficient for the state to fulfil its duties. The distribution of powers in the US between the federal state and the member states is undertaken according to the usual approach for federal states. The federal state is competent for those subject matters that it is explicitly assigned in the constitution, while the member states remain responsible for all those areas that are not explicitly transferred to the responsibility of the federal state.[222] From the perspective of constitutional law, the people grant certain powers to the federal state and, at the same time, take away these powers from the member states. Thus, the federal state can make use only of those competences that it is

---

218 *Ibid.* (Vedder does not consider the conclusion of an international agreement as an effective use of an internal power (*effet-utile*), but in formal and qualitative terms as something else, at 120, 128, 132); Kay Hailbronner, in: Kay Hailbronner, Eckart Klein, Siegfried Magiera and Peter-Christian Müller-Graff, eds. (1995), *EUV/EGV* (*EU Treaty/EC Treaty*), vol. 2, (Cologne, Berlin, Bonn, Munich: Carl Heymanns Verlag) Article 228 EC at para 14 [Hailbronner].

219 Bleckmann, European Law, *supra* note 75 at para 1074ff.; see in particular Manfred Zuleeg (1969), *Das Recht der Europäischen Gemeinschaften im innerstaatlichen Bereich* (*The Law of the European Communities in the National Sphere*), (Cologne, Berlin, Bonn, Munich: Carl Heymanns Verlag) [Zuleeg, EC Law in the National Sphere].

220 Vedder, Treaty-making Power, *supra* note 69 at 132.

221 Arnim, *supra* note 203 at 216.

222 Hans Kruse, 'Implied Powers und Implied Limitations' (Implied Powers and Implied Limitations), (1953/54) *Archiv des Völkerrechts* (*Archives of Intl. L.*) 169 at 172 [Kruse]; Hans-Jürgen Rabe (1963), *Das Verordnungsrecht der Europäischen Wirtschaftsgemeinschaft* (*The Legislation Law of the European Economic Community*), (Hamburg: Kommissionsverlag Ludwig Appel) at 138 [Rabe].

granted by the people.[223] The general rule is that the member states are responsible for the regulation of a subject matter, unless the constitution foresees an action of the federal state. Since the federal state's powers are an exception, they should normally be applied, from a methodological perspective, in a restrictive manner. However, the American constitution does not only provide for the explicit powers of the federal state, but also contains another, very particular provision in Article I, sec. 8, cl. 18. This clause, called 'sweeping clause,' says the following:

> The congress shall have all powers to make all laws which shall be necessary and proper for carrying into execution the foregoing powers, and all other powers vested by this Constitution in the Government of the USA or in any department thereof.

The sweeping clause was the starting point for the foundation of the implied-powers theory. It was argued that the federal state must have those implied powers that it needed to be able to make full and effective use of its explicitly granted powers. As a result the federal state would be enabled to regulate subject matters that originally were not covered by its explicit competences.[224] Despite this extension of the federal state's responsibility, the implied-powers theory did not provide for new powers, as a link to explicit powers laid down in the constitution must always exist.[225] Since the implied-powers theory was developed with respect to federal states, it must be scrutinized, whether it can be also applicable with respect to EC law. Due to the principle of limited attribution pursuant to Article 5(1) EC one can already draw a first analogy between federal states and the Community. Similarly to a federal state that receives its powers from the people, the Community is also assigned only a limited number of powers by the Member States. Still, some scholars refused to acknowledge the applicability of the implied-powers theory in EC law due to the existence of Article 308 EC.[226] It was argued that the lack of a required power, which is the condition for an application of the implied-powers theory, is the typical situation that triggers the application of Article 308 EC. The acknowledgement of the implied-powers theory would lead to a circumvention of Article 308 EC and the necessity to unanimous decisions that are foreseen in this provision. Article 308 EC would be taken away an own area of application and the effet-utile of this provision would be nullified.[227] Contrary to the inflexible and limited powers of federal states

---

223 Kruse, *ibid.*

224 Vedder, Treaty-making Power, *supra* note 69 at 116; Gert Nicolaysen, 'Zur Théorie von den implied powers in den Europäischen Gemeinschaften' (The Theory of Implied Powers in the European Communities), (1966) *Europarecht (Journal of European Law)*, 129, 131 [Nicolaysen].

225 Kruse, *supra* note 222 at 176; Norbert Achterberg, 'Die Annex-Kompetenz' (The Annex Power) (1966) *Die öffentliche Verwaltung* (*J. of Public Admin.*) 695 at 696.

226 Dörr, *supra* note 194 at 41; Rabe, *supra* note 222 at 151; Ernst Wohlfahrt, in: Ernst Wohlfahrt, Ulrich Everling, Hans Joachim Glaesner and Rudolf Sprung, eds. (1960), *Die Europäische Wirtschaftsgemeinschaft* (*The European Economic Community*) (Berlin, Frankfurt: Verlag Franz Vahlen GmbH); (Wohlfahrt does not exclude the implied-powers theory completely in EC law, but argues that, if it shall be applicable, due to Article 308 EC, it can only have a very limited scope, Article 228 at para 9 and Article 235 at para 7.)

227 Rabe, *ibid.* at 156.

and thus their need for implied powers, the EC Treaty contains Article 308 EC, which avoids a similar rigidity and grants the necessary flexibility to the EC, when it makes use of its competences.[228] Rabe described Article 308 EC as a 'means to canalize implied competences' that itself takes in the implied-powers theory.[229] According to his opinion, Article 308 EC was the 'sweeping clause of the EC Treaty'.[230] At this point Rabe contradicts himself. If Article 308 EC is supposed to be the sweeping clause of the EC Treaty then it cannot include the implied-powers theory at the same time. The sweeping clause was the pre-requisite for the emergence of implied powers, but did not provide for any implied powers itself. But also equating implied powers and Article 308 EC is not correct. From a methodological perspective implied powers are the result of a teleological interpretation of explicit powers, but are not original powers themselves.[231] Article 308 EC on the other hand is an independent competence.[232] Furthermore, implied powers and Article 308 EC differ in respect of their objectives. While the first ones serve as a completion to existing explicit competences, Article 308 EC requires that other powers are not available.[233] Rabe further argues that, if the implied-powers theory should be applicable in EC law, then Article 308 EC should read '... the EC Treaty has not provided the necessary explicit or implicit powers'.[234] According to Rabe such an understanding of Article 308 EC is, however, not possible and to justify his opinion, he refers to Article 211 Rec. 2 EC, where it is said that the Commission is only responsible to undertake a measure, if this is explicitly foreseen in the EC Treaty. Article 211 Rec. 2 EC says:

> To ensure the proper functioning and development of the common market, the Commission shall:
>
> ...
>
> - formulate recommendations or deliver opinions on matters dealt with in this Treaty, if it expressly so provides or if the Commission considers it necessary, ...

Based on this provision Rabe concludes that the EC Treaty only contains explicit powers and cannot provide for further implied powers.[235] This reasoning cannot be followed. Firstly, Article 308 EC contains a vertical division of powers between the Community and the Member States, while Article 211 Rec. 2 EC provides for a horizontal division of powers among the institutions within the Community. Secondly, Article 211 Rec. 2 EC also grants the Commission the power to formulate recommendations or deliver opinions 'if the Commission considers it necessary'. This possibility is in contrast to the alternative presented by Rabe. Thereby no reference

228 *Ibid.* at 152.
229 *Ibid.* at 157.
230 *Ibid.*
231 Nicolaysen, *supra* note 224 at 131; Böhm, *supra* note 96 at 153, 155, 183.
232 Böhm *ibid.* at 189.
233 *Ibid.*; Oppermann, European Law, *supra* note 196 at para 527.
234 Rabe, *supra* note 222 at 155.
235 *Ibid.* at 156.

is made to explicit regulations in the EC Treaty any more. On the contrary, due to this second alternative it could be argued that the EC Treaty provides for other than only explicit powers. Rabe's understanding that the EC Treaty only contains explicit powers is therefore rejected. From the indifference of the wording of Article 308 EC to the powers contained in the EC Treaty it can be further concluded that implied powers are part of those powers Article 308 EC refers to.[236] This means that existing implied powers can exclude the applicability of Article 308 EC.[237] Contrary to Rabe the implied-powers theory and Article 308 EC cannot be regarded as being identical. Article 308 EC is only applicable, if an action of the Community is necessary to attain an objective of the Community and the EC Treaty does not provide for the necessary explicit or implicit powers.[238]

Although the notion that Article 308 EC precludes an application of the implied-powers theory is rejected, it must be still answered, whether implied powers can exist in the EC Treaty. It can be concurred with Rabe insofar, as he describes Article 308 EC as the 'sweeping clause of the EC Treaty'. The existence of this provision indicates that the signatory states knew that the EC Treaty could be too rigid and intended to include an element that would provide for more flexibility. From this perspective it would be in accordance with the intention of the signatory states, if the implied-powers theory, as another means providing flexibility, would be also applicable in EC law.[239] In addition, it shall be further regarded that in many jurisdictions the existence of implied powers is acknowledged, partially even without that a provision comparable to the sweeping clauses exists.[240] For example, in Germany it is undisputed today that the federal state has implied powers, although the Grundgesetz [Basic Law], the constitution of Germany, does not contain any provision that is comparable to the sweeping clause.[241] If Germany's Basic Law is strictly followed, implied powers of the federal state would even have to be rejected

---

236 Dörr, *supra* note 194 at 40; Schwartz, in: Groeben/Schwarze, Article 308, *supra* note 77 at para 67.

237 Dörr, *ibid.*; Schwartz, *ibid.* at para 69; Böhm, *supra* note 96 at 182, 190; Vedder, Treaty-making Power, *supra* note 69 at 150.

238 Nicolaysen, *supra* note 224 at 136.

239 Krück, International Agreements, *supra* note 194 at 34; Heribert Franz Köck, 'Die "implied powers" der Europäischen Gemeinschaften als Anwendungsfall der "implied powers" interntaionaler Organisationen überhaupt' (The 'Implied Powers' of the European Community as an Application of 'Implied Powers' of International Organizations), in: Karl-Heinz Böckstiegel, Hans-Ernst Folz, Jörg Manfred Mössner and Karl Zemanek eds. (1988), *Völkerrecht, Recht der Internationalen Organisation, Weltwirtschaftsrecht – Festschrift für Ignaz Seidl-Hohenveldern (Law of Nations, Law of International Organizations, World Economy Law – Liber Amicorum Honouring Ignaz Seidl-Hohenveldern)*, (Cologne: Heymann Verlag), at 279, at 294.

240 Böhm, *supra* note 96 at 132.

241 Christoph Degenhart (2001), *Staatsrecht I – Staatsorganisationsrecht (Constitutional Law I – State Organization Law)*, 17th edn. (Heidelberg, Germany: C. F. Müller Verlag) at paras 133ff.; Philip Kunig (2003), in: Ingo von Münch and Philip Kunig, eds., *Grundgesetz-Kommentar* (*Basic Law Commentary*), vol. 3, 5th edn. (Munich: Verlag C.H. Beck) Article 70 at paras 22ff.

due to Articles 30, 70, 83 of the Basic Law. These provisions stipulate that generally the member states of Germany shall be competent unless the Basic Law provides for a different regulation. But implied powers are not only common in federal states. They are also acknowledged in international law. The International Court of Justice has granted implied powers to the United Nations already on two occasions.[242] If one takes into account that the European Community, as a supranational organization, is a construct between a federal state and an international organization and considering that implied powers are available in states as well as in international organizations, it is only consequent to also acknowledge implied powers in EC law.[243] Finally, it is stressed that the implied-powers theory is, in fact, only part of the general rules of legal interpretation.[244] By means of a teleological interpretation, further implied powers are derived from an explicit power.[245] Since the means of a teleological interpretation is not limited to explicit powers contained in constitutions of federal state, implied powers are not a phenomenon that can only occur in federal states.[246] Thus, there is no reason to deny the admissibility of the implied-powers theory in EC law.[247]

After the legitimacy of implied powers in EC law was acknowledged, it is necessary to determine their scope. As already mentioned, implied powers can be applied only to complete explicit powers. The question is, whether by referring to implied powers one is limited to the measures foreseen in the corresponding explicit power. This is affirmed by one school of thought, which argues that the limitation of the Community to particular measures is a sort of separation of powers.[248] This means that, if pursuant to an explicit power the Community can adopt only directives, the respective implied power can be also used only to enact directives.[249] As far as this school refers to directives, it can be concurred with. A different result must be achieved though, if an explicit power foresees the adoption of a regulation. Since compared to a regulation, directives or recommendations have a less invasive effect *vis-à-vis* the Member States, there is no persuasive reason, why, in such a situation, the Community should not be also able to enact a directive or make a recommendation. Thereby, it is not decisive, whether these measures are

---

242 Rabe, *supra* note 222 at 147; Vedder, Treaty-making power, *supra* note 69 at 99.

243 Vedder, *ibid.* at 103.

244 Kruse, *supra* note 222 at 170; Martin Bullinger, 'Ungeschriebene Kompetenzen im Bundesstaat' (Implied Powers in Federal States) (1971) *Archiv des öffentlichen Rechts* (*Archives of Public L.*) 237 at 247; Erich Küchenhoff, 'Ausdrückliches, stillschweigendes und ungeschriebenes Recht' (Explicit, Implicit and Unwritten Law) (1957) *Archiv des öffentlichen Rechts* (*Archives of Public L.*) 413 at 425.

245 Fuß, *supra* note 88 at 240; Kruse, *ibid.* at 171; Joachim Becker (1976), *Die Anwendbarkeit der Theorie von den implied powers im Recht der Europäischen Gemeinschaft* (*The Application of the Implied-Powers Theory in EC Law*), (Münster, Germany) at 30f.

246 Becker, *ibid.* at 46.

247 *Ibid.*; Carl Friedrich Ophüls, 'Quellen und Aufbau des Europäischen Wirtschaftsrechts' (Sources and Structure of European Commerical Law) (1963) *Neue Juristische Wochenschrift* (*New Weekly L. Rev.*) 1697 at 1699 [Ophüls].

248 Becker, *ibid.* at 88ff.

249 *Ibid.* at 86ff.

based on an explicit or implied power.[250] Still, difficulties in this respect arise, when it comes to the question, whether the conclusion of international agreements can be based on implied powers, when such a measure is not foreseen in the explicit competence. According to one opinion, it shall not be possible to have recourse to the implied-powers theory when it comes to treaty-making powers, because of the independence of internal and external measures. According to this school treaty-making powers are not variations of internal competences. The conclusion of an international agreement is an independent activity that does not serve to complete explicit internal powers.[251] Another school of thought considers the implied-powers theory as a means to extend explicit powers to other areas that originally do not fall into the scope of these explicit powers. The extension of explicit powers allowing for the adoption of internal measures to external activities is not perceived as such an expansion to other areas, but merely as a procedural widening of an explicit internal power to other measures.[252] According to this opinion, the extension of an internal competence to a treaty-making power is not part of the implied-powers theory, but an extensive application of an internal power. This is considered as legitimate, if the Community needs a corresponding treaty-making power, to make efficient use of an internal power.[253] This second opinion is refused on the ground that, if a treaty-making power was already part of an internal competence, then there would be no need to grant explicit external competences to the EC like it is done in for example Articles 133(3) or 310 EC. The first school is also rejected as an extension of the implied-powers theory to treaty-making powers can be justified through a systematical and teleological interpretation of Article 300 EC. When Article 300 EC was originally included in the EC Treaty only two other provisions, Article 133 and Article 310 EC, provided for a treaty-making power of the Community. Since Article 133 EC contains own procedural rules on the conclusion of international agreements, Article 300 EC would have been limited to treaties concluded pursuant to Article 310 EC. Under these circumstances it would have been more consequent to include a corresponding procedural rule in Article 310 EC as it was done in Article 133 EC, instead of setting up a whole Article for that purpose. Therefore, other treaty-making powers must have existed next to Article 310 EC in the EC Treaty to which Article 300 EC would have been applicable.[254] Another reason for the existence of treaty-making powers based on the implied-powers theory is the fact that the Member States have explicitly granted a legal personality to the EC in Article 281 EC. The main feature of the legal personality of an international organization or, in this case, a supranational organization is its ability to conclude international agreements.[255] If the treaty-making power of the Community was

250 Manfred Zuleeg, 'Die Kompetenzen der Europäischen Gemeinschaften gegenüber den Mitgliedstaaten', (The Powers of the European Communities towards the Member States), (1971) *Jahrbuch des öffentlichen Rechts* (*Yearbook of Public L.*) 1 at 14.

251 Arnold, *supra* note 198 at 430.

252 Vedder, Treaty-making Power, *supra* note 148 at 116.

253 *Ibid.* (Vedder basically adopts the formulation used in Opinion 1/76, at 117f.).

254 Krück, International Agreements, *supra* note 194 at 35.

255 Christian Tomuschat, 'Völkerrechtliche Grundlagen der Drittlandsbeziehungen der EG', (Legal Basis in International Law for the External Relations of the EC), in: Meinhard

limited to the subject matters included in Articles 133, 310 EC, a separate granting of legal personality in Article 281 EC would have been dispensable. The ability to conclude international contracts in the areas covered by Articles 133, 310 EC could be directly derived from these two provisions. The fact that Article 281 EC separately provides for legal personality of the Community is another reason, to acknowledge further treaty-making powers of the EC. Since, when Article 281 EC was included in the EC Treaty, Articles 133, 310 EC were the only provisions providing for a treaty-making power, any further treaty-making powers must have been implied ones.[256] For these reasons it is submitted that the implied-powers theory can be applied to derive implied treaty-making powers from explicit competences set out in the EC Treaty. Having said this, it is this author's opinion that the ERTA judgement was based on the implied-powers theory.

After having concluded that the ERTA judgement is based on the implied-powers theory, it can be scrutinized in the following, whether an extensive application of the ERTA judgement is possible. As was described above, it shall be examined whether the ERTA judgement could be applied in a manner that would allow to grant the EC the treaty-making powers not only for those subject matters that are part of an EC regulation, but to extend these treaty-making powers to such aspects that are closely related with those regulated subject matters although they themselves are not included in any EC act. As far as the implied-powers theory is concerned it could be argued that, since this theory is an exception to the explicit distribution of powers between federal states and their member states, it is necessary, from a methodological perspective, to apply it in a restrictive manner only. Hence, the same would be true for the ERTA judgement and an extension of the ERTA judgement to areas, for which the EC has not adopted any regulations, could be therefore not in accordance with the implied-powers theory. On the other hand, when applying the implied-powers theory in EC law, two aspects must be taken into regard. Firstly, the implied-powers theory was developed with respect to federal states and was thus influenced by this system. One significant difference between federal states and the European Community is that the constitutions of federal states usually provide for a limited set of powers that aims at fixing the *status quo* of the relationship between the federal state and its member states. The EC Treaty, on the other hand, is structured in a very dynamic manner allowing for a lot of flexibility and continuous development. Indications for this dynamic nature can be found in, for example, the preamble of

Hilf and Christian Tomuschat eds. (1991), *EG und Drittstaatsbeziehungen nach 1992 (EC and External Relations after 1992)* (Baden-Baden, Germany: Nomos-Verlagsgesellschaft), at 139, 142f. [Tomuschat, External Relations of the EC].

256 Krück, International Agreements, *supra* note 194 (Krück has a very pro-European attitude in respect of this question and suggests that the Community, irrespective of the internal powers, shall be granted concurrent treaty-making powers for all areas to which the objectives of the EC extend. He justifies his position with the legal personality of the Community, its objectives and a missing contrary intention of the signatory states of the EC Treaty, at 41f. According to my opinion such an extensive approach shall be rejected, because it is held that implied powers can be only derived from primary law. Their scope must be determined based on their wording, objective and structural incorporation in the EC Treaty and cannot be abstractly and generally fixed.).

the EC Treaty, where reference is made to 'an ever closer union among the peoples of Europe' or in the preamble of the EU Treaty that speaks of 'further steps to be taken in order to advance European integration'.[257] Generally, preambles included in international treaties are attributed little legal significance and are considered as soft law. They do not provide for any direct legal obligations.[258] However, preambles can be used to determine the historical background of an international agreement and can give an indication on the intentions of the signatory states. Furthermore, preambles contain the objectives that are pursued with an international treaty. The knowledge of the intentions, the historical background and the objectives can be helpful, when the provisions of an agreement are construed. With respect to the EU- and EC Treaty, where no other material on the legislative history is available, the preambles are therefore of great significance.[259] Additionally, the objectives of an international agreement can be referred to for the creation of customary law and when provisions of the treaty provide for discretionary decisions. Then the objectives set out in the preamble can have a guiding function, how the discretionary powers shall be used and how customary law may be created.[260] If the Community is based on a principle of continuous and advancing integration, it would contravene this principle when the scope of activities of the Community is restricted through a too narrow application of the Community's powers.[261] The fact that the Community operates in a much more flexible and dynamic framework has the consequence that principles applicable within the stricter framework of a federal state cannot be transferred without any amendments to EC law.[262] Secondly, it shall be borne in mind that the problems dealt with here concern treaty-making powers. This may require a particular approach and a differentiated application of the implied-powers theory. Within federal states treaty-making powers are usually regulated in the federal constitution and are assigned to the federal state. In Germany, for example,

257 Ophüls, *supra* note 247 at 1699; Fernand Schockweiler, 'Zur Kontrolle der Zuständigkeitsgrenzen der Gemeinschaft' (The Control of the Limits of the Competences of the Community) (1996) *Europarecht* (*Europ. L. Rev.*) 123 at 126; Albert Bleckmann, 'Teleologie und dynamische Auslegung des Europäischen Gemeinschaftsrechts' (Teleological and Dynamic Interpretation of European Community Law) (1979) *Europarecht* (*Europ. L. Rev.*) 239 at 256; Hans Peter Ipsen (1972), *Europäisches Gemeinschaftsrecht* (*European Community Law*), (Tübingen, Germany: J.C.B. Mohr (Paul Siebeck)) at chapter 5, note 72.

258 Meinhard Hilf and Eckhard Pache, in: Eberhard Grabitz and Meinhard Hilf eds., (2004), *Das Recht der Europäischen Union* (*The Law of the European Union*), looseleaf (Munich: C.H. Beck, January, 8th supplement), Preamble EU Treaty, at para 7 [Hilf/Pache, in: Grabitz/Hilf, Preamble].

259 Manfred Zuleeg, in: Hans von der Groeben and Jürgen Schwarze, eds. (2003), *Vertrag über die Europäische Union und Vertrag zur Gründung der Europäischen Gemeinschaft* (*Treaty on the European Union and Treaty establishing the European Community*), vol. 1, 6th edn. (Baden-Baden, Germany: Nomos Verlagsgesellschaft), Preamble EU Treaty, at para 3 [Zuleeg, in: Groeben/Schwarze, EU Preamble]; Hilf/Pache, *ibid.* at para 9.

260 Hilf/Pache, *ibid.* at para 6.

261 Krück, International Agreements, *supra* note 194 at 34.

262 Helmut Lecheler, 'Die Pflege der auswärtigen Beziehungen in der Europäischen Union' (The Fostering of External Relations in the European Union), (1994) *Archiv des öffentlichen Rechts* (*Archives of Public L.*) 1 at 13.

the member states only have a treaty-making power, if the federal state grants its explicit permission for any external activity pursuant to Article 32(3) Basic Law. The question, whether the federal state can have a treaty-making power based on the implied-powers theory does not even arise. Here is a difference with respect to the European Community. Contrary to a federal state, the Community was granted only few treaty-making powers. Since this has proven as insufficient for the Community to fulfil its obligations, additional implicit treaty-making powers of the EC have been established in the ERTA judgement and in Opinion 1/76. This resulted in a situation that is significantly different from the one existing in federal states: the treaty-making powers are divided between the Community and the Member States. For this reason, it could be also necessary to apply the implied-powers theory with respect to treaty-making powers, and thus the ERTA judgement, in a different, maybe more extensive manner. So far treaty-making powers pursuant to the ERTA judgement are only granted, if this is deemed necessary to make efficient use and to protect explicit powers laid down in the EC Treaty. This understanding of the ERTA judgement limits the scope of the implied-powers theory to the effectuation of explicit powers and to the protection of primary law, which is also in accordance with its original understanding. However, it is proposed to apply the implied-powers theory, and thus the ERTA judgement, more extensively. The main focus shall not be limited to explicit powers of the EC Treaty, but attention shall be also given to the effectuation of treaty-making powers themselves. For this purpose, the scope of a treaty-making power pursuant to the ERTA judgement should not only depend on the content and scope of primary law and on an existing EC regulation, but be also determined by the necessity of an effective use of the respective treaty-making power. The implied-powers theory shall be applied in a manner that the emerging treaty-making powers pursuant to the ERTA judgement can be also used efficiently. The existing situation – that treaty-making powers based on the ERTA judgement are themselves fragmented and incomplete – causes new difficulties and shall be avoided. Therefore, when treaty-making powers pursuant to the ERTA judgement are established, the following approach is suggested. Firstly, the interpretation of an explicit power laid down in the EC Treaty must provide for the possibility to derive an implied treaty-making power there from. Secondly, only if the EC has also adopted additional secondary law based on this competence, the Community will also have a corresponding implied and exclusive treaty-making power for the subject matter dealt with in the enacted secondary law. So far, no differences exist with the principles established in the ERTA judgement. It is now additionally suggested that these implied and exclusive treaty-making powers shall not depend on, and be strictly limited to, the subject matters included in secondary law, but shall also extend to such subjects, whose inclusion in an international agreement is indispensable, even if the Community has not adopted corresponding secondary law with respect to this subject. The only condition that must be fulfilled is that the so extended implied treaty-making power is still covered by the explicit power laid down in the EC Treaty. Insofar it was already previously conceded that Article 80(2) EC provides for a treaty-making power with respect to comprehensive air service agreements that include the granting of air traffic rights.

Concluding, it is therefore submitted, that as a result of an extensive application of the implied-powers theory and thus the ERTA decision, the EC shall also have the external power to conclude air service agreements on the exchange of the third and fourth freedom, when the fifth freedom is also granted in the same agreement. This is necessary, because the power to grant the fifth freedom could not be effectively exercised without a corresponding treaty-making power on the granting of the third and fourth freedom.

Finally it shall be also scrutinized, whether the EC could be competent for the granting of the first two air traffic rights. Although the first and second freedom are covered by the Chicago Convention and the Transit Agreement to a great extent, it must be taken into regard that some countries have not signed, in particular, the Transit Agreement. Therefore, the granting of the first two freedoms in an air service agreement could become necessary. Furthermore, considering that air traffic rights are usually granted on a reciprocal basis, it is also important to notice that the EC is party to neither the Chicago Convention nor the Transit Agreement. Therefore, if the Community is involved in negotiations on the conclusion of an air service agreement, it might be asked by third countries to grant the first and second freedom in an air service agreement.

The Community has not adopted any secondary law that could directly serve as the basis for the emergence of treaty-making powers pursuant to the ERTA judgement for the granting of the first and second freedom. Like in the previous situation, the only possibility for establishing an implicit treaty-making power of the Community with respect to the first and second freedom, would be an extensive application of the ERTA judgement. The first two air traffic rights are only significant for international transit flights that do not include any stop-overs for commercial purposes. These rights are exercised *vis-à-vis* a third country, while air transportation services between two other countries are undertaken. If, on the other hand, an air carrier is granted the third, fourth and fifth freedom by one state, it is implicitly permitted to enter the airspace and to land on an airport within this respective state. Once it has landed, the air carrier must be also given the permission to take care of any technical problems that could exist and that might preclude it from making use of its granted air traffic rights. Insofar the third, fourth and fifth freedom implicitly also include the rights usually granted with the first two freedoms, if a landing for commercial purposes is not undertaken. Hence, unlike in the previous case, the granting of the first and second freedom does not seem indispensable for the use of the fifth freedom. On the other hand, the technical freedoms, and in particular the first freedom, are insofar of great significance, as they need to be obtained to be able to offer efficient international air transportation services. An utilization of the third, fourth and fifth freedom by an air carrier also depends on the granting of the first two freedoms by other states not involved in the third, fourth and fifth freedom traffic. For example, if the Community grants the third, fourth and fifth freedom to for example Australia, the latter once will also need to obtain, at least, the first freedom by the states located between Australia and the EC. Australian air carriers could offer no air transportation services, if those states would reject to grant the necessary transit rights to Australia. It is for this reason that, despite the multilateral granting of the first and second freedom in the Transit Agreement, usually these two air traffic rights are additionally included in bilateral air service agreements. By granting each other the first two air traffic rights, the contracting states have more legal security in case the other party should decide to denounce the Transit Agreement.

If one assesses bilateral air service agreements on a case-by-case basis, one could come to the conclusion that the Community does not require the treaty-making power with respect to the technical freedoms, to effectively grant the fifth freedom. An air service agreement with a state (state A) that provides for an exchange of the third, fourth and fifth freedom can be viable, even if first and second freedom are not included in this agreement. On a bilateral basis the mere granting of the commercial freedoms is sufficient for air transportation services to be installed between the two contracting states. Thus, at first glance a similar link that exists between the third and fourth freedom on one side and the fifth freedom on the other side, does not seem to exist with respect to the first, second and fifth freedom. The first two air traffic rights become only important, if another air service agreement is concluded with another state (state B) that is located behind state A. To be able to fly to state B it would be beneficial for the air carriers of the contracting state, if they had transit rights through the airspace of state A (first freedom) or if they could do stop-overs in state A for technical reasons (second freedom). However, the granting of these two freedoms by state A is not always absolutely necessary for offering air transportation services to state B. Much depends also on the location of a state. A country like Australia does not need many transit rights by other countries, if it wants the establishment air transportation services between its territory and other countries in the South-East Asian area. On the other hand, in a country like Switzerland no air carrier would be able to offer any air transportation services into and out of Switzerland, if the latter one was not granted the first freedom by its neighbouring countries. At this point it is important to take into account that international air transportation is not based on single bilateral air service agreements, but that these agreements form a system. The rights granted in one air service agreement by state A may be indispensable for air transportation services offered to state B pursuant to the rights and conditions included in another and totally independent air service agreement concluded with state B. Therefore, very often air service agreements are also concluded by having regard to the other existing air service agreements the respective countries have concluded with other states. Under these circumstances it could happen that a country negotiating with the European Community may insist on being granted the first and second freedom. For example, Australia and the USA conclude an 'open-skies' agreement. Additionally, Australia concludes another similarly liberal agreement that provides for fifth freedom rights with for example Egypt. To provide fifth freedom traffic between Egypt and the US, an Australian air carrier could very likely have to make a stop-over in the Community for refuelling before continuing its flight to the US. Here the second freedom would be indispensable for the exercise of the rights granted to Australia by Egypt and the US in the other two air service agreements. Hence, although with respect to a single bilateral air service agreement no link may exist between the use of a fifth freedom and the first and second air traffic right, such a link comes into existence, when the entirety of all bilateral air service agreements of a state are taken into regard. For this reason, and by applying the ERTA judgement in an extensive manner as set out above, the Community shall be also granted the treaty-making power with respect to the first and second freedom. However, this competence is linked with the treaty-making power regarding the fifth freedom and therefore can be only exercised with the latter one.

As far as the remaining seventh and eighth freedom are concerned, due to their little significance for international air transportation, a detailed examination with respect to these two air traffic rights is not necessary. Since the exercise of the eight freedom would be limited to the territory of a single Member State anyway, the respective Member State must have the treaty-making power with respect of the granting of this air traffic right. The exercise of the seventh freedom within the EC would mean that an air carrier from a non-EC state could offer intra-Community air transportation services. Hence, the reasons required to grant the European Community the treaty-making power with respect to the fifth freedom are also applicable with regard to the seventh freedom.[263] As for the sixth freedom, since this air traffic right is usually not regulated in bilateral air service agreements, it is also not necessary to further elaborate, who (theoretically) might have the respective treaty-making power.

Distribution of treaty-making powers with respect to air service agreements between the EC and the Member States

Taking into account the 'open-skies' judgement and the conclusions reached above, it will be examined in this chapter, how and by whom air service agreements will be concluded in the future.

The starting point of the following analysis shall be the proposal brought forward by many scholars in the aftermath of the 'open-skies' judgement that Member States and the European Community will need to act jointly and future air service agreements would have to be concluded as mixed agreements.[264] Mixed agreements are usually concluded, if neither the Member States nor the European Community have all the necessary treaty-making powers to conclude an international agreement alone.[265] A mixed agreement is therefore only admissible, if the subject matter of an agreement falls under the responsibility of the Community and the Member States, in other words, when each institution needs the other one to complete its competence deficit.[266] This also means that a mixed agreement is not admissible, if the European Community can adopt an international agreement on its own.[267] The reason for this is that, since an international agreement must be ratified by all parties involved, the Community would make itself dependent on the unanimous acceptance of the agreement by all Member States involved in that treaty as a party. If a treaty can be concluded without the participation of the Member States, the Community's

263 Compare above at 55 with respect to the fifth freedom.

264 See above at 42ff.

265 Schweitzer/Hummer, *supra* note 102 at para 670f.; Streinz, European Law, *supra* note 73 at para 428.

266 Oppermann, European Law, *supra* note 196 at para 1, 712; Geiger, *supra* note 102 at 977; Arnold, *supra* note 198 at 420; Stein, *Mixed Agreement*, *supra* note 203 at 59f., 158.

267 Stein, *ibid.* at 150.

dependence is unnecessary.[268] Although the involvement of the Member States in an international agreement can be beneficial for the acceptance of such a treaty, it can also bring disadvantages. The Member States could use their position to request a concession from the Community in return for their approval to the agreement that eventually does not need to be linked to the subject matter of the agreements.[269] It is even possible that powerful non-EC countries exert pressure on smaller and less powerful Member States on a bilateral basis to force them to adopt a certain position in the negotiations. This would enable non-Member States to indirectly influence the proceedings within the EC.[270] In this respect Hahn even reports that, in the past, Member States and third countries have formed alliances against the Commission and other Member States.[271] Although, when it referring to mixed agreements, in its Ruling 1/78 the ECJ imposed a duty on the Member States and the Community to cooperate and work together, in reality the Member States may be driven by their own national interests.[272] Such a dividing approach by a third country *vis-à-vis* the European Community could be witnessed recently and with respect to air transportation. In a statement, Deputy Assistant Secretary Byerly, when commenting on the EC Council's refusal of the proposed US–EC air service agreement, said that 'the Transport Council's rejection of that agreement was indisputably a defeat for Ireland' and that he considered it odd 'that those countries, like Ireland and Spain, that did not negotiate bilateral Open Skies agreements with the United States in the past should today be held hostage by those in Europe who now adopt an "everything or nothing" approach. This form of apartheid doesn't seem fair to me.'[273] If third countries can succeed in driving a wedge into the EC on a specific subject matter, one can easily imagine that mixed agreements can be abused by the Member States to put through own national interests. For these reasons some authors describe mixed agreements as 'an evil that needs to be eliminated.'[274] Sometimes the Member States

268 Karl-Matthias Meessen, 'Das Abkommen von Lomé als gemischter Vertrag', (The Treaty of Lomé as a Mixed Agreement) (1980) *Europarecht* (*Europ. L. Rev.*) 36 at 40 [Meessen]; Geiger, *supra* note 102 at 978; Stein, *ibid.* at 171f.; Vedder, in: Grabitz/Hilf, Article 133, *supra* note 61 (pursuant to Vedder mixed agreements lead to a 'renaissance of unanimity' since the conclusion of these agreements requires the approval of every single Member State involved, at para 55).

269 Hahn, in: Calliess/Ruffert, Article 133, *supra* note 65 (Hahn describes this development as the 'balkanisation of exclusive treaty-making powers of the Community', at para 120); Byerly, Open-Skies Talks with EU, *supra* note 161.

270 Horst Günter Krenzler. and Hermann da Fonseca-Wollheim, 'Die Reichweite der gemeinsamen Handelspolitik nach dem Vertrag von Amsterdam – eine Debatte ohne Ende?' (The Scope of the Common Commercial Policy after the Treaty of Amsterdam – an Endless Debate?) (1998) *Europarecht* (*Europ. L. Rev.*) 223 at 229 [Krenzler and Fonseca-Wollheim].

271 Hahn, in: Calliess/Ruffert, Article 133, *supra* note 65 at para 16, see even more explicitly Hahn in the 1st edition of this Commentary, Article 133 at para 17.

272 ECJ, *Ruling 1/78*, [1978] E.C.R. 2151, 2181 [Ruling 1/78]; Opinion 1/94, *supra* note 64 at I-5420.

273 Byerly, Negotiations, *supra* note 172.

274 For example, Tomuschat, External Relations of the EC, *supra* note 255 at 146; *contra* Werbke, *supra* note 99 (although in respect of the distribution of powers Werbke

also argue that an international agreement, which the Commission intends to conclude for the European Community, includes a subject matter that does not fall into the responsibility of the EC. It also happens that clauses are introduced in an international agreement, whose subject matter falls into the responsibility of the Member States. The intention behind this idea is to deny a sole treaty-making power of the Community and to force a participation of the Member States leading to the conclusion of mixed agreements.[275] This again allows the Member States to influence the agreement, and to maintain and strengthen their identity and position on the international stage.[276] But even if the Member States are right and an envisaged agreement indeed extends to an area not covered by the treaty-making power of the Community, this should not automatically lead to a participation of the Member States.[277] The ECJ has already decided that auxiliary provisions are not relevant for the classification of an international agreement and that its scope shall be determined by the main subject matters regulated in that agreement.[278] Considering that such auxiliary clauses might be also left out of an agreement and are only included to facilitate the implementation of the treaty, it is stressed and concurred by the ECJ that the EC should be able to conclude international agreements all alone and without any participation of the Member States; even if the agreement contains clauses that normally would fall into the responsibility of the Member States, if the clauses are only of a subordinate nature.[279] An exception to this general rule should be only made if the contracting non-EC country demands the involvement of the Member States despite the sole treaty-making power of the Community. The reasons for such a request could be that either the European Community is not accepted as a subject of international law or that the Member States shall be obliged next to the EC to ensure the implementation of the treaty.[280] Although the European Community is accepted by a multitude of nations as a subject of international law, doubts may still exist in respect of the Community's ability as far as the effective implementation of the respective agreement is concerned. Here, the Member States' participation serves only as a backup to safeguard the third country's rights. Although pursuant to Article 307(7) EC, the Member States are obliged to respect the treaties concluded

---

describes mixed agreements quite harshly as 'bastards,' he still favours them over the dubious ERTA doctrine, at 2107).

275 Krenzler and Fonseca-Wollheim, *supra* note 270 at 229; Hahn, in: Calliess/Ruffert, Article 133, *supra* note 65 at paras 16, 119.

276 Tomuschat, External Relations of the EC, *supra* note 255 at 146.

277 Peter W. Gilsdorf, 'Die Außenkompetenzen der EG im Wandel – Eine kritische Auseinandersetzung mit Praxis und Rechtsprechung', (The External Powers of the EC Changing – a Critical Discussion with Practice and the Jurisprudence) (1996) *Europarecht* (*Europ. L. Rev.*) 145 at 146 [Gilsdorf].

278 Stein, Mixed Agreements, *supra* note 203 (Stein also proposes that the decision who is responsible for a certain area shall not be based on the single clauses, but on the major regulation contained in an agreement, at 151); Opinion 1/78, *supra* note 62 at 2917f.

279 Stein, *ibid.* at 151.

280 Albert Bleckmann, 'Der gemischte Vertrag im Europarecht', (The Mixed Agreement in European Law) (1976) *Europarecht* (*Europ. L. Rev.*) 301 at 302; Arnold, *supra* note 198 at 424.

by the Community anyway, and can be obliged by the EC to implement them under their national laws, the third countries cannot refer themselves to Article 307(7) EC to ensure an effective implementation of the agreements they conclude with the EC as they are not party to the EC Treaty. However, generally there is no reason to grant the Member States any influence in external matters, if an area falls entirely or mostly into the responsibility of the Community. In addition, the Community can negotiate more flexibly and may be able to achieve better results, if it does not need to have regard to the special interests of every single Member State. Furthermore, a sole representation of the Community in the conclusion of international treaties has the advantage that the identity of the European Union on the international stage is strengthened, which is also one of the objectives set out in Article 2 Rec. 2 EU. Although, technically speaking, the conclusion of an air service agreement would be undertaken by the European Community and not by the European Union, the first one is pursuant to Article 1 EU part of the latter one and hence its activities have a direct impact on the European Union, and on its perception among the general public and on the international stage. Hence, an unnecessary participation of the Member States in the negotiations and conclusions of international agreements will weaken the Community. Therefore, a sole activity of the EC is always preferable over a joint action of the Member States and the Community.[281] While the inclusion of an auxiliary clause in an international agreement cannot render it into a mixed agreement, the same is not true vice versa. An international agreement that falls into the responsibility of the Member States, which contains a clause that falls into the responsibility of the Community, even if it is only a subordinated clause, will always require a participation of the EC and the conclusion of a mixed agreement. Due to the fact that the ERTA judgement is part of customary EC law,[282] a different approach is necessary. Therefore, whenever a subject matter is regulated in an international agreement that was already the subject of secondary law, the Community has an exclusive treaty-making power and must be involved as a party in such an international agreement, notwithstanding that the respective clause may have only a subordinate significance.

Another difficulty of mixed agreements exists in respect of their legal nature and when it comes to their implementation. One possibility is to divide mixed agreements into a Community and a national part. This solution leads to a variety of problems. In particular, the national part would not be subject to the jurisdiction of the ECJ. When it comes to the implementation of the treaty it could therefore result in uncertainty, fragmentation of the law and disputes with the contracting third country.[283] For these reasons another school argues that mixed agreements must be considered in their entirety as EC law, to which the principles of EC law, which include, for example, direct applicability, primacy of Community law, and so on are applicable.[284] As far as the implementation of mixed agreements is concerned, this shall be accomplished

281 Opinion 1/94, *supra* note 64 at 5420.

282 See above at 54.

283 Gilsdorf, *supra* note 277 at 146f.

284 Bleckmann, Mixed Agreements, *supra* note 280 at 310; Bleckmann, European Law, *supra* note 75 at para 1402.

either by the Member States or by the European Community. Thereby and as far as the outcome is concerned, it shall make no difference, which institution finally is responsible.[285] Mixed agreements shall be applied uniformly by the Member States and the Community.[286] The ECJ remarked in this respect that the conclusion and implementation of mixed agreements should be undertaken in close cooperation.[287] The second solution has the advantage that it provides for legal clarity and a coherent and uniform interpretation of the respective international agreement by only one court in the European Community, namely the European Court of Justice. Furthermore, it shall be borne in mind that the Community and the Member States act jointly as one party *vis-à-vis* the contracting non-EC state. Although the EC and the Member States each act only within their own responsibility, they also support and complement each other. While the contracting third country is pursuing its interests, the Member States and the Community pursue a common interest. Thus, Member States and EC do not compete against each other, but cooperate and work together. Although consisting of two institutions, they constitute one uniform party to the contract. It would be contradictory, if the Community and the Member States are acknowledged as one uniform party during the negotiations and the conclusion of an international agreement, but go separate ways when it comes to the subsequent implementation of the agreement. Furthermore, doubts can be raised, whether a partition of an international agreement into a national and a community part, as it is suggested by the first school of thought, is always possible. The various provisions of a treaty can be so inter-related that a split of the responsibilities for the different provisions may not be feasible. As an example one may consider the General Agreement on Trade in Services. This treaty contains four kinds of services. In Opinion 1/94, the ECJ concluded that the Community has the treaty-making power with respect to one of these four services, while the remaining three remained in the competence of Member States. If, on the implementation level, one would split this treaty into a national and a community part, it could result in different interpretations of the agreement, depending on which service is concerned and whether it falls into the responsibility of the Community or the Member States.]It is therefore submitted, that with respect to the legal nature and implementation of mixed agreement, they shall be classified, in their entirety, as EC law.

The conclusion of air service agreements as mixed agreements would be legitimate, if neither the Member States nor the Community could contract on this subject matter alone. It was already concluded that the Community has the treaty-making power with respect to fares that are applicable on intra-Community flights, regulations regarding computerized reservation systems, provisions on the allocation of slots and the granting of the first five freedoms if the fifth freedom is included in an air service agreement. In the following, the major provisions usually contained in air service agreements and their significance shall be examined. Having regard to the proposal that auxiliary clauses cannot transform an agreement into a mixed agreement, irrespective of the fact that the Community may not have the

285 Bleckmann, *European Law*, *ibid.* at para 1405.

286 Gilsdorf, *supra* note 277 at 157.

287 Opinion 1/94, *supra* note 64 at 5420; Ruling 1/78, *supra* note 272 at 2181.

power for the subject matters contained in such clauses, it will be assessed, whether a clause has only a subordinate significance or whether it is a core element of air service agreements. Based on this evaluation it can be then scrutinized, in whose responsibility the main regulations contained in air service agreements fall and thus, whether these agreements need to be concluded as mixed agreements or whether the Community can maybe act alone.

EXCHANGE OF AIR TRAFFIC RIGHTS

The first aspect regulated in air service agreements, is the granting of air traffic rights between the contracting parties. The Chicago Convention and the Transit Agreement usually limit this to the third, fourth and fifth freedom. If air service agreements are concluded with countries that have not signed the Chicago Convention and/or the Transit Agreement, it is also necessary to negotiate the first two freedoms.[288] The bargaining power of the states involved depends very much on their size and location. It could have been witnessed in the past that countries located at the geographical outskirts, for example Japan or Australia, have a stronger position that they can use in the negotiations compared to smaller and centrally located countries when it comes to the granting of the fifth freedom.[289] If a state that is surrounded by several countries refuses to grant the fifth freedom, the other state can simply contact one of the neighbouring countries to obtain this freedom. The refusal of the fifth freedom by a state located on the geographical outskirts, on the other hand, cannot be compensated that easily by contacting a neighbouring country.

DETERMINATION OF ROUTES

Another very important issue dealt with in bilateral air service agreements is the determination of the destinations, where the granted air traffic rights can be exercised. Thereby it can happen that the routes are exactly pre-determined by fixing the airport of departure in the registration state of an airline, the airports for stop-overs and the destination airports in the host state.[290] However, route plans are usually not directly included in air service agreements any more, but are negotiated via diplomatic correspondence between the two contracting states and annexed to the air service agreements.[291] In the very liberal 'open-skies' agreements, routes are not fixed at all and the airlines are completely allowed to decide where they want to fly.

---

288 See above at 7.

289 Schwenk/Giemulla, *supra* note 13 at 641; Günther Becher, 'Luftverkehrspolitische Entwicklungstendenzen in Asien im Spannungsfeld zwischen Liberalismus und Dirigismus' (Air Traffic Policy Development Tendencies in Asia between Liberalization and State Control) (1979) *Zeitschrift für Luft – und Weltraumrecht* (*Air and Space L. Rev.*) 317 at 323 [Becher].

290 Giemulla, in: Giemulla/Schmid, *supra* note 4, § 21 at para 9.

291 Schwenk/Giemulla, *supra* note 13 at 632.

DESIGNATION OF AIR CARRIERS

Air service agreements regularly contain two clauses that regulate what kind of air carriers (nationality clause) and how many air carriers (designation clause) can make use of the granted air traffic rights. The nationality clause prescribes in a qualitative manner that only such air carriers can make use of the air traffic rights that are controlled and majority owned by the state that is granted these rights, by its citizens or by both combined. The designation clause determines in a quantitative manner, how many air carriers can be designated under an air service agreement at maximum. Three variations are possible: a single designation clause, a multiple designation clause with a maximum number of possible designations and an unlimited multiple designation clause.[292]

CAPACITIES AND FREQUENCIES

Most agreements further contain regulations concerning the available capacities that can be used by the designated airlines on the routes agreed upon. This can be accomplished by determining the size and type of aircrafts to be used in terms of available seats and cargo space. Thereby some typical clauses exist that are included in air service agreements depending on the interests and objectives of the contracting parties. Pursuant to the Bermuda-clause[293] the capacities are based on the anticipated demand for the respective air transportation services. The exact fixing of the capacities is left to the airlines, which can decide, based on their assessment of the demand, how much capacity they want to offer on a route. This approach grants a great deal of flexibility to the airlines. However, a restriction results from the possibility of each of the contracting states, to ask for consultations with the other country on a change of the existing regulation and to demand an amendment to the capacities offered, if one of the states is unsatisfied with the share its national airline has in the total amount of passengers or cargo transported between the two countries. The flexibility the Bermuda-clause offers to the airlines on the one hand, is set off by the possibility of an *ex post* control and restrictions by the contracting states. An even more restrictive and less flexible way to regulate capacities is the pre-determination clause,[294] which fixes the available capacities in advance. This approach is often chosen, if the airlines of the contracting states differ very much in their size and economic power. It allows for the state with the weaker airline, to secure a certain share in the total traffic available on routes between the two states for its air carrier. The advantage of this solution is equal treatment of the airlines involved and the avoidance of excess capacities. The disadvantage is that any expansion of the airlines is precluded, innovation hampered and new routes are not developed. The opposite of this approach is the free-determination clause,[295] which does not foresee

292 *Ibid*. at 638.
293 *Ibid*. at 643f.
294 *Ibid*. at 644.
295 *Ibid*. at 644f.

any limitations on capacities. The states involved leave it to the airlines to decide how often, when and with what aircrafts they want to offer their air transportation services. Hence, full confidence is given to the self-regulation of the market. The states refrain from imposing any limitations on the air carriers, apart from ensuring fair and undistorted competition. This solution requires similarity of the designated air carriers in terms of size and economic power and an equally high amount of traffic in both states. This approach is mostly used in air service agreements between the Member States and North-American countries. The markets between these two regions provide for a high demand for air transportation services and the operating airlines are, from an economic and competitive perspective, quite equal.[296]

When it comes to frequencies, air service agreements can be again either liberal or restrictive. While liberal agreements leave it to the airlines to decide, when they want to offer their services, restrictive agreements determine the exact day or even time of a day when a service shall be provided.[297]

TARIFFS

To achieve an equal treatment of their air carriers and to enable them to operate at a level playing field, states sometimes also fix the tariffs the air carriers can apply. The advantage is that the airlines are guaranteed an income at a stable level. The downside is the preclusion of any possible competition from the start. The idea behind this approach is that only those airlines are considered as able to offer safe and reliable air transportation services that have sufficient income and that found themselves in sound economic conditions.[298] Competition, on the other hand, is considered as leading to lower fares and thus less income. Air carriers will be able to compensate this only by reducing their costs. There is a fear that a reduction of the air carriers' expenditures may occur at the expense of safety. Apart from this safety-related aspect, states have also an interest in protecting their airlines from being pushed out of a market due to competition.[299] This is in particular true, when states themselves own airlines and would incur direct losses in case their airlines have to terminate services and eventually go bankrupt. But even if airlines are not state owned and are run by private investors, states have an interest in protecting their national air carriers for political and economical reasons.[300] Firstly, the air transport industry is of outmost

296 *Ibid.* at 645.

297 See overview in Schwenk/Giemulla, *ibid.* at 643ff.

298 *Ibid.* at 649.

299 *Ibid.* at 650; see also above at 6.

300 See EC, Council Resolution on problems caused by congestion and crisis situations in air traffic in Europe of 17 November 1995 [1995] O.J. C 317/1 (second recital: 'Considering that air transport has a key role to play in economic development and strengthening social cohesion, providing swift and efficient links between regions and in particular in making peripheral and island regions accessible.'); Prodromos Dagtoglou, 'Der Luftverkehr und die Europäische Gemeinschaft – Eine Bestandsaufnahme' (Air Transportation and the European Community) in Roland Bieber and Dietmar Nickel, eds. (1981), *Gedächnisschrift für Christoph Sasse – Das Europa der zweiten Generation* (*Liber Amicorum Christoph Sasse – The Europe*

importance for the development of a state's economy.[301] It allows for the export and import of goods, or brings tourists into a country. According to the World Bank 5 per cent of the World Gross Domestic Product can be attributed to aviation's impact on the economy (equals $1.4 trillion).[302] Secondly, air transportation is also part of a state's infrastructure. The biggest and most outstanding advantage it offers in this respect is the possibility to cover wide distances within a short period and enables the citizens of a state to participate in the global economy.[303] In addition to the infrastructural significance and its assisting function for a state's economy, air transportation is an important industry itself. It offers many jobs, and serves as a consumer for services and goods from other industries. As far as the European Community is concerned, the Commission reported that air carriers account directly for about 1.5 per cent of the European Gross Domestic Product. The sector employs about three million people, while the 30 major air carriers alone have 350,000 employees.[304] Another very important function of the air transportation sector is its contribution to the security of a nation. States usually conclude contracts with commercial airlines that allow the states to make use of the transportation capacities of the airlines in times of emergency and, in particular, during warfare. This transportation capacity is called 'civil reserve fleet' or 'strategic reserve'.[305] For example, in 2003, during the second Iraq War, the US made use of its civil reserve fleet to transport military personnel and military equipment to the Middle East.[306] In 2005, many airlines offered their aircrafts to provide emergency airlifts or to transport goods to the areas affected by the various natural disasters, which occurred that year (for example, the tsunami in Asia and hurricane Katrina in New Orleans, USA). Thereby it is remarkable that not only the national airlines of the affected countries were engaged in these emergency activities, but also air carriers from other states volunteered. With respect to security, aviation can be attributed another indirect feature since 11 September 2001. It is the possibility to abuse aircrafts as weapons. Of course, this possibility must not be used by a civilized nation. Still, it is a possible aspect of how aircrafts can be misused that also influences national air legislation and fosters control over airlines and aircrafts by the states. Another characteristic of airlines is its significance in the social and political sphere. Since the advent of international aviation, airlines are perceived as representatives of their countries, as a kind of ambassadors that

---

of the Second Generation) vol. 1 (Baden-Baden, Germany: Nomos Verlagsgesellschaft) 539 at 549 [Dagtoglou].

301 Schwenk/Giemulla, *supra* note 13 at 623, 627.

302 Mineta, Foreign Ownership, *supra* note 171; see also Shane, Liberalization, *supra* note 174 (e.g. a single daily Boeing 777 roundtrip between the US and China will produce a total annual benefit of $158 million to the US economy).

303 Dagtoglou, *supra* note 300 at 548; Schwenk/Giemulla, *supra* note 13 at 623; Mineta, US/EU Talks, *supra* note 166.

304 Commission, Staff Working Document, *supra* note 138 at 2.

305 Paas Hans-Werner, 'Rechtsgrundlage des Charterluftverkehrs' (Legal Basis for non-Scheduled Air Transportation), (1972) *Zeitschrift für Luft- und Weltraumrecht* (*Air and Space L. Rev.*) 55 at 63 [Paas].

306 See *Flug Revue* (*Flight Magazine*) (March 2003), at 27; Byerly, Comprehensive Accord, *supra* note 165.

carry the flag of their registration states around the world.[307] Through the worldwide operating aircrafts states can be present everywhere.[308] That is also why the big and internationally operating air carriers are referred to as 'flag carriers'.[309] Furthermore, the establishment of permanent air transportation services between two countries and in particular between their capitals is of a great political significance. The existence of air transportation services is considered as proof that the relations between the two respective countries are intact and that both nations are bound together through an 'air line'. Apart from the political function, national air carriers also have a social significance. Airlines are attributed the ability to strengthen national pride and national identity.[310] In particular when citizens are abroad and away from their home countries, the sight of a familiar entity like an air carrier can enhance their patriotic feelings.[311] Furthermore, the pride and self-confidence of a nation and its citizens can be improved through the existence of a national airline, because of the high degree of technological sophistication that goes along with it.[312] Having an own national air carrier shows that a country has the ability to run a very complicated and technologically demanding sort of business. The progress of airlines is often perceived as a measure for the success and progress of the whole nation.[313] Finally, the last feature of aviation is its ability to bring people from around the world together. It therefore strengthens cultural exchange, tolerance and understanding. Aviation can help to overcome prejudices and thus foster a peaceful living together. For all these particularities of the air transportation industry previously mentioned, many states consider the existence of a national air carrier as very important for the existence of the state itself and want to protect the industry from competition and the negative effects it may bring.[314] Seldom do the states allow for competition to take place in this industry. An exception hereto has been the development in the US air transportation sector, where competition was introduced very early. Since the 1990s full competition with respect to air transportation services also exists in the European Community.[315]

If a restrictive approach is chosen and the applicable tariffs are regulated, they are not directly determined by the respective states, but the airlines are at first free

---

307 'Lufthansa übernimmt Swiss' (Lufthansa Takes over Swiss), *Handelsblatt* (*Trade Gazette*) (23 March 2005) at 1.

308 Schwenk/Giemulla, *supra* note 13 at 609; Werner Guldimann, 'Luftverkehrspolitik' (Air Traffic Policy) (1993) *Zeitschrift für Luft – und Weltraumrecht* (*Air and Space L. Rev.*) 369 at 371.

309 'Eine Frage der Ehre' (A Question of Honour), *Handelsblatt* (*Trade Gazette*) (11 March 2005) at 10 (with respect to the history of Swiss and its predecessor Swissair it is mentioned that 'the Swiss love their airline because it made their little country big in the world') [Question of honour].

310 Dagtoglou, *supra* note 300 at 549.

311 Question of honour, *supra* note 309.

312 *Ibid.*; 'Lufthansa übernimmt Swiss' (Lufthansa Takes Over Swiss), *Financial Times Deutschland* (*Financial Times Germany*) (23 March 2005) at 1.

313 Dagtoglou, *supra* note 300 at 549.

314 Becher, *supra* note 289 at 318.

315 Schwenk/Giemulla, *supra* note 13 at 637.

to set the fares they want to apply.[316] However, in a second step the selected tariffs must be approved by the states, which, for example under an air service agreement pursuant to Bermuda I standards, will have regard to that:

i) the expenditures of the airlines are covered
ii) a sufficient profit is achieved
iii) the airlines have taken account of the competitive situation in a market and the available traffic, and
iv) the interests of the consumers are observed.[317]

In very restrictive air service agreements, the tariffs must be approved by both states concerned (single-disapproval-rule).[318] When the US started to liberalize their air transportation market, they forbade their national airlines to participate in any consultation procedures on tariffs. Since this approach was rejected by most other countries as they wanted to retain control on the applicable fares, a compromise was found, originally only applied to air service agreements concluded between the US and third countries, according to which states were only entitled to have control over those tariffs that were applicable for flights out of their territory (country-of-origin-rule).[319] If a more liberal approach is chosen, tariffs are only notified by the air carriers with the authorities of the two states and become automatically valid, if they are not objected by both authorities (double-disapproval-rule). An objection can be only raised, if an airline has a dominating position in the market that it abuses, if the tariffs are discriminating or if they are part of predatory behaviour.[320]

USE OF A STATE'S AIRSPACE, NECESSARY DOCUMENTS AND COMPLIANCE WITH IMMIGRATION PROCEDURES

Usually a provision included in air service agreements prescribes that national regulations on the entrance into, flight through and *exit* from the contracting states' airspace must be followed by the airlines. In another clause it is determined, which documents must be always carried along by the air carriers and to which extent the states accept documents issued by the other party. Finally, it is also stipulated that national rules of the contracting states in respect of immigration, customs and quarantine must be followed.[321]

---

316 *Ibid.* at 648.

317 *Ibid.*

318 *Ibid.* (Schwenk/Giemulla call it 'double approval rule', at 654).

319 *Ibid.*

320 *Ibid.* at 655.

321 See e.g., Air Transport Agreement Between Germany and the USA of 7 July 1955 (as amended on 25 April 1989, on 23 May 1996 and on 10 October 2000), online: <http://luftrecht-online.de/regelwerke/d-us-e.htm> (date accessed: 2 November 2005), Articles 5 and 6 [German-US air service agreement].

FEES AND CUSTOMS

A provision regularly prescribes that fees, which are imposed for the use of airports and other facilities as well as air navigation services, must be non-discriminatory. Regularly it is also agreed upon that fuel and other fluids, spare parts and goods that are intended for consumption on board of the aircraft are exempted from any customs.[322]

TRANSFER OF INCOME AND THE ESTABLISHMENT OF FACILITIES

Usually a clause is included that permits the free transfer of income realized in the host state to the registration state of the airline. Another provision grants the right to the airlines to establish and maintain facilities and to employ a staff in the host state to facilitate the provision of air transportation services.[323]

SAFETY AND SECURITY

In the newer air service agreements, regulations are included that contain the minimum requirements on safety and security on airports and in the air transportation sector that shall be met by the contracting states. Thereby reference is made to the regulations adopted by ICAO. It is also laid down which consequences arise, if one country does not fulfil these minimum requirements.[324]

FINAL PROVISIONS

A final clause provides for a regulation on consultations and arbitration/ mediation.[325]

* * *

After giving this overview on the content of air service agreements, it will be discussed in the following, which of the provisions contained in air service agreements have only an auxiliary function and which form the core element of such an agreement. For this purpose one can have recourse to the legal source of all air service agreements. As mentioned above, Article 6 Chicago Convention is considered as the basis for the existence of bilateral air service agreements. This provision stipulates:

> No scheduled international air service may be operated over or into the territory of a contracting State, except with the special permission or other authorization of that State, and in accordance with the terms of such permission or authorization.

---

322 See e.g., *ibid.*, Articles 7 and 7bis.
323 See e.g., *ibid.*, Article 9.
324 See e.g., *ibid.*, Article 11.
325 See e.g., *ibid.*, Article 13.

Pursuant to this regulation, states can prescribe the conditions under which international scheduled air transportation to and from their territories is conducted. From this point of view, all provisions referring to air transportation contained in an air service agreement would have to be considered as the core of such a treaty, that is, provisions regarding capacities, frequencies, and tariffs, designation, entrance to and exit from a state's airspace, granting of air traffic rights, the documents to be carried along and the determination of routes. Auxiliary provisions would be the ones referring to the establishment of facilities and employment of staff, the equal treatment in respect of fees, the transfer of income and the exemption of fluids, spare parts and goods from customs. These latter provisions are only indirectly related to air transportation and rather regulate the remote economic aspects of the air transportation services offered. In addition to Article 6 Chicago Convention, one shall also have regard to the principle set out in Article 1 Chicago Convention. Article 1 stipulates:

> The contracting States recognize that every State has complete and exclusive sovereignty over that airspace above its territory.

As indicated above this provision renders the state's sovereignty over its airspace as the dominating and most important principle of public air law.[326] All further provisions and regulations refer to or are based on this principle. sovereignty means, firstly, that a state has authority over its citizens and all persons that find themselves on its territory (internal sovereignty). Secondly, sovereignty means that a state is not subject to any other higher authority, but that it subjects itself only to the consensually agreed rules of international law.[327] Although Article 6 Chicago Convention is widely considered as the legal basis for the conclusion of bilateral air service agreements, the real basis for these treaties is Article 1 Chicago Convention. Article 6 Chicago Convention has a declaratory function and only confirms, what can be already derived from all other provisions of the Chicago Convention and, in particular, *e contrario* from Article 1 in connection with Article 5 Chicago Convention; namely that the use of a state's airspace by air carriers engaged in international scheduled air transportation requires a permission of the respective state. The possibility of a state to limit its own sovereignty is not due to Article 6 Chicago Convention, but a general principle of international law.[328] When concluding an air service agreement, the contracting countries agree to the use of their airspace and thus to a limitation of their sovereignty pursuant to the conditions set out in the agreement. The permission to use the territory of a state is not limited to air transportation, but is quite common in international law. It is usually granted in international agreements and commonly referred to as *servitut*.[329] The means by which countries allow for the use of their airspace and limit their sovereignty in public international air law is the granting of air traffic rights. While all other provisions contained in bilateral air service

---

326 See above at 2.

327 Verdross/Simma, *supra* note 2 at 29f.; Ipsen, Intl. Law, *supra* note 28 at 310, para 67.

328 Verdross/Simma, *ibid.* at 638; Ipsen, *ibid.* at 913, para 36.

329 Ipsen, *ibid.* at 314f., paras 77ff.

agreements further regulate the carrying out of international air transportation services, it is the granting of air traffic rights that enables these services at all. The exchange of air traffic rights is indispensable for international air transportation. The sole granting of air traffic rights is sufficient for the establishment of international air transportation services. The other clauses of air service agreements, for example regarding capacities, frequencies, fares, and so on, are insignificant without the granting of air traffic rights and serve only as a limitation and further substantiation of the granted air traffic rights. They do not provide for any independent rights and thus are only a subordinate nature. The core element of every air service agreement is the granting of air traffic rights.[330]

With respect to the conclusion of air service agreements as mixed agreements this means that the institution that has the treaty-making power for the granting of air traffic rights, is responsible for the conclusion of air service agreements as a whole. It was concluded above that the European Community shall be responsible for granting air traffic rights to third countries if the respective air service agreement includes the fifth freedom. In such a case the Community has the treaty-making power to conclude an air service agreement with the third country. Only if a third country does not request the fifth freedom, the Member States shall be still able to conclude an air service agreement alone. In such a case, the Community would not have to participate in the air service agreement, unless the agreement also contains provisions on subject matters that fall into the responsibility of the EC, like for example provisions on computerized reservation systems. Although such clauses may be only of a subordinated importance, they would still trigger a necessary participation of the EC as a result of the ERTA judgement and lead to the conclusion of a mixed agreement.[331] A mixed agreement would be also necessary, if a third country requested fifth freedom traffic rights and eight freedom traffic rights with respect to a specific Member State.

CONCLUSION

Unlike in the 'open-skies' judgement, the previous analysis has shown that, with respect to the conclusion of air service agreements, the European Community has far reaching treaty-making powers. Although no direct treaty-making powers for these agreements exist in primary law, such competences can be derived from the application of the ERTA judgement. Thereby it is this author's opinion that the ERTA judgement shall be applied in a more extensive manner as it was done in the past. The Community shall not only be granted treaty-making powers exclusively for such subject matters that it has included in a regulation before, but the treaty-making powers shall have such an extend that they enable the Community to make

330 See also Byerly, Open-Skies Talks with EU, *supra* note 161; Assistant Deputy Secretary Jeffrey Shane US Department of Transportation, 'US Official comments on EU "open skies" ruling' of 8 November 2002, online: US Mission to the EU <http://www.useu.be/Categories/Transportation/Nov0602USEUOpenSkies.html> (date accessed: 2 June 2005) [Shane, US Comments on ECJ Ruling].

331 See above at 82.

efficient use these competences. Having also regard to the reasoning of the European Court of Justice in Opinion 1/94 and the 'open-skies' judgement, a more extensive understanding and application of the ERTA judgement seems necessary to remedy the discontinuation of the ECJ's theory of implied treaty-making powers pursuant to Opinion 1/76 that the latter one seems to have given up. Unlike with respect to most scholars, it is this author's opinion that future air service agreements do not need to be concluded as mixed agreements. Whether an international agreement can be concluded by the Community alone or whether it requires a coordinated and joint activity of the EC and the Member States depends on the content on the agreement. Thereby the treaty-making power for the whole agreement depends on, which institution is responsible for the core elements included in an international agreement. In case of air service agreements, the core element is the granting of air traffic rights. In this respect it was concluded that the Community is responsible for the first five freedoms provided that the fifth freedom is included in the air service agreement. The Member States can only conclude air service agreements, if they do not grant the fifth freedom or if the eight freedom (cabotage) with respect to their territory is included in an air service agreement. If, in addition to the eight freedom the fifth freedom shall be also part of the agreement, a mixed agreement needs to be concluded in which the Member States and the Community shall both participate. A mixed agreement would be also necessary, if, notwithstanding that the fifth freedom might not be included, such subject matters would be part of the air service agreements that are regulated in secondary law, even if these clauses may be only of a subordinated nature. Due to the ERTA judgement the Community has the treaty-making power for all subject matters that are dealt with in an EC regulation, irrespective of the significance this subject matter is attributed in an international agreement.

PART TWO
The Distribution of Air Traffic Rights

Chapter 3

Background

Before the problem of the distribution of air traffic rights is further examined, another part of the 'open-skies' judgement shall be discussed first, because, as a result of this ruling, new difficulties with respect to the allocation of air traffic rights can be expected that will have to be considered.

THE 'OPEN-SKIES' JUDGEMENT AND THE INFRINGEMENT OF THE RIGHT OF ESTABLISHMENT

SUBMISSIONS OF THE PARTIES

One of the consequences of the establishment of a common market for air transportation services is that EC law becomes applicable to a larger extent. Therefore, the market freedoms laid down in the EC Treaty may be applicable to Community air carriers. There are five market freedoms:

i) the right to free movement of goods;
ii) the right to free movement of workers;
iii) the right to provide services;
iv) the right of free movement of capital, and
v) the right of establishment.

In the following attention will be given to the right of establishment set out in Articles 43, 48 EC. Article 43 EC stipulates:

> Within the framework of the provisions set out below, restrictions on the freedom of establishment of nationals of a Member State in the territory of another Member State shall be prohibited. Such prohibition shall also apply to restrictions on the setting-up of agencies, branches or subsidiaries by nationals of any Member State established in the territory of any Member State.
>
> Freedom of establishment shall include the right to take up and pursue activities as self-employed persons, and to set up and manage undertakings, in particular companies or firms within the meaning of the second paragraph of Article 48, under the conditions laid down for its own nationals by the law of the country where such establishment is effected, subject to the provisions of the chapter relating to capital.

Article 48 EC says:

> Companies or firms formed in accordance with the law of a Member State and having their registered office, central administration or principal place of business within the

Community shall [for the purpose of this Chapter] be treated in the same way as natural persons who are nationals of Member States.
'Companies or firms' means companies or firms constituted under civil or commercial law, including cooperative societies, and other legal persons governed by public or private law, save for those which are non-profit-making.

Hence, the right of establishment precludes the Member States from imposing any restrictions on nationals of other Member States, if they intend to establish a company, an agency, a branch or a subsidiary in the territory of a Member State, whose nationality they do not have. According to Article 43(2) EC Member States are obliged to treat nationals of other Member States making use of the right of establishment like they treat their nationals.[332] Pursuant to Article 48 EC the right of establishment is applicable not only to natural persons, but also to companies.

The Commission accused the Member States of infringing the right of establishment by having included a nationality clause in their 'open-skies' agreements. From the Commission's perspective the nationality clause results in a discrimination of air carriers that establish an undertaking in another Member State and which are not majority owned by this latter Member State itself, by its citizens or by both combined. As already described above, when states conclude an air service agreement, they grant each other air traffic rights. These rights are then given to air carriers, which can make use of them. The allocation of air traffic rights is undertaken independently by each of the contracting states. They alone decide, which air carrier is granted air traffic rights. After the allocation, both states inform each other the air carriers they have allocated the air traffic rights. However, the designated air carriers still cannot start operating between the two countries. They first need to obtain some additional permissions from the state into which it wants to operate. At this stage, the nationality clause becomes important. Airlines that are majority owned by the contracting states or by their citizens usually cannot be refused the necessary permissions unless they do not comply with specific aviation laws. If, on the other hand, a state designates an air carrier that is not majority owned by this state or by its citizens, the other state is entitled by the nationality clause to deny the necessary permits to such an air carrier. Usually a nationality clause might look like this:[333]

1. Each contracting party shall have the right to designate as many airlines as it wishes to conduct international air transportation in accordance with this Agreement and to withdraw or alter such designations. Such designations shall be transmitted to the other contracting party in writing through diplomatic channels, and shall identify which type of air service specified in the Route Schedule the airline is authorized to conduct

332 Open-Skies, *supra* note 105 at I-9911.

333 German-US Air Service Agreement, *supra* note 321, Articles 3 and 4.

2. On receipt of such a designation, and of applications from the designated airline, in the form and manner prescribed for operating authorizations and technical permissions, the other contracting party shall grant appropriate authorizations and permissions with minimal procedural delay, provided that:
   a) Substantial ownership and effective control of that airline are vested in the contracting party designating the airline, nationals of the contracting party (which may include natural or legal persons), or both ....

3. Either contracting party may revoke, suspend, or limit the operating authorizations or technical permissions of an airline designated by the other contracting party where:
   a) Substantial ownership and effective control of that airline are not vested in the other contracting party, the other contracting party's nationals (which may include natural or legal persons), or both ....

With respect to EC law the inclusion of a nationality clause in an air service agreement means that a Community air carrier cannot make fully use of the right of establishment since, as far as designation for international routes is concerned, the danger exists that its designation is not accepted by a third country. Therefore, Member States would always designate their national air carriers. And even if the Member States were willing to designate all Community air carriers under their 'open-skies' agreements, the USA could always refuse to grant the necessary permissions to those Community air carriers that were not majority owned by the designating Member State or its citizens. According to the Commission this results in discrimination based on nationality and violates the right of establishment pursuant to Articles 43, 48 EC. The Commission argued that the Member States have directly caused this discrimination by including the nationality clause in their air service agreements. The respective Member States disagreed and, firstly, argued that the right of establishment was not applicable in the present situation at all. On the one hand, air transportation services did not fall in the scope of the right of establishment, but under the freedom of services subject to the limitations imposed by secondary law based on Article 80(2) EC. On the other hand, a link to the European Community, which is always necessary, if the market freedoms shall be applicable, was lacking, because the matter concerned the relation of the Member States and another non-Member State. But, even if the Court should affirm the applicability of EC law in this case, the Member States denied that they were responsible for any possible discrimination. It were the USA only that might cause a discrimination by denying the necessary permissions to Community air carriers, which were not majority owned by the designating Member State or by its citizens. Additionally, the Member States argued that, in any case, a possibly existing discrimination would be justified for reasons of public interests pursuant to Article 46 EC. Article 46 EC prescribes:

> The provisions of this chapter and measures taken in pursuance thereof shall not prejudice the applicability of provisions laid down by law, regulations or administrative action providing for special treatment for foreign nationals on grounds of public policy, public security or public health ....

The Member States explained that they must have the possibility to refuse permissions to airlines that are designated by the US, but which are not owned and controlled by the US or by its citizens. For reasons of reciprocity, the same privilege must be also granted to the US. Therefore, the nationality clause was indispensable in the air service agreements due to public interest.

FINDINGS OF THE EUROPEAN COURT OF JUSTICE

The Advocate General fully agreed with the Commission and, in particular, rejected the Member States' argument with respect to a justification based on Article 46 EC, explaining that this provision, since it contained an exception to the right of establishment, which is a fundamental principle of EC law, must be applied in a very restrictive manner. Its scope must be limited to those measures that were absolutely necessary to protect the public's interests. The selected measure must provide for the least restrictive limitation of the right of establishment possible. From his point of view the Member States failed to prove that a less restrictive solution than the one provided by the nationality clause was not available.[334]

The European Court of Justice concurred with the Commission and with its Advocate General and affirmed an infringement of the right of establishment pursuant to Articles 43, 48 EC for reasons of discrimination based on nationality. The judges refused the notion that the discrimination was only caused by the USA emphasizing that, by legally refusing an air carrier that had been designated by a Member State, the USA only made use of a right granted by the Member States. The mere inclusion of the nationality clause in the air service agreements was already to be considered as discriminatory. The argument of the Member States that such discrimination was justified pursuant to Article 46 EC was also dismissed. Article 46 EC must be applied in a very restrictive manner and is only applicable, if a real, present and substantial danger existed that might affect interests of the general public. Furthermore, there must be a link between the danger and the discrimination. These conditions were not fulfilled for several reasons. Firstly, the nationality clause allowed to refuse the granting of necessary permissions for other reasons than solely those of public interests. Secondly, a link between the discrimination of Community air carriers and the danger caused by airlines designated by the US was missing. Hence, since Article 46 EC was not applicable, the Court of Justice affirmed an infringement of the right of establishment by the Member States.

DISCUSSION

CONSEQUENCES OF THE 'OPEN-SKIES' JUDGEMENT FOR EXISTING DESIGNATIONS AND AIR SERVICE AGREEMENTS

With respect to the treaty-making powers of the EC *vis-à-vis* air service agreements, the European Court of Justice decided that the Community was partially competent

334 Opinion of AG Tizzano, *supra* note 81 at I-9482.

for the conclusion of air service agreement. According to this author's opinion as concluded in the previous chapter the EC could even have the treaty-making power for the majority of the Member States' air service agreements. Furthermore, the ECJ ruled that the Member States have infringed the right of establishment by including the nationality clause in their air service agreements. To answer the question which consequences the 'open-skies' judgement may have for the existing designations and current air service agreements of the Member States, it is necessary to have regard to Articles 307(1) and (2) EC.

Respect for international agreements of the Member States pursuant to Article 307(1) EC

Pursuant to Article 307(1) EC international agreements concluded before 1 January 1958 or the accession to the European Community by the respective Member States, are not affected by EC law. Article 307(1) EC says:

> The rights and obligations arising from agreements concluded before 1 January 1958 or, for acceding States, before the date of their accession, between one or more Member States on the one hand, and one or more third countries on the other, shall not be affected by the provisions of this Treaty.

In the 'open-skies' judgement, the involved Member States argued that the disputed air service agreements the ECJ had to decide upon, had been concluded before the establishment of the European Community or before the accession of the respective Member States to it respectively and therefore a breach of the EC Treaty was precluded based on Article 307(1) EC. Although, the concerned air service agreements were amended after their conclusion, in particular during the 1990s, the Member States held the opinion that they were still to be considered as the original agreements concluded before the establishment of the Community or the accession of some Member States to it. Therefore, Article 307(1) EC should apply (if at all) then only the undertaken amendments could be assessed in relation to the provisions of the EC Treaty. The Commission disagreed and argued that the former air service agreements of the Member States had been amended to such an extent that, in fact, they had been transformed into wholly new 'open-skies' air service agreements. Since those modifications were undertaken in the aftermath of the establishment of the EC or the accession to it respectively, Article 307(1) EC was inapplicable. Therefore, it should be possible to assess all clauses of the respective 'open-skies' agreements in relation to the provisions of Community law, irrespective as to whether an amendment had taken place or whether clauses were transferred without any modifications from the original agreements. Advocate General Tizzano was of the opinion that an infringement of the Community's treaty-making power has been caused only by those clauses of the air service agreements that had been introduced during the various revisions in the 1990s. In respect of these new clauses Article 307(1) EC was inapplicable.[335] Furthermore, Advocate General Tizzano argued that, in the case an infringement of EC law was based on a newly arisen treaty-making powers

335 *Ibid.* at I-9473ff., 9482ff.

of the Community, the Member States neither could nor should take any measures. The only solution in such a situation was the conclusion of a new agreement by the Community and/or the termination of their existing agreements by the Member States. The disadvantage of the latter solution was the rise of a legal gap.[336] Tizzano therefore concluded that, as far as an infringement of newly established treaty-making powers pursuant to the ERTA judgement was concerned, Article 307 EC as a whole could not be applied. When it came to the nationality clause, the Advocate General held the opinion that this provision had been transferred from the previous air service agreements without any amendments and thus Article 307(1) EC could be applicable. On the other hand, he added that it should be taken into regard that this clause might have not been formally amended, but its scope of applicability and its significance within the respective air service agreements has changed substantially since its original inclusion in these agreements, due to the additional rights that were included in the new 'open-skies' agreements. Since the amendments occurred in the 1990s, Article 307(1) EC would be not applicable. A particular situation existed in respect of Great Britain, which had concluded a wholly new air service agreement with the USA, after its accession to the European Community. In respect of this new agreement, including the nationality clause it contained, Article 307(1) EC was definitely inapplicable.[337] The European Court of Justice mostly concurred with the Advocate General and adopted his reasoning.

To determine the applicability of Article 307(1) EC, the European Court of Justice, the Advocate General and the Commission in their deliberations only had regard to the time, when the disputed air service agreements had been concluded or amended respectively. Thereby they did not consider sufficiently the objective of and legal principle behind Article 307(1) EC. This provision entitles the Member States to refuse the performance of their duties under the EC Treaty, if this would result in a breach of other duties they have committed to under previous agreements concluded with non-Member States.[338] Although according to its wording the provision seems to be addressed to the Member States only, it also obliges the institutions of the Community not to demand any actions from the Member States that would require

336 *Ibid.* at I-9476.

337 *Ibid.* at I-9476, 9482ff.

338 *Ibid.*, ECJ, *ex parte Evans Medical Ltd. and MacFarlan Smith Ltd. The Queen v Secretary of State for the Home Department*, C-324/93, [1995] E.C.R. I-563 at 607 [Evans Medical and MacFarlan Smith]; ECJ, *Office National de l'Emploi v Madleine Minne*, C-13/93, [1993] E.C.R. I-371 at 384 [Minne]; ECJ, *ex parte Centro-Com Srl v HM Treasury and Bank of England The Queen*, C-124/95, [1997] E.C.R. I-81 at 130 [Centro-Com]; ECJ, *T. Port GmbH and Co. KG v Hauptzollamt Hamburg-Jonas*, C-364/95 and C-365/95, [1998] E.C.R. I-1023 at 1051f. [T. Port]; Ernst-Ulrich Petersmann and Christoph Spennemann, in: Hans von der Groeben and Jürgen Schwarze, eds. (2004), *Vertrag über die Europäische Union und Vertrag zur Gründung der Europäischen Gemeinschaft* (*Treaty on the European Union and Treaty establishing the European Community*), vol. 4, 6th ed. (Baden-Baden, Germany: Nomos Verlagsgesellschaft) Article 307 EC at para 12 [Petersmann/Spennemann, in: Groeben/Schwarze, Article 307].

them to breach their former agreements.[339] The objective of Article 307(1) EC is to ensure that the Member States do not release themselves from former obligations they have undertaken towards other, non-EC states.[340] This provision protects the rights of third countries and confirms the principle of *pacta sunt servanda*, a core element of international law.[341] Due to Article 307(1) EC the European integration is describes as 'being in accordance with international law'.[342] Therefore, the reference to 'rights arising from agreements' in Article 307(1) EC means the rights of the third countries only, while 'obligations arising from agreements' refers to the obligations of the Member States.[343] Consequently Article 307 (1) EC preserves the obligations of the Member States towards third countries. At the same time it limits the exercise of the corresponding rights of the Member States granted by third countries, if this would result in an infringement of their duties flowing from EC law.[344] Thus, if Member States are authorized, but not obliged, to exercise a right under an international agreement, they shall not make use of this right, if it is not in accordance with EC law.[345] This understanding of Article 307(1) EC also complies with another principle of international law, according to which a country *ipso facto* gives up its former rights, if it has accepted new obligations that are incompatible with these rights it has obtained under a former agreement.[346]

For Article 307(1) EC to be applicable it is necessary that an agreement was concluded before 1 January 1958 or before the accession of the respective Member States to the European Community.[347] Difficulties arise, if international agreements with non-EC countries were concluded after these dates and if, at that time, the Community was lacking any treaty-making powers for the respective subject matters.

---

339 B. Beutler, A. Epiney, R. Bieber and M. Haag, eds. (2001), *Die Europäische Union*, (*The European Union*), 5th ed. (Baden-Baden, Germany: Nomos Verlagsgesellschaft) at para 1208 [Beutler/Epiney/Bieber/Haag]; Hans Krück, in: Jürgen Schwarze, ed. (2000), *EU-Kommentar* (*EU Commentary*), (Baden-Baden, Germany: Nomos Verlagsgesellschaft) Article 307 EC at para 4 [Krück, in: Schwarze, Article 307]; ECJ, *Attorney General v Juan C. Burgoa*, 812/79, [1980] E.C.R. 2787 at 2803 [Burgoa].

340 Evans Medical and MacFarlan Smith, *supra* note 338 at 605f.; Centro-Com, *supra* note 338 at 130; T-Port, *supra* note 338 at 1051.

341 26 Vienna Convention on the Law of the Treaties [Vienna Convention]; Beutler/Epiney/Bieber/Haag, *supra* note 339 *at* para 1208; Dierk Booß, in: Carl-Otto Lenz and Klaus Borchardt eds. (2003), *EU- und EG-Vertrag* (*EU and EC Treaty*), 3rd ed. (Cologne: Bundesanzeiger) Article 307 EC at para 1[Booß, in: Lenz/Borchardt, Article 307].

342 Petersmann/Spennemann, in: Groeben/Schwarze, Article 307, *supra* note 338 at paras 1, 4.

343 *Ibid.* at para 4; ECJ, *Commission v Italy*, 10/61, [1962] E.C.R. 5 at 22 [Commission v Italy]; Krück, in: Schwarze, Article 307, *supra* note 307 at para 5.

344 Petersmann/Spennemann, *ibid.*

345 *Ibid.*; Evans Medical and MacFarlan Smith, *supra* note 338 at 607; Centro-Com, *supra* note 338 at 130; Kirsten Schmalenbach, in: Christian Calliess and Matthias Ruffert eds., *Kommentar zum EU-Vertrag und EG-Vertrag* (*Commentary to the EU Treaty and EC Treaty*), 2nd edn. (2002) (Neuwied, Kriftel, Germany: Luchterhand Verlag) Article 307 EC at para 6 [Schmalenbach, in: Calliess/Ruffert, Article 307].

346 *Commission v Italy*, *supra* note 343 at 22; Schmalenbach, *ibid.*

347 Krück, in: Schwarze, Article 307, *supra* note 339 at para 6.

In such a case it seem appropriate to apply Article 307(1) EC analogously. Firstly, such an approach respects the principle of *pacta sunt servanda*. Secondly, it seems impossible to foresee which treaty-making powers, in particular if they are established based on the ERTA judgement, may arise on the part of the Community one day and lead to the incompatibility of an international agreement with EC law.[348] It would be a disproportionate limitation of the Member States' treaty-making powers, if they were precluded from concluding international agreements solely for the reason that the responsibility for such an agreement could fall into the responsibility of the Community one day. However, an analogous application of Article 307(1) EC shall also not be possible, if the emergence of a particular treaty-making power of the Community could have been expected.[349] According to Vedder such an emergence of treaty-making powers of the Community is only unforeseeable in case of Article 308 EC, while in all other situations, even with respect to implied powers, the emergence of treaty-making powers of the Community shall be expected. To justify his opinion Vedder has recourse to Article 10(2) EC and argues that already before treaty-making powers emerge, the Member States are obliged not to prejudge and endanger the Community's exercise of its powers with own international agreements. If the Member States nevertheless want to conclude international agreements, they shall include clauses in their treaties that ensure the compatibility with EC law. One such possibility would be the inclusion of a provision that would allow for the termination of the treaty in case the Community becomes responsible for the respective subject matter.[350] Vedder's opinion can be countered by arguing that apart from Article 308 EC there are other provisions in the EC Treaty, which grant a similar discretion to the institutions of the Community in respect of the exercise of their powers. One of such provisions similar to Article 308 EC is Article 80(2) EC.[351] Thus, at least, with respect to Article 80(2) EC even Vedder would have to affirm the possibility for an analogue application of Article 307(1) EC to air service agreements concluded by the Member States after 1 January 1958 or after their accession to the European Community respectively. In the 'open-skies' cases an additional difficulty arose from the fact that many Member States had concluded their respective air service agreements with the US already before 1 January 1958 or their accession to the EC, but that these treaties were modified in the aftermath and partially even after the common air transportation market has been fully established. While the Member States argued that Article 307(1) EC should be applied to the original treaties and only

348 Rudolf Bernhardt, 'Die Europäische Gemeinschaft als neuer Rechtsträger im Geflecht der traditionellen zwischenstaatlichen Rechtsbeziehungen', (The European Community as a New Legal Entity in the Context of Traditional Bilateral Relations) (1983) *Europarecht* (*Europ. L. Rev.*) 199 at 205 [Bernhardt]; Petersmann/Spennemann, in: Groeben/Schwarze, Article 307, *supra* note 338 at para 6.

349 Petersmann/Spennemann, *ibid.*; Christoph Vedder, in: Eberhard Grabitz and Meinhard Hilf, eds. (1986), *Das Recht der Europäischen Union* (*The Law of the European Union*), looseleaf (Munich: Verlag C.H. Beck, May, 2nd supplement) Article 234 EC at para 21 [Vedder, in: Grabitz/Hilf, Article 234].

350 Vedder, *ibid.*

351 Petersmann/Spennemann, in: Groeben/Schwarze, Article 307, *supra* note 338 at para 6.

the modifications undertaken to these agreements could be assessed in relation to EC law, the Commission held the opinion that the amendments were so extensive that *de facto* new agreements were concluded, which should be fully assessed in relation to the provisions of Community law. The ECJ concurred with the Commission and denied the applicability of Article 307(1) EC as a whole.[352] Since some of the modifications occurred after the beginning of the liberalization process of the air transportation market by the EC, the emergence of implicit treaty-making powers on the part of the Community was, under these circumstances, also foreseeable and thus an analogous application of Article 307(1) EC must be also rejected.

The question is now, what consequences arise from the fact that the Member States have firstly violated the EC Treaty by concluding air service agreements, for which the treaty-making power partially falls into the sphere of the Community and secondly infringed the right of establishment by including the nationality clause in their air service agreements. Firstly, it can be stressed that it is of no relevance for the effectiveness of an international agreement concluded by the Member States, whether Article 307(1) EC is directly, analogously or not at all applicable. Pursuant to Article 46 Vienna Convention the agreement remains in force anyway. Whether a Member State can invoke Article 307(1) EC is only relevant for the question, whether it has infringed EC law or whether its actions were justified.[353] Secondly, and as explained above, Member States cannot have recourse to Article 307(1) EC, if they are authorized by an international agreement to adopt certain measures, but if they are not obliged to do so and if these measures would infringe EC law. In respect of the air service agreements concluded by the Member States with third countries, the first ones are obliged to grant the necessary permissions to the air carriers designated by the third country to be able to make use of the rights agreed upon in the respective air service agreement. At the same time, the Member States are entitled to designate their own air carriers, which are granted the same permissions by the third country. The sole designation of air carriers and the establishment of air transportation services between the Member States and third countries do not by itself lead to an infringement of EC law. Difficulties arise only from the application of the nationality clause and the condition included therein that the designated air carriers must be majority owned by the designating country and/or its citizens. This results in an infringement of the right of establishment. Thereby is shall be underlined that the designation of exclusively national air carriers is a right and not an obligation, granted to the Member States in the air service agreements. If the Member States make use of this right, they infringe EC law and, as set out above, the Member States must refrain from undertaking such a measure. Pursuant to Article 307(1) EC they are precluded from continuing to designate solely national air carriers. Since the nationality clause is part of all air service agreements, the Member States would have to revoke the designations of their national air carriers towards all countries with

352 Open-Skies, *supra* note 105 at I-9890f.

353 Bernhardt, *supra* note 348 at 206f.; Vedder, in: Grabitz/Hilf, Article 234, *supra* note 349 at para 20; see also Oppermann, European Law, *supra* note 196 at paras 608f., 1721; Bleckmann, European Law, *supra* note 75 at paras 1407, 1420; Meessen, *supra* note 268 at 49.

whom they have concluded air service agreements. Although the ECJ has conceded that Article 307(1) EC was not applicable on the 'open-skies' agreements, the legal consequences of Article 307(1) EC remain applicable. If the Member States are obliged to refrain from designating solely national air carrier and to revoke existing designations that contravene EC law, if Article 307(1) EC would be directly or analogously applicable, this consequence must be even more valid, if the provision is not applicable at all. The result of such an approach would be that most Community air carriers would have to limit their operations to intra-Community flights, while international flights between the EC and third countries would be offered by air carriers from third countries. The benefit of this solution would be that, from a legal perspective, all Community air carriers were treated equally and an infringement of EC law was avoided. From an economic point of view, this solution would lead to serious competitive disadvantages and financial damages of the European aviation industry. As a first consequence, excess capacities would emerge, if air carriers were limited to operate solely on intra-Community routes. Additionally, the fleets of many Community air carriers are structured to provide international air transport services. Due to their size aircrafts intended to operate on long-haul routes would have to be laid up for reasons of profitability. This would not only have a negative impact on the air carriers themselves, but also on aircraft manufacturers and aviation insurers. The investments undertaken in the fleets by the airlines might not be compensated through the reduced operations, eventually finally leading to bankruptcies of air carriers and thus less competition in the common air transport market in the EC. The only entities benefiting from such a situation would be the air carriers from non-EC states. They could obtain a monopoly or oligopoly on the international routes between their countries and the EC. Depending on the regulations included in the respective air service agreements, these airlines could be entitled to freely fix fares and conditions of carriage. Under these circumstances one can expect that tariffs would increase. Even if this scenario would last 'only' for one year, as air service agreements usually contain a clause that allows for a termination of these treaties with a notice period of one year, the damages for the European aviation industry would be tremendous. Air carriers from third countries could use their advantage to strengthen their position in the market *vis-à-vis* Community air carriers. When the latter ones would start to operate on international routes again, they would enter into competition with other airlines from a weakened position. For these reasons, an obligation of the Member States to stop designating national air carriers and to revoke former designations would be contrary to the attainment of the objectives of the EC Treaty in the area of aviation. The goals mentioned in Article 2 EC to promote a harmonious, balanced and sustainable development of economic activities, a high level of employment, a sustainable and non-inflationary growth, a high degree of competitiveness and convergence of economic performance, and the raising of the standard of living and quality of life would be endangered. To avoid such a development, a teleological reduction of Article 307(1) EC in accordance with the objectives of the EC shall be undertaken. Therefore, if a measure normally required by Article 307(1) EC results in that goals of the EC would be impaired, Article 307(1) EC must be applied restrictively and any of such measures must be omitted. Concluding, it is submitted that, based on Article 307(1) EC, it is neither directly nor analogously possible to derive an obligation of the Member States to stop designating national airlines or to revoke existing designations.

Duty of the Member States to amend their air service agreements pursuant to Article 307(2) EC

According to Article 307(2) EC the Member States could be obliged to take all appropriate steps to eliminate all incompatibilities of their existing air service agreements with EC law. Article 307(2) EC says:

> To the extent that such agreements are not compatible with this Treaty, the Member State or States concerned shall take all appropriate steps to eliminate the incompatibilities established. Member States shall, where necessary, assist each other to this end and shall, where appropriate, adopt a common attitude.

In the 'open-skies' judgement the Commission argued alternatively, in case that the ECJ would have held Article 307(1) EC as applicable, that the Member States have infringed their obligations under the EC Treaty in so far as they have not undertaken all necessary measures pursuant to Article 307(2) EC to rescind those provisions of the air service agreements that existed before the establishment of the EC or before the accession of a Member State to it, and that were incompatible with EC law. As far as the amendments to the agreements after the establishment of the EC or the accession to it were concerned, the Commission maintained the position that they should be fully assessed in relation to the EC Treaty. As far as an infringement of the right of establishment was concerned, the Advocate General agreed with the Commission.[354] According to his point of view, the Member States were obliged to make more efforts to eliminate the incompatibility of the nationality clause with EC law. Even a termination or at least serious considerations of a termination as *ultima ratio* could have been part of such efforts. The ECJ decided in an *obiter dictum* that Article 307 EC applies only to treaties that were concluded before the EC Treaty entered into force, but not to those modifications that were made to existing agreements after the respective Member States have adopted the EC Treaty.[355] The Court of Justice further argued that the past amendments undertaken in respect of existing air service agreements were, in parts, quite substantial. But even where only slight changes to former provisions had taken place or where provisions had been adopted without any changes, these provisions obtained a new and greater significance due to the other modified clauses as well as the developments of air transportation law in the European Community. Hence, even the unmodified clauses had become more important than they had been before, when they had been first included in the original air service agreements. Under these circumstances one would have to conclude that the Member States have concluded wholly new air service agreements to which Article 307 EC was not applicable at all.

In general, the ECJ's understanding of Article 307 EC is in accordance with the wording of the provision. As a result the Member States would not be obliged by

354 Opinion of AG Tizzano, *supra* note 81 at I-9484ff.

355 Open-Skies, *supra* note 105 at I-9891.

Article 307(2) EC to take back amendments undertaken to their air service agreements after 1 January 1958 or after their accession to the European Community. However, such a conclusion seems quite contradictory. If the Member States are obliged to eliminate incompatibilities of their former agreements and EC law, this obligation must be affirmed even more for such treaties or amendments to such treaties that the Member States have concluded or undertaken after 1 January 1958 or their accession to the EC. Otherwise the Member States could deliberately violate the EC's treaty-making powers and conclude international agreements that are not in accordance with EC law. The objective of Article 307(2) EC is to foster the primacy of EC law and to ensure a uniform application of the rules of the Community. If these goals can be impaired by the Member States through the conclusion of international agreements with third countries, there must be a possibility to impose obligations on the Member States to omit or to take back such measures. Although such a duty could be derived from Article 10 EC, it seems more appropriate to have recourse to the more special Article 307(2) EC. Through an application of this provision by analogy the Member States would not only be obliged to take all appropriate steps to eliminate incompatibilities of their old agreements, but also of agreements that they have concluded after 1 January 1958 or their accession to the Community. Insofar it is not relevant, whether incompatibilities between the air service agreements of the Member States and EC law result from the original treaties or from the amendments made hereto. The Member States must eliminate all existing incompatibilities resulting from their international agreements by taking appropriate steps. Due to the reference to 'appropriate steps,' only activities are allowed that are in accordance with international law.[356] This encompasses, firstly, an interpretation of the international agreement that is in accordance with EC law, secondly re-negotiations to adapt the agreement to EC regulations or, if this should fail, thirdly a termination of the treaty.[357] One school of thought proposes that appropriate steps pursuant to Article 307(2) EC also includes the possibility for the Member States to refrain from making use of the rights they were granted in the agreement.[358] However, according to this author's opinion, this solution is already available under Article 307(1) EC and, in any case, as was shown above, is not an appropriate measure to remedy the infringement of the right of establishment and the treaty-making powers of the EC by the air service agreements. Therefore, as far as the infringement of the right of establishment caused by the nationality clause is concerned, the Member States are requested to achieve an amendment of this clause to ensure its compatibility with

356 Krück, in: Schwarze, Article 307, *supra* note 339 at para 10.

357 *Ibid.* at paras 11, 12; Minne, *supra* note 338 at 384; Schmalenbach, in: Calliess/Ruffert, Article 307, *supra* note 345 at para 11; Petersmann/Spennemann, in: Groeben/Schwarze, Article 307, *supra* note 338 at para 8; Vedder, in: Grabitz/Hilf, Article 234, *supra* note 349 at para 9; ECJ, *Ministère public and Direction du travail and de l'emploi v Levy*, C-158/91, [1993] E.C.R. I-4287 at 4305 [Levy]; Gunnar Schuster and Peter-Tobias Stoll, 'Gemeinschaftskompetenz und Altverträge mit Drittstaaten' (Competences of the Community and International Agreements with Third Countries) (1996) *Recht der internationalen Wirtschaft* (*L. of Intl. Commerce*) 89 at 93 [Schuster and Stoll]; Commission, Aviation Relation with Russia, *supra* note 176 at 9.

358 Schuster and Stoll, *ibid.*; Petersmann/Spennemann, *ibid.*

EC law. Should this fail, then the respective air service agreements would have to be terminated. For practical reasons, terminations of air service agreements should not be undertaken by all Member States at the same time, as the negotiations and conclusions of new agreements can take a while. To avoid legal uncertainty for the air carriers and to guarantee continuity in international air transportation between the EC and third countries, a coordinated and gradual approach by the Community and the Member States is recommended, when the air service agreements of the Member States are amended.[359]

CONSEQUENCES OF THE 'OPEN-SKIES' JUDGEMENT FOR THE ALLOCATION OF AIR TRAFFIC RIGHTS

Although as far as the violation of the right of establishment through the nationality clause is concerned the judgement was less disputed than with respect to the question of treaty-making powers of the EC in the area of air transportation, in particular, Wassenbergh did not hold back with criticism. He strongly disagreed with the judge's reasoning and accused the Court of acting like a world court instead of like an institution of regional organization, because it had decided on a matter that partially took place outside the territory of Community. Thereby the ECJ extended the right of establishment beyond its scope and applied it on other, non-EC countries.[360] From the perspective of EC law, the judges exceeded their powers and acted *ultra vires*, thereby infringing the sovereignty of the Member States.[361] He further blamed the ECJ for having limited its considerations to EC law only and not having taken into account the international character of civil aviation. To reach a correct decision, the judges should have had regard to international law, when applying and construing EC law.[362] Contrary to the Court, it was Wassenbergh's conviction that the Member States had no obligation under EC law to treat Community air carriers like national airlines, in particular, when it came to the question of designation under air service agreements.[363] From his perspective the nationality clause was fully in accordance with Community law.[364] Apart from the European Court of Justice Wassenbergh also criticized the Commission for having accused the Member States in the 'open-skies' cases for their discrimination of air carriers that were registered in other Member States. He considered this as hypocritical since the Community and the Commission were discriminating air carriers based on their nationality themselves by reserving the access to the common air transportation market only to Community air carriers, which were majority owned by the Member States or by the citizens of the European Union.[365] Wassenbergh's criticism that the European Court of Justice

359 See also Council, Conclusions, *supra* note 159.

360 Wassenbergh, Mandate, *supra* note 150 at 139, 142; Wassenbergh, ECJ Decision, *supra* note 146 at 22.

361 Wassenbergh, ECJ Decision, *ibid.* 24.

362 *Ibid.* at 21.

363 Wassenbergh, Mandate, *supra* note 150 at 139.

364 *Ibid.* at 139, 142.

365 Wassenbergh, ECJ Decision, *supra* note 146 at 23f.

based its decision merely on Community law is insofar quite surprising, because, as an institution of the European Community, the Court can base its judgements only on EC law. International law and international agreements can be taken into considerations only, if the EC has adopted such obligations and if they have become part of Community law. The Chicago Convention, the Transit and the Transport Agreements, for instance, are not applicable in the Community as the EC has not signed these treaties. Furthermore, the judges did not demand from the Member States to violate their obligations under international law, but only concluded that some of these agreements were incompatible with EC law. Any measures of the Member States to remedy these violations, maybe even through a denunciation of the respective treaties, must be undertaken in accordance with international law.[366] Wassenbergh's opinion that the judges have infringed the sovereignty of the Member States by a too extensive application and interpretation of Article 43 EC must be also rejected. Firstly, Article 43 EC was not applied and construed in a particularly extensive manner, but in accordance with the usual practice and the common understanding of the scope of the right of establishment as it was previously already done in other situations as well. Secondly, if the judgement indeed results in a limitation of the Member States' sovereignty, the reason for this is not the judgement itself, but dates back to when the Member States had concluded the EC Treaty; when they not only included the right of establishment, but also vested the Court of Justice with the power to construe and apply EC law. Under these circumstances the Member States' sovereignty was not limited by the ECJ, but voluntarily transferred to the EC by the Member States themselves. Finally, Wassenbergh's comparison between the discrimination of Community air carriers by the Member States and the discrimination of airlines from third countries by the Community must be also rejected as both situations are not comparable. While the EC Treaty forbids the discrimination of citizens and entities of one Member State by another Member State, no similar prohibition exists, according to which airlines from third countries shall not be treated differently than Community air carriers when it comes to access to the common air transportation market.

Only shortly after the 'open-skies' judgement was delivered, the Commission announced its intention to focus on the amendment of existing air service agreements of the Member States and the modification of the nationality clause. This clause is contained in all air service agreements basically in the same standardized version and it precludes the Community air carriers from taking advantage of the right of establishment to the full extent.[367] Therefore, the judgement relating to the nationality clause does not only affect the few 'open-skies' agreements concluded between some Member States and the United States of America, but is also applicable in respect of all air service agreements concluded between the Member States and third countries.[368] To overcome this situation, the Commission proposed a new sort of Community nationality clause that should substitute the existing nationality clause in the Member States' air service agreements and which would enable the latter ones

366 Krück, in: Schwarze, Article 307, *supra* note 339 at para 10.

367 Commission, Communication, *supra* note 127 at 3.

368 See also Wassenbergh, ECJ Decision, *supra* note 146 at 23.

to designate all Community air carriers under their air service agreements. Pursuant to this clause it would be sufficient, if an air carrier was majority owned by a Member State or by citizens of the EC.[369] In this respect the Commission was also granted a horizontal mandate by the Council on 5 June 2003 to negotiate with third countries the substitution of the nationality clause with the Community nationality clause.[370] So far the Commission was able to conclude 15 horizontal agreements for the Community with Chile, Georgia, Lebanon, Azerbaijan, Croatia, Bulgaria, Romania, Albania, Bosnia-Herzegovina, Macedonia, Singapore, Morocco, Ukraine, Australia and New Zealand.[371] The conclusion of an air service agreement by the Community and a third state under the horizontal mandate has no impact on the number of air traffic rights. The latter ones are still mainly agreed upon by the Member States and the third countries in their respective air service agreements. The substitution of the nationality clause by a Community nationality clause only results in that the Member States have a greater choice among airlines they can consider for designation, the number of air carriers they can designate still remains subject to the provision of the existing agreements of the Member States and the third countries.[372]

The re-negotiation of existing air service agreements for the substitution of the traditional nationality clause with a Community nationality clause is an important first step to allow the Community air carriers to make use of the right of establishment. However, attention shall be also paid to the subsequent step following the conclusion

---

369 Commission, Communication, *supra* note 127 at 8ff.; see the new Community nationality clause in the German model air service agreement, online: Luftrecht-Online <http://www. luftrecht-online. de/regelwerke/pdf/e-muster.pdf> (date accessed: 2 November 2005), Article 3 (2):

(2) The Contracting Party granting these rights shall, subject to the provisions of Article 10 (1) of this Agreement, give without delay the said authorization to operate the international air services under the following conditions:

(a) in the case of an airline designated by the Federal Republic of Germany:

(i) it is established in the territory of the Federal Republic of Germany under the Treaty establishing the European Community and has a valid Operating Licence in accordance with European Community Law;

(ii) effective regulatory control of the airline is exercised and maintained by the European Community Member State responsible for issuing its Air Operator's Certificate and the relevant aeronautical authority is clearly identified in the designation; and

(iii) the airline is owned and shall continue to be owned directly or through majority ownership by European Community Member States and/or nationals of European Community Member States, or by other States listed in Annex 1 and/or nationals of such other states, and shall at all times be effectively controlled by such states and/or such nationals … [*German model air service agreement*].

370 See also above at 45.

371 Commission, Aviation Policy Agenda, *supra* note 137 at 7; EC, Commission, Press Release, EU aviation agreements with Australia and New Zealand, IP/05/783 of 23 June 2005, online: European Commission, Press Releases <http://www.europa.eu.int/rapid/pressReleasesAction.do?reference=IP/05/783andformat=HTMLandaged=1andlanguage=ENandguiLanguage=en> (date accessed: 26 October 2005).

372 Commission, Staff Working Document, *supra* note 138 at 7.

of an air service agreement, namely the designation of air carriers. It must be borne in mind that the exchange of air traffic rights occurs only among the states involved. The airlines do not participate, at least not formally, in the proceedings at this stage. Only in a second step, the received air traffic rights are distributed by the respective states among air carriers. Apart from the nationality clause that sets out the qualitative criteria an air carrier has to fulfil to be designated, every air service agreement also contains a quantitative regulation on the total number of air carriers a state can designate. A country can be allowed to designate only one airline (single designation) or a multitude of airlines (multiple designation). Multiple designation clauses can further foresee numerical limitations or allow for an unlimited number of airlines to be designated.[373] Very often the question of designation gives rise to conflicts between the countries involved. While a state with only one national airline will very likely opt for a single designation clause to limit the competitive pressure on its air carrier, a state with several national air carriers will favour a multiple designation clause to open the respective market for as many of its national airlines as possible. Due to the situation in the European Community with its multitude of air carriers and the interest of the Member States that their national airlines are granted access to as many markets as possible, it will be necessary for the Member States and/or the Community, when future air service agreements are concluded with non-EC states, to arrange for the inclusion of a multiple designation clause with as many designations as possible.[374] Limitations in respect of possible designations would not only preclude an effective use of the right of establishment by Community air carriers, but could also cause frictions among the Member States, if they had the impression that their national air carriers are not sufficiently designated and discriminated compared to other air carriers. In the worst case, aviation could become a subject of disputes among the Member States. Therefore, the Commission even argued that air service agreements without a multiple designation clause are not acceptable to the EC and the Member States.[375] Certainly, to enable the Community air carriers to make use of the right of establishment, it is first necessary to substitute the nationality clause with a Community nationality clause. Still, only little progress would be made, if due to a single or limited multiple designation clause, only few Community air carriers could be designated and make use of the rights granted in an air service agreement. Insofar the Commission is right with its demand for the inclusion of multiple designation clauses. This aspect was also included in Article 3 of Regulation 847/2004. Pursuant to this provision, Member States are not permitted to amend their air service agreements, if the number of possible air carriers to be designated under the new agreement would be lower compared to the old agreement. However, this approach may be too selfish. The EC must also take into account that other, non-EC countries are pursuing their own interests in air transportation matters and may not be willing to accept a multiple designation clause by any means. Every state shall have the

---

373 Schwenk/Giemulla, *supra* note 13 at 638.

374 Wassenbergh Mandate, *supra* note 150 at 140f., 145f.

375 Commission, Communication, *supra* note 127 at 14.

equal opportunity to develop its own air transportation industry.[376] This also includes the possibility to structure air service agreements according to the actual needs of the respective aviation industry. The fact that the number of Community air carriers that can be designated under an air service agreement could be limited leads to a problem that neither the Community nor any of the Member States had to deal with before to a similar extent. In the future, it could emerge that more air carriers apply for air traffic rights than there are available. So far, the Member States allocated available air traffic rights only among their national airlines. This task was not very difficult as most Member States had one or at most only a few airlines that were able to offer international air transportation services. This situation could change significantly in the future as a result of the substitution of the nationality clause with the Community nationality clause. The first question to be answered is whether every Community air carrier will be entitled to apply to a Member State for being designated under each of the Member States air service agreements and whether the Member State will be obliged to consider applications of every Community air carrier. The relevant EC Regulation here is 47/2004, adopted in the aftermath of the 'open-skies' judgement, which contains a clause on the distribution of air traffic rights. Article 5.[377] It says:

> Where a Member State concludes an agreement, or amendment to an agreement or its Annexes, that provide for limitations on the use of traffic rights or the number of Community air carriers eligible to be designated to take advantage of traffic rights, that Member State shall ensure a distribution of traffic rights among eligible Community air carriers on the basis of a non-discriminatory and transparent procedure.

It is remarkable that this provision speaks of 'Community air carriers eligible to be designated' and of a distribution of air traffic rights among 'eligible Community air carriers'. It seems that not all Community air carriers will have the right to be designated and to be granted air traffic rights under the respective air service agreements of the Member States. If all Community air carriers would be eligible for designation, the term 'eligible' could be omitted. Thus, there must be an additional condition that needs to be fulfilled by the Community air carriers to be designated. An indication under which circumstances Community air carriers can be designated, can be found in Article 2 of Regulation 847/2004.[378] This provision says:

> In so far as air carriers and other interested parties are to be involved in the negotiations referred to in Article 1, Member States shall treat equally all Community air carriers with an establishment on their respective territories to which the treaty applies.

According to Article 2 only those Community air carriers are expected to have an interest in an air service agreement that have an establishment in the territory of

---

376 See also the preamble of the Chicago Convention: 'Therefore, the undersigned governments having agreed on certain principles and arrangements in order that international civil aviation may be developed in a safe and orderly manner and that international air transport services may be established on the basis of equality of opportunity and operate soundly and economically.'

377 Regulation 847/2004, *supra* note 179.

378 *Ibid.*

the respective Member State. This means *e contrario* that all other Community air carriers not having an establishment in the respective Member State do not have any interest in such an air service agreement. The reason for the lack of any interest must necessarily lie in the fact that the air service agreement will not provide for any rights for these Community air carriers. Hence, only those Community air carriers are eligible for designation under an air service agreement by the Member States that have an establishment in the respective Member State. This understanding is further supported by the third recital of the preamble of Regulation 847/2004, which stipulates:

> The Court has also clarified the right of Community air carriers to benefit from the right of establishment within the Community, including the right to non-discriminatory market access.

This statement discloses that one of the reasons for the adoption of Regulation 847/2004 was the promotion of the right of establishment for Community air carriers. Therefore, only those Community air carriers can have the benefit of being designated under an air service agreement by a Member State that have made use of the right of establishment and established an undertaking in that respective Member State. Accordingly Germany, for example, has amended the nationality clause of its model air service agreement to include also those Community air carriers that are 'established in the territory of the Federal Republic of Germany'.[379] Of course, this does not mean that Article 5 of Regulation 847/2004 has to be construed according to the clauses included in Germany's model air service agreement. However, taking into account that Germany has amended its model air service agreement in the same month as Regulation 847/2004 was adopted, the assumption is that this amendment occurred to comply with that regulation.

The next question to be answered is, what kind of establishment must exist for a Community air carrier in a Member State to be designated by the latter one under its air service agreements. To start with, one has to have recourse to the right of establishment laid down in Article 43 EC. According to a definition set up by the European Court of Justice, an establishment is acknowledged, if it enables a temporally unlimited exercise of a commercial activity through the use of a fixed facility in another Member State.[380] EC law itself provides for two kinds of establishments. The primary establishment refers to the place, where the headquarters of a company are located. This can be at the statutory seat of a company or where the management takes its decisions.[381] Secondary establishments are defined in Article 43(1) s.2 EC and include agencies, branches and subsidiaries. While agencies and branches are legally and economically dependent on the primary establishment, subsidiaries, established in other Member States, are legally autonomous and only economically dependent

---

379 German model air service agreement, *supra* note 369 at Article 3 (2) (a).

380 Alexander Scheuer, in: Carl-Otto Lenz and Klaus Borchardt eds. (2003), *EU- und EG-Vertrag* (*EU and EC Treaty*), 3rd edn., (Cologne: Bundesanzeiger), Article 43 EC at para 2 [Scheuer, in: Lenz/Borchardt, Article 43].

381 Bleckmann, European Law, *supra* note 75 at para 1,594.

on the primary establishment.[382] The main functions of agencies and branches are the support of the sales and marketing department, the promotion of the company's products and services and the carrying out of various administrative matters.[383] Other common secondary establishments are supporting facilities like construction facilities, storehouses or distribution centres. Generally, one can say that by founding secondary establishments, a company transfers some parts of its infrastructure to another Member State.[384] While the determination of a primary establishment is not very complicated, difficulties can arise with respect to secondary establishments. Thereby the additional difficulty can arise to distinguish between the foundation of permanent establishments and the mere use of the right of free movement of services pursuant to Article 49 EC. The difference between an establishment and the mere provision of services is that an establishment is of a permanent nature, while the provision of services is usually temporally limited. A company making use of the right of establishment decides to be part of the commercial life of a Member State in a permanent manner and usually will shift parts of the company's infrastructure to the respective Member State. Regularly the company will establish some sort of facilities and in particular buildings in the Member State.[385] These facilities must enable a company to become actively involved in the business life of the respective Member State where such a facility is located. Whether this is the case must be decided by having regard to the function of the facility and what kind of transactions it is supposed to facilitate.[386] A business address, a bank account and a phone line with an answering machine are regularly not sufficient to affirm the existence of an establishment.[387] The foundation of an establishment is a process that cannot be reversed easily and it is this permanent nature of an integration into the business life of a Member State that is the core element of the right of establishment.[388] A natural person or entity-making use of its right to free movement of services, on the other hand, does not have the intention of beoming permanently installed and involved in the business life of another Member State. Its participation in another Member State's

382 Jürgen Tiedje and Peter Troberg, in: Hans von der Groeben and Jürgen Schwarze, eds. (2003), *Vertrag über die Europäische Union und Vertrag zur Gründung der Europäischen Gemeinschaft* (*Treaty on the European Union and Treaty establishing the European Community*), vol. 1, 6th edn., (Baden-Baden, Germany: Nomos Verlagsgesellschaft), Article 43 EC Treaty at paras 35f. [Tiedje/Troberg, in: Groeben/Schwarze, Article 43].

383 *Ibid.* at para 36.

384 *Ibid.* at para 33.

385 Jürgen Böhmer, in: Christian Calliess and Matthias Ruffert, eds. (2002), *Kommentar zu EU-Vertrag und EG-Vertrag* (*Commentary to the EU Treaty and EC Treaty*), 2nd edn. (Neuwied, Kriftel, Germany: Luchterhand Verlag) Article 43 EC at para 11 [Böhmer, in: Calliess/Ruffert, Article 43].

386 Tiedje/Troberg, in: Groeben/Schwarze, Article 43, *supra* note 382 at para 9.

387 *Ibid.* at para 8.

388 Scheuer, in: Lenz/Borchardt, Article 43, *supra* note 380 at para 2; Tiedje/Troberg, *ibid.* paras 11, 13; Albrecht Randelzhofer and Ulrich Forsthoff in: Eberhard Grabitz and Meinhard Hilf, eds. (May 2001), *Das Recht der Europäischen Union* (*The Law of the European Union*), looseleaf (Munich: Verlag C.H. Beck, 18th supplement), Article 43 EC at paras 22ff. [Randelzhofer/Forsthoff, in: Grabitz/Hilf, Article 43].

economy is only a temporal one. Usually there is no transfer of the service provider's infrastructure to another Member State.[389] However, sometimes service providers also have some facilities in the Member State, where they offer their services. Such facilities might be necessary for the services to be provided.[390] Hence, although the existence of a permanent facility is a strong indication for an involvement with the economy and the business life of a Member State, it is not sufficient to automatically assume that a person or entity makes use of the right of establishment. Additional aspects shall be also considered, like for example the scope of the use of the facility or the centre of an economic activity.[391] This is particularly true when activities are undertaken that do not require the existence of a permanent establishment. In such a case, the question, whether the right of establishment is applicable, is affirmed, if the respective person or entity conducts its commercial activities frequently, continuously or temporally unlimited in a Member State[392] or, even more, if significant portions of its entire business operations are conducted in a Member State.[393] Under these circumstances the right of establishment can be applicable, despite the lack of any existing facility of a company in a Member State.

After giving this short overview on the right of establishment and the different kinds of establishments possible under EC law, it can be discussed in the following, under which circumstances Community air carriers could be able to apply for designation to a Member State. As was concluded above, an airline is only eligible for designation, if it has an establishment in the respective Member State. Thereby it shall not matter what kind of undertaking was established, primary and secondary establishments are to be treated equally. Firstly it can be admitted that Community air carriers are eligible for designation by those Member States, where they have their primary establishment. More difficulties may arise when it comes to secondary establishments. Usually, all airlines offering services to, from and within a Member State will, at least, have an office at the airport, where they operate. This can be also only a check-in counter. The question is, whether this kind of facility is sufficient to affirm the existence of a secondary establishment. Taking into account the previous description and definition of establishments, this could be affirmed, as check-in counters are of high significance when it comes to manage the air transportation services offered by an airline. The employees working there organize the transfer of the baggage to the aircraft and issue the necessary boarding passes with the seat assignments for the flights to the passengers. On the other hand, it must be also regarded that nowadays the check-in procedure can be also accomplished without any assistance of the employees of the airline offering the air transportation service.

---

389 Scheuer, *ibid.*; Tiedje/Troberg, *ibid.* para 11.

390 Böhmer, in: Calliess/Ruffert, Article 43, *supra* note 385 at para 13.

391 Tiedje/Troberg, in: Groeben/Schwarze, Article 43, *supra* note 382 at paras 1, 7, 12.

392 Tiedje/Troberg, *ibid.* para 14; Böhmer, in: Calliess/Ruffert, Article 43, *supra* note 385 at para 14; Bleckmann, European Law, *supra* note 75 (Bleckmann even draws the conclusion that no facility is necessary in another Member State in order to make use of the right of establishment, at paras 1600f.).

393 Scheuer, in: Lenz/Borchardt, Article 43, *supra* note 380 at para 2; Tiedje/Troberg, in: Groeben/Schwarze, Article 43, *supra* note 382 at para 13.

Firstly, many airlines do not conduct check-ins on their own, but conclude agreements with other airlines that take care of that. This is in particular very common within the various airline alliances. Secondly it becomes more and more popular that passengers print out their boarding passes and submit their baggage themselves at self-check-in machines. Some airlines provide even for the possibility to obtain boarding passes before the arrival on the airport from the Internet. These developments show that check-in counters are not indispensable for managing air transportation services. Another aspect that could be raised against check-in counters as a form of secondary establishments is the fact that these facilities existing at the airports are established in a very standardized form, and can be transferred from one airline to another without any complicated and cumbersome reconstructions being necessary. This contravenes the idea that check-in counters represent a permanent establishment of an air carrier. Therefore, check-in counters at airports shall not be regarded as a secondary establishment of an airline. What kind of undertaking must be established to constitute a secondary establishment of an air carrier according to the EC Treaty is difficult to answer. Complex facilities like maintenance hangars would certainly be acknowledged as a secondary establishment. But also ticket sales offices could be regarded as secondary establishments, if they are of a more permanent nature and cannot be abandoned easily by the airlines. Unlike check-in counters on the airport, such sale centres are more customized and require the airline to invest a considerable amount of money in establishing such offices. According to one opinion it shall be sufficient for the acknowledgement of secondary establishments, if an airline hires sales agents that represent the air carrier in a Member State instead of establishing such ticket sales offices on their own.[394] This opinion must be rejected. Comparable to check-in counters at airports, in such a case the bond with the Member State is not strong enough and the involvement of the air carrier in the business life of a Member State can be terminated quite easily without any considerable losses for the airlines due to prior investments. The possibility for an airline to leave a market easily shows the lack of confidence and intention to become permanently involved in the business life of a Member State. Hence, if the secondary establishment of an air carrier shall be justified with the existence of a permanent facility, this facility must be of a considerable size and importance for the operations. In this respect the following definition is interesting:

> Establishment in the territory of a Member State implies the effective and real exercise of air transport activity through stable arrangements; the legal form of such an establishment, whether a branch or a subsidiary with a legal personality, should not be the determining factor in this respect. When an undertaking is established on the territory of several Member States, as defined by the Treaty, it should ensure, in order to avoid any circumvention of national law, that each of the establishments fulfils the obligations which may, in accordance with Community law, be imposed by the national law applicable to its activities.[395]

394 Tiedje/Troberg, *ibid.* at paras 44f.

395 Commission, Staff Working Document, *supra* note 138 at 8.

Similarly it is set out in recital 10 of the preamble to Regulation 847/2004 that undertakings established pursuant to the EC Treaty, should ensure the fulfilment of such obligations that are imposed by the national law in respect of air transportation activities. Thereby it is understood that establishments on the territory of a Member State imply the effective and real exercise of air transportation activities through stable arrangements. These conditions are deemed necessary to avoid any circumvention of national laws.[396] However, as was explained above, the existence of a company's infrastructure is only one aspect that can serve as an indication for the foundation of an establishment. Some industries do not require many facilities to be able to offer their products or services. This is in particular true for the air transportation sector. Aircrafts are the only means and the basis for the provision of air transportation services. Due to their mobility they can be shifted quite easily from one place to another one. Airlines do not need to have any establishment in another State at all to be able to offer their services. They only need to have an aircraft. Tickets can be bought on the phone or on the Internet, which renders ticket sales centres in the Member States unnecessary. Crews can be chartered and maintenance work outsourced. Despite the lack of any fixed facilities, an airline can still have an establishment in a Member State. But in such a case it must be proven that the airline is very active in the respective Member State and that it generates a certain amount of traffic and income in that country. How high this threshold must be, cannot be generalized, but it shall clearly constitute a special market for the airline. A market that is, compared to the other markets of the airline, of a higher significance. Concluding, it is submitted, that, to be designated, Community air carriers need to have an establishment in the respective Member State. Such an establishment can be either based on a facility that sufficiently shows the air carrier's intention to be permanently involved in the business life of a Member State or the Community air carrier must conduct a considerable amount of operations in a Member State compared with its entire operations.

Currently, there are more than 130 air carriers existing in the European Community.[397] If it is assumed that in every Member State there is only one air carrier capable of offering international air transportation services and that such an air carrier has an establishment in every Member State, the Member States potentially may have to deal with 25 air carriers applying for designation. There are only few markets worldwide in which all these airlines could operate profitably. Apart from the transatlantic route between the European Community and North-America and some other routes to Asia, all other markets can support only a limited number of airlines. Therefore, the difficulty, resulting from the introduction of a Community nationality clause combined with a limited designation clause, will be the allocation of air traffic rights.[398] The Commission is aware of this problem, but has not presented any solution yet. The new Regulation 847/2004 only contains an obligation of the Member States in Article 5 to allocate air traffic rights in a transparent and non-

---

396 Regulation 847/2004, *supra* note 179 at 7, recital 10.

397 Commission, Staff Working Document, *supra* note 138 at 2.

398 See also Wassenbergh, ECJ Decision, *supra* note 146 at 23; Wassenbergh, Open-Skies, *supra* note 122 at 175; Giemulla/Schmid/Mölls, *supra* note 42 at para 151.

discriminatory procedure. Other criteria and principles that should be applied are not suggested. Therefore, it will be undertaken below to develop a procedure for the distribution of air traffic rights among Community air carriers in the EC, whereby particular attention shall be given to the allocation of air traffic rights for passenger and passenger/cargo air transportation services. Air traffic rights for all-cargo operations will not be taken into regard due to the particularities that exist for this kind of operations, mainly resulting from the use of seventh freedom rights that are granted more frequently with regard to all-cargo services.

Chapter 4

Administrative Power to Distribute Air Traffic Rights

The air traffic rights granted by non-EC states must be distributed among eligible Community air carriers. In the following chapter it will be assessed, who shall be responsible to execute this duty. Three solutions seem possible:

i) a distribution by the Member States;
ii) a distribution solely by the European Community; or
iii) a mixed distribution by the European Community and the Member States.

Before any further analysis is undertaken, what is to be understood by the term 'distribution of air traffic rights' shall be defined . On the one hand, the distribution of air traffic rights can occur on the legislative level by prescribing the applicable procedure in an act, but leaving the administrative decision about the allocation of air traffic rights that is the execution of this act to a different institution. In this chapter the term 'distribution of air traffic rights' is understood in the latter sense as the 'administrative decision of a single case', that is, the direct allocation of air traffic rights to air carriers. Since air service agreements, to which the European Community will be a party, are to be qualified as EC law, the question of the implementation of these treaties can be subsumed under the general problem of who is responsible for the administration and execution of EC law: the EC itself or the Member States.

DISTRIBUTION BY THE MEMBER STATES

So far, international air traffic rights have been distributed by the Member States. Even under current EC law, it is foreseen that the Member States are in charge of the distribution of air traffic rights.[399] As long as air service agreements are concluded solely by the Member States without any participation of the European Community, there is no reason at hand, why the Member States should not be also exclusively responsible for the distribution of the obtained air traffic rights. Even if the fifth freedom is included in an air service agreement and thus triggers a participation of the Community in the negotiations and in the conclusion of the agreement, the situation for the distribution of air traffic rights remains the same. As long as the respective air traffic rights use are limited with respect to one Member States only that is, when departure and arrival points in the EC are limited to the territory of one

399 See Regulation 847/2004, *supra* note 179, Article 5.

Member State there is no reason for the Community to be involved in the allocation procedure and it should remain in the exclusive responsibility of the respective Member State.

However, the situation may change, if an air service agreement is not concluded between the EC and one Member State on one side and a non-EC state on the other side, but when several Member States and the Community are parties to the agreement. Due to the fact that the nationality clause contained in the air service agreements of the Member States will have to be amended and substituted by a Community nationality clause,[400] it would also make sense to conclude future air service agreements on a multilateral basis. So far, the nationality clause was the reason, why air service agreements were concluded in a bilateral setting. With the amendment of that clause, it is not necessary any more to limit the number of the participating states, at least not on the side of the European Community. If, in the future, a Member State will be able to designate air carriers registered in other Member States, then there is also no reason, why specific Member States shall be excluded from participating in the respective air service agreement. On the contrary, to obtain a coherent system of international air transportation, it is preferable that the EC and all Member States conclude future air service agreements with non-EC states on a multilateral basis. This would help reducing the patchwork of air service agreements with different rights and obligations that each Member State has concluded with third countries. Not only is this advantageous from a legal perspective, since it would establish a single legal regime between the EC, the Member States and the respective non-EC country, but it would also establish a level playing field for the operating airlines. From a legal point of view, it would not matter any more from and to which destination within the EC an air carrier would operate since, the conditions of operation would be the same for all airlines. Airlines could improve services and profits through management activities and their success would be less dependent on the air service agreements their registration states were able to conclude with other countries. The downside of preference for a single regime concluded between the EC and the Member States with non-EC countries is the difficulty arising with respect to the distribution of air traffic rights granted by the non-EC states. Due to the fact that all, or at least a multitude, of Member States would participate in such an agreement, the allocation procedure could become very complicated. Only if a very liberal 'open-skies' air service agreement providing for an unlimited multiple designation clause and unlimited air traffic rights is concluded, the Member States would be still capable to directly distribute air traffic rights to the airlines. If there are no on the operations in the air service agreement, then there is also no reason at hand to introduce any restrictions in the allocation process. Thus, in such a situation, all Member States should be given the permission to freely allocate air traffic rights, of course in accordance with EC law. A different situation would emerge, if the number of possible designations and/or air traffic rights is limited. A limitation of air traffic rights can occur in two different ways. Either the number of flights in a certain period is limited (limitation of the frequency) or a maximum amount of passenger or cargo capacity that can be offered in a certain period is prescribed (limitation of capacity).

---

400 See insofar the German model service agreement, *supra* note 369, Article 3 (2).

Capacity restrictions are usually either included by requiring the use of a specific aircraft or by granting a quota of passenger seats or cargo space available in a certain period. The two aspects, frequency and capacity, can be also combined and both foresee restrictions. If, in such a situation, every Member State is given the right to distribute air traffic rights and designate air carriers independently and on its own, it could happen that the maximum number of air carriers that can be designated and/or of air traffic rights is exceeded and the respective air service agreement is violated. To prevent such a situation, a different scheme needs to be established, where a control of the total use of designations and/or of air traffic rights is guaranteed. This could be accomplished through a co-operation between the European Community and the Member States or by centralizing the distribution of air traffic rights with one institution. In the latter case it would be most likely to empower the EC to carry out this task. In any case, an independent responsibility of the Member States does not seem feasible in this setting. Which approach should be finally taken, will be examined in the following.

DISTRIBUTION BY THE EUROPEAN COMMUNITY

For the reasons set out above, the Member States will not be able to distribute air traffic rights, if a multilateral air service agreement is concluded, to which several Member States and the EC are parties to. It should be therefore scrutinized, whether in such a situation the European Community could distribute the air traffic rights granted by the non-EC state. For this purpose it needs to be examined, whether the EC has the necessary competences.

ADMINISTRATIVE POWER OF THE EUROPEAN COMMUNITY - PRINCIPLE OF LIMITED ATTRIBUTION

Generally, numerous authors and even some bodies of the European Community take the position that the Community is responsible for legislative measures, while the execution of EC law is left to the Member States.[401] Indeed, in most

401 Stefanie Schreiber (1997), *Verwaltungskompetenzen der Europäischen Gemeinschaft*, (*Administrative Powers of the European Community*), (Baden-Baden, Germany: Nomos Verlagsgesellschaft,) at 39, 59 [Schreiber]; Joachim Suerbaum (1998), *Die Kompetenzverteilung beim Verwaltungsvollzug des Europäischen Gemeinschaftsrechts in Deutschland* (*The Distribution of Powers in the Execution of EC Law*), (Berlin: Duncker and Humblot) at 113f. [Suerbaum]; Oppermann, European Law, *supra* note 196 at paras 637, 640; Anton Klösters (1994), *Kompetenzen der EG-Kommission im innerstaatlichen Vollzug von Gemeinschaftsrecht* (*The Powers of the EC Commission in the National Implementation of Community Law*), (Cologne, Berlin, Bonn, Munich: Carl Heymanns Verlag) at 12, 17 [Klösters]; Manfred Zuleeg (2004), *Der rechtliche Zusammenhalt in der Europäischen Union* (*The Legal Unity in the European Union*), (Baden-Baden, Germany: Nomos Verlagsgesellschaft) at 70 [Zuleeg, Legal Unity]; Manfred Zuleeg (1994), 'Deutsches und europäisches Verwaltungsrecht – wechselseitige Einwirkungen' (German and European Administration Law – Interrelated Impacts) in: *Veröffentlichungen der Vereinigung der*

situations the EC does not execute its regulations on its own, but leaves this task to the Member States.[402] Still, that does not mean that the Community itself cannot execute EC law at all.[403] Unlike many constitutions, the EC Treaty does not contain a specific provision on the distribution of executive competences.[404] The distinction between legislative, executive and judicial powers, which is very common in many constitutions, is not made to the same extent in the EC Treaty. Only the judicial power is clearly assigned to the European Court of Justice. Legislative and executive functions are very often mixed and interrelated. Although the Commission is generally regarded as the executive body of the EC, it is also possible that pursuant to Article 202(3) EC the Council becomes involved in administrative matters.[405] The same is true for legislation, which usually falls into the responsibility of the Council, but which sometimes can be also dealt with by the Commission.[406] Still, as can be seen in Article 202(3) and Article 211(4) EC, the Commission shall be generally responsible for the execution of EC law and the Council's involvement in this area should remain an exception.[407] With respect to Articles 202(3), 211(4) EC, it shall be stressed that these two provisions do not distribute executive powers between the EC and the Member States, but that they only establish a horizontal distribution of powers among the bodies within the Community. Neither the Council nor the Commission can refer to these two Articles to obtain administrative competences and such powers must be derived from other provisions.[408]

---

deutschen Staatsrechtslehrer (*Publications of the Association of German Scholars for Constitutional L.*), vol. 53, (Berlin, New York: Walter de Gruyter) 154 at 158; Hans D. Jarass, 'Die Kompetenzverteilung zwischen der Europäischen Gemeinschaft und den Mitgliedstaaten' (The Distribution of Powers between the European Community and the Member States) (1996) 121 *Archiv des Öffentlichen Rechts* (*Archives of Public L.*), 173 at 182 [Jarass]; see also the 43rd declaration of the Member States to the Treaty of Amsterdam stipulating that the execution of EC law should be generally accomplished by the Member States.

402 Schreiber, *ibid.* at 39; Klösters, *ibid.* at 10, 20; Eberhard Schmidt-Aßmann (2004), *Das allgemeine Verwaltungsrecht als Ordnungsidee* (*General Administration Law as Means of Structuring*), 2nd ed., (Berlin, Heidelberg, New York: Springer) at 382, 385 [Schmidt-Aßmann]; Jürgen Schwarze (1988), *Europäisches Verwaltungsrecht* (*European Administration Law*), vol. 1, (Baden-Baden, Germany: Nomos Verlagsgesellschaft) at 9 [Schwarze, *Administration Law*].

403 But compare Rupert Stettner, in: Manfred A. Dauses, ed. (2004), *Handbuch des EU-Wirtschaftsrechts* (*Manual of EU Commercial Law*), vol. 1, looseleaf (Munich: Verlag C.H. Beck), at B.III (due to the principle of limited attribution, Stettner argues for the existence of a principle of institutional and procedural autonomy of the Member States with respect to execution of EC laws, at para 9).

404 Schreiber, *supra* note 401 at 49, 67; Schmidt-Aßmann, *supra* note 402 at 379.

405 Schreiber, *ibid.* at 42; Schwarze, Administration Law, *supra* note 402 at 23.

406 Schreiber, *ibid.* at 20; Oppermann, European Law, *supra* note 196 at para 357.

407 Oppermann, *ibid.* para 657; Schmidt-Aßmann, *supra* note 402 at 379.

408 Schreiber, *supra* note 401 at 71; Helmut Schmitt von Sydow (1980), *Institutionen der erweiterten Europäischen Gemeinschaften – Die Kommission* (*Institutions of the Extended European Communities – The Commission*), (Baden-Baden, Germany: Nomos Verlagsgesellschaft) at 67 [Sydow, Commission]; Reimer von Borries, 'Verwaltungskompetenzen der Europäischen Gemeinschaft' (Administrative Powers of

Throughout the EC Treaty one can find up to 60 provisions that explicitly grant executive powers to the European Commission.[409] Most of them relate to competition law, state aid, the common fishery policy, the common agriculture market, the common trade policy, the management of some EC funds and the internal organization of the Community.[410] In the area of the common transport policy, Article 75(4) and Article 76(2) EC also provide for a direct and explicit administrative power of the Community. In all these areas, where the Commission is granted explicit administrative powers, it can act without that any additional regulations needed to be adopted to vest the Commission with the necessary executive powers. It is disputed, whether the EC can have additional executive powers, apart from those explicitly stipulated in the EC Treaty. Schwarze mentions that the few administrative powers of the EC are spread throughout the EC Treaty in a manner that they leave the impression to be there only by chance.[411] His statement suggests that any additional administrative powers of the Community were never foreseen by the Member States, when they signed the EC Treaty. Consequently he argues that, within the scope of the EC Treaty, the Community has only legislative, but no administrative competences. The principle of limited attribution pursuant to Article 5(1) EC would foreclose the possibility to extend the Community's administrative powers beyond the areas that are explicitly foreseen in the EC Treaty.[412] Referring to the regulation contained in Article 211(4) EC, Schmitt von Sydow concludes that the Commission shall be only empowered to adopt additional regulations for the execution of EC law, but that it is not vested with any other administrative powers.[413] It is further suggested by Klösters that, if administrative powers are not explicitly set out in the EC Treaty, one will have to resort to the implied-powers theory, Article 308 EC or Articles 94, 95 EC.[414] Apart from these provisions, administrative powers could not be based on any other provisions of the EC Treaty and in particular not on

---

the European Community), in: Ole Due, Marcus Lutter and Jürgen Schwarze, eds. (1995), *Festschrift für Everling* (*Liber Amicorum Everling*), vol. I, (Baden-Baden, Germany: Nomos Verlagsgesellschaft) 127 at 132 [Borries].

409 Helmut Schmitt von Sydow, in: Hans von der Groeben and Jürgen Schwarze, eds. (2004), *Vertrag über die Europäische Union und Vertrag zur Gründung der Europäischen Gemeinschaft* (*Treaty on the European Union and Treaty establishing the European Community*), 6th ed., (Baden-Baden, Germany: Nomos Verlagsgesellschaft), Article 211 at para 66.

410 Schreiber, *supra* note 401 at 29; Suerbaum, *supra* note 401 at 112; Klösters, *supra* note 401 at 8; Schwarze, Administration Law, *supra* note 402 at 27f.; Wolfgang Kahl, 'Hat die EG die Kompetenz zur Regelung des Allgemeinen Verwaltungsrechts?' (Does the EC has the Power to Adopt a Regulation on General Administration Law?) (1996) *Neue Zeitschrift für Verwaltungsrecht* (*New J. of Admin. L.*) 865 at 866 [Kahl]; Albert Bleckmann, 'Zur Auflage im europäischen Beihilfenrecht' (The Condition in European Anti-Dumping Law) (2004) *Neue Zeitschrift für Verwaltungsrecht* (*New J. of Admin. L.*) 11 at 13 [Bleckmann, Anti-Dumping]; see also Article 147 EC for the administration of the Social Fund.

411 Schwarze, *ibid.* at 38; similar Suerbaum, *ibid.* at 113f.

412 Schwarze, *ibid.* at 39; similar Suerbaum, *ibid.* at 103.

413 Sydow, Commission, *supra* note 408 at 67; Similar Klösters, *supra* note 401 at 43.

414 Klösters, *ibid.* at 14, 47f.; Schwarze, Administration Law, *supra* note 402 at 48; Borries, *supra* note 408 at 132, 137.

those, which generally allow the Community to regulate a certain subject matter.[415] In my opinion these positions are too restrictive and there is no provision in the EC Treaty, which would require such a restriction of the Community's powers. Pursuant to Article 202(3) and Article 211(4) EC, the Commission is vested with executive powers to implement the regulations adopted by the Council. However, this cannot mean that the Commission is only limited to the adoption of regulations to implement the legislative acts adopted by the Council. Neither the wording of these two provisions nor any other provision contained in the EC Treaty require a limitation of the Commission's administrative powers to the activities foreseen under Articles 202(3), 211(4) EC. The only limitation of the Commission's executive powers could be derived from a declaration of the Member States made to the Treaty of Amsterdam, whereas the execution of EC law shall remain in the responsibility of the Member States.[416] According to Article 31(2) Vienna Convention on the Law of Treaties, when construing international agreements, annexes and any instruments, made by one or several parties in connection with the conclusion of a treaty, shall be considered, when determining the context of a treaty. However, the declaration does not absolutely forbid any executive measures of the Community, but only stipulates that in general the Member States shall be responsible. This does not preclude that the Community can also implement EC law itself. Furthermore, there is also no need to limit the administrative powers of the EC to the extent Klösters and Schwarze propose to do, since the principles contained in Article 5 EC guarantee a modest and appropriate approach of the Community when discharging these powers. The past experience and the Community's limited involvement in administrative matters show that the principles contained in the EC Treaty preclude an extensive activity of the EC when directly executing EC law. The fact that EC law is mostly executed by Member States is not because of a fixed pre-determination of administrative powers explicitly foreseen in the EC Treaty, but is due to the application of general principles included in the EC Treaty.[417] Jarass even concludes that the EC has far reaching executive competences, which it simply does not use sufficiently.[418]

The EC Treaty does not make a distinction between the different powers of the Community. Administrative powers can be derived from the EC Treaty in the same manner as legislative competences. According to the principle of limited attribution set out in Article 5(1) EC, the Community can have only those administrative powers, which are laid down in the EC Treaty.[419] Without the necessary delegation of competences in the EC Treaty, the Community will not be able to undertake any administrative activities. Article 5(1) EC says:

> The Community shall act within the limits of the powers conferred upon it by this Treaty and of the objectives assign to it therein.

---

415 Klösters, *ibid.* at 24f.; Schwarze, *ibid.* at 48.

416 Declaration 43 to the Treaty of Amsterdam.

417 Klösters, *supra* note 401 at 12.

418 Jarass, *supra* note 401 at 183.

419 Schreiber, *supra* note 401 at 54; Suerbaum, *supra* note 401 at 95; Oppermann, European Law, *supra* note 196 at paras 517, 636; Zuleeg, Legal Unity, *supra* note 401 at 61; Schmidt-Aßmann, *supra* note 402 at 382; Kahl, *supra* note 410 at 866.

An administrative power of the EC can be laid down explicitly in the EC Treaty, as described above, or the Community can have the power to adopt secondary law, in which it includes administrative competences for itself.[420] Thereby it is important to note that the fact that the EC has a legislative power for a specific subject matter, does not imply automatically that it also has the corresponding administrative competence.[421] Whether a provision of the EC Treaty provides for executive powers to the EC or whether it only grants legislative powers, must be examined independently.[422] Apart from explicit executive powers laid down in the EC Treaty and executive powers based on secondary law, it is acknowledged that the European Community can also be vested with administrative powers based on the implied-powers theory and on Article 308 EC.[423] The distinction, whether an administrative power is explicitly set out in the EC Treaty, granted to the EC by secondary law or whether this is based on the implied-powers theory or Article 308 EC Treaty is of great importance. The conditions under which an administrative power can be discharged may vary greatly, depending on which provision it is based. Administrative powers, based on the implied-powers theory, can be only discharges, if the EC requires these executive powers to ensure an effective application of EC law that may not be guaranteed, if the Member States are in charge of executing the respective EC law.[424] Thereby, an assignment of administrative powers to the EC, based on the implied-powers theory, must be only of little significance compared to the remaining regulation.[425] Even more than with respect to legislative competences, the implied-powers theory must be applied with caution, when establishing administrative powers of the Community. In particular the more restrictive school of thought that believes that the execution of EC law falls into the responsibility of the Member States, holds the opinion that an unlimited extension of the executive powers of the EC would contravene the principle of limited attribution and is therefore forbidden under current EC law.[426] Still, if the execution of EC law by the Member States cannot guarantee a coherent and uniform application of EC law, the Community is entitled to adopt measures prescribing the administrative procedure. This competence may be based on either Article 308 EC or on the implied-powers theory.[427] However, it is not sufficient that vesting the EC with administrative powers would solely result in a more efficient application of EC law to justify a competence of the Community.[428] The fact that in some federal states, like for example Germany, the legislative and the executive

420 Schreiber, *ibid.* at 56, 69; Suerbaum, *ibid.* at 110, 116; Oppermann, *ibid.* para 361; Zuleeg, *ibid.* at 70; Bleckmann, Anti-Dumping, *supra* note 410 at 13.

421 Schreiber, *ibid.* at 59; Suerbaum, *ibid.* at 110.

422 Schreiber, *ibid.* at 56.

423 *Ibid.* at 137, 142; Klösters, *supra* note 401 at 13f.; Schmidt-Aßmann, *supra* note 402 at 382; Schwarze, Administration Law, *supra* note 402 at 46f.

424 Suerbaum, *supra* note 401 at 103; Klösters, *ibid.* at 22f.; Kahl, *supra* note 410 at 867; contra Zuleeg, EC Law in the National Sphere, *supra* note 219 (according to Zuleeg, the implied-powers theory cannot be applied and only Art. 308 EC is applicable, at 221).

425 Schreiber, *supra* note 401 at 143.

426 See e.g. Schwarze, Administration Law, *supra* note 402 at 48.

427 Schwarze, *ibid.* at 50.

428 Suerbaum, *supra* note 401 at 102.

powers are divided between the federal state and its member states, and that the federal state is mostly responsible for legislation while the member states are vested with the power to execute federal laws shows that it is not necessary to concentrate administrative and legislative functions within one institution.[429]

For the distribution of air traffic rights Article 80(2) EC could be the appropriate provision, on which an administrative competence of the EC could be based. So far, the Member States have fulfilled most administrative duties with respect to transportation matters.[430] Von Borries argues that, in the field of transportation, Articles 70, 71 EC foresee a legislative function of the Community only and do not grant any administrative powers. Although the term 'common transport policy' as it is used in Article 70 EC does not preclude administrative activities *per se*, Article 71 EC limits the Council to adopt rules. The term 'rules' can be understood as the adoption of generally applicable regulations and not decisions of single cases in administrative procedures. According to von Borries the same shall be true for sea and air transportation and thus administrative powers in the transportation sector could be only established through an application of the implied-powers theory or by having recourse to Article 308 EC.[431] This is even more valid, if one takes into account that Article 80(2) EC also makes reference to 'provisions'. Again the use of the term 'provisions' could indicate that only the adoption of general rules is possible and thus Article 80(2) EC provides for a legislative competence only.[432] Finally, Article 80(2) EC prescribes that the procedural provisions of Article 71 EC shall apply, which leads to the application of the procedure set out in Article 251 EC. Article 251(1) EC contains regulations on the procedure for the adoption of 'acts'. Here again one could argue that acts are to be understood as generally applicable rules. On the other hand, one has to have regard to the systematical context, in which Article 251 EC is placed. The provision is laid down in the same chapter as Article 249 EC. The latter provision describes all possible measures that can be taken by the EC. Next to generally applicable directives and regulations, the EC can also adopt non-binding recommendations and opinions and binding decisions. According to its definition, decisions can be limited to a certain number of addressees. Thus, it can be also used to issue binding regulations of a single case, which is the typical scenario of an administrative decision.[433] Furthermore, Article 253 EC refers to regulations, directives and decisions as 'such acts'. Hence, the term 'act' does not only refer to generally applicable rules, but also to decisions, which are usually issued in administrative procedures. Finally, Article 254(1) EC stipulates that regulations, directives and decisions can be adopted according to the procedure set out in Article 251 EC. For these reasons, it must be concluded that Article 80(2) EC, by referring to the procedure of Articles 71, 251 EC, allows not only for the adoption of regulations and directives, which are legislative acts, but also for the

429 *Ibid.* at 103; Borries, *supra* note 408 at 129.
430 Schreiber, *supra* note 401 at 29.
431 Borries, *supra* note 408 at 140.
432 Schreiber, *supra* note 401 at 101f.
433 Jarass, *supra* note 401 at 182.

issue of decisions, which are the typical instrument in administrative proceedings.[434] As far as the horizontal distribution of powers with respect to air transport matters is concerned, at first, the appropriate institution for taking such decisions would be the Council pursuant to Article 80(2) EC. But due to Article 202(3) and Article 211(4) EC, the Council shall transfer this right to adopt decisions to the Commission. This is also implicitly included in Article 80(2) EC, where the Council is given the power to decide upon the procedure how provisions, in this case decisions, for air transportation may be laid down. Thus, Article 80(2) EC provides for the possibility of the EC to issue administrative decisions on the distribution of air traffic rights.

PRINCIPLE OF SUBSIDIARITY

Overview

As mentioned above, the general principles of the EC Treaty that are applicable, if legislative measures are adopted, are as well valid, if administrative regulations are enacted.[435] One principle that is of great significance in this respect, is the principle of subsidiarity set out in Article 5(2) EC.[436] Article 5(2) EC says:

> In areas which do not fall within its exclusive competence, the Community shall take action, in accordance with the principle of subsidiarity, only if and in so far as the objectives of the proposed action cannot be sufficiently achieved by the Member States, and can therefore, by reason of the scale or effects of the proposed action, be better achieved by the Community.

---

434 Compare Zuleeg, Legal Unity, *supra* note 401 at 70; Jarass, *ibid.* 182.

435 Zuleeg, Legal Unity, *ibid. at* 82; Borries, *supra* note 408 at 146; Gerd Langguth, in: Carl-Otto Lenz and Klaus Borchardt eds. (2003), *EU- und EG-Vertrag* (*EU and EC Treaty*), 3rd edn., (Cologne: Bundesanzeiger), Article 5 at para 22 [Langguth, in: Lenz/Borchardt, Article 5]; Peter M. Schmidhuber and Gerhard Hitzler, 'Die Verankerung des Subsidiaritätsprinzips im EWG-Vertrag – ein wichtiger Schritt auf dem Weg zu einer föderalen Verfassung der Europäischen Gemeinschaft' (The Inclusion of the Principle of Subsidiarity in the EC Treaty an Important Step towards a Federal Constitution of the European Community) (1992) *Neue Zeitschrift für Verwaltungsrecht* (*New J. of Admin. L.*) 720 at 721 [Schmidhuber and Hitzler, Principle of Subsidiarity]; Günter Hirsch, 'Das Subsidiaritätsprinzip' (The Principle of Subsidiarity), in: Reinhard Bötcher, Götz Hueck and Burkhard Jähnke, eds. (1996), *Festschrift für Walter Odersky zum 65. Geburtstag* (*Liber Amicorum Walter Oderksy to his 65th Birthday*), (Berlin, New York: Walter de Gruyter) 197 at 200 [Hirsch].

436 Thomas Goppel, Die Bedeutung des Subsidiaritätsprinzips' (The Meaning of the Principle of Subsidiarity), (1993) *Europäische Zeitschrift für Wirtschaftsrecht* (*Europ. J. of Commercial L.*) 367 at 368 [Goppel]; Wernhard Möschel, 'Subsidiaritätsprinzip und europäisches Kartellrecht' (The Principle of Subsidiarity and European Anti-Trust Law) (1995) *Neue Juristische Wochenschrift* (*New Weekly L. Rev.*) at 281 [Möschel]; Gerhard Konow, 'Zum Subsidiaritätsprinzip des Vertrages von Maastricht' (The Principle of Subsidiarity in the Treaty of Maastricht) (1993) *Die öffentliche Verwaltung* (*J. of Public Admin.*) 405 at 410 [Konow].

This principle is neither a power itself, nor does it distribute powers between Member States and the Community.[437] Whether the Community can adopt measures depends on the application of the principle of limited attribution in connection with the various powers set out in the EC Treaty. The principle of subsidiarity is additionally applied to asses, whether and how the EC can regulate a subject matter, for which it has a concurrent or parallel competence.[438] The objective of the principle of subsidiarity is to strengthen the national identity of the Member States and to preclude a total shift of powers from the Member States to the EC.[439] Still, the principle of subsidiarity is not supposed to reduce existing competences of the Community, if necessary measures need to be adopted. In this respect attention shall be also given to Article 10 EC, which obliges the Member States to support the Community in its efforts and to Article 6(4) EU, which stipulates that the European Union, of which the European Community is a part, shall provide itself with the necessary means to achieve its objectives and to carry out its policies.[440] These two provisions show that, despite the existence of the principle of subsidiarity, the Community should still have all necessary powers that it needs to be able to discharge its duties. When concluding the Treaty of Amsterdam, the Member States have added a protocol, in which they have expressed their opinion, on how the principle of subsidiarity is supposed to be applied.[441] Pursuant to Article 311 EC this protocol is part of the EC Treaty and therefore must be regarded, when the principle of subsidiarity is applied. Protocol 30, Section 3 of the respective protocol stipulates that the principle of subsidiarity is a dynamic concept and should allow for an increase of EC competences, if this is necessary and a reduction, if an activity of the EC is dispensable. The application of the principle of subsidiarity should help striking a balance between the desire of the Member States to preserve their national identities and the need of the Community to be able to efficiently fulfil its duties. As it is with respect to the principle of limited attribution, some authors, arguing that administration is a matter that genuinely falls into the responsibility of the Member States, suggest that the principle of subsidiarity needed to be applied more strictly, when it comes to administrative issues compared to legislative acts.[442] Again such an opinion and a stricter application of the principle of subsidiarity must be refused. As it is the case with the principle of limited attribution, the scope of the principle of subsidiarity does not depend on whether a legislative or an executive measure is undertaken. There is no provision in the EC Treaty that assigns administrative responsibilities primarily to the Member States and that would require a stricter application of the principle of subsidiarity. If the impression arises that Member States are responsible for the execution of EC law, this results from the application

437 Hirsch, *supra* note 435 at 205; Möschel, *ibid.* at 281.

438 Schreiber, *supra* note 401 at 57.

439 Oppermann, European Law, *supra* note 196 at para 522; Hirsch, *supra* note 435 at 204, note 22; Stefan Pieper, 'Subsidiaritätsprinzip – Strukturprinzip der Europäischen Union' (The Principle of Subsidiarity – Structural Principal of the European Union) (1993) *Deutsche Verwaltungsblätter* (*German J. of Admin. L.*) 705 at 710 [Pieper].

440 Oppermann, *ibid.* at para 522.

441 See Protocol No. 30 to the Treaty of Amsterdam.

442 Schreiber, *supra* note 401 at 58.

of the various principles set out in the EC Treaty and in particular of the principle of subsidiarity.[443] But it is wrong to set up the premise that Member States are *de jure* responsible for the execution of EC law and that the Community must bring forward additional reasons, why it should have the administrative power for a subject matter. An administrative competence to execute EC law by the Community is generally possible and will be excluded pursuant to Article 5(2) EC only, if the Member States can execute EC law in a sufficient manner.[444]

Concurrent power

The first prerequisite for an application of the principle of subsidiarity is that a matter does not fall into the exclusive competence of the European Community. Where this is the case, the Member States cannot regulate a subject matter anyway and thus it does not make sense to apply the principle of subsidiarity. In such a situation, the measures undertaken by the Community are only subject to the principle of proportionality stipulated in Article 5(3) EC.[445] In general, because of the principle of limited attribution, the Member States are responsible for all subject matters, unless the Community is given the power to undertake a particular activity.[446] Under these circumstances, only few exclusive powers of the European Community exist and the EC will have mostly only concurrent and parallel competences.[447] Some authors even hold the opinion that the Community has no exclusive competences at all, but only concurrent powers.[448] Such an approach clearly contradicts the wording of Article 5(2) EC and must be therefore refused. Which powers are exclusively assigned to the EC is difficult to assess. The EC Treaty does not contain any definition or catalogue of exclusive powers of the European Community.[449] Therefore, it is necessary to analyse each power according to its wording, objective and structure, to determine, whether the Community shall be exclusively competent and thereby should exclude any corresponding activity of the Member States.[450] According to von

443 Borries, *supra* note 408 at 127, 130.

444 Armin von Bogdandy and Martin Nettesheim, in: Eberhard Grabitz and Meinhard Hilf, eds. (September 2004), *Das Recht der Europäischen Union* (*The Law of the EU*), looseleaf (Munich: Verlag C.H. Beck, 7th supplement), Article 3b at para 33 [Bogdandy/Nettesheim, in: Grabitz/Hilf, Article 3b].

445 Schreiber, *supra* note 401 at 64.

446 Zuleeg, Legal Unity, *supra* note 401 at 61.

447 Oppermann, European Law, *supra* note 196 at para 518; Bogdandy/Nettesheim, in: Grabitz/Hilf, Article 3b, *supra* note 444 at para 15.

448 See Christian Calliess, 'Der Schlüsselbegriff der "ausschließlichen Zuständigkeit" im Subsidiaritätsprinzip des Article 3b II EGV' (The Core Element of the 'Exclusive Competence' in the Principle of Subsidiarity stipulated in Article 3b II EC Treaty), (1995) *Europäische Zeitschrift für Wirtschaftsrecht* (*Europ. J. of Commercial L.*) 693 at 694 [Calliess, Exclusive Competence]; Clemens Stewing, 'Das Subsidiaritätsprinzip als Kompetenzverteilungsregel im Europäischen Recht' (The Application of the Principle of Subsidiarity to Distribute Powers in European Law) (1992) *Deutsche Verwaltungsblätter* (*German J. of Admin. L.*) 1516 at 1517 [Stewing].

449 Konow, *supra* note 436 at 407; Calliess, *ibid.* at 693.

450 Compare Konow, *ibid.*; Calliess, *ibid.* at 696.

Bogdandy and Nettesheim it is not possible to evaluate, whether a power is exclusive, concurrent or parallel by having regard to the wording of a provision. The evaluation must be rather based on the general structural principles of the EC Treaty, whereby a coherent and effective application of EC law and an appropriate approach to solve legal problems shall be of greatest significance.[451] Von Bogdandy's and Nettesheim's suggestion is too broad and too vague. It does not give any useful guidance by itself, how provisions contained in the EC Treaty providing for competences of the EC shall be construed. It is agreed that only the wording of a power might not be sufficient to determine, whether a power is exclusive, concurrent or parallel. For this purpose further aspects must be taken into account. The principles von Bogdandy and Nettesheim mention can be also applied, but the scope of a provision must be first determined by using the common instruments of legal interpretation. Therefore, it is suggested that the more precise a provision providing for a power of the EC is and the more a subject matter can be regulated by the European Community, the more likely is an exclusive power of the EC to be acknowledged.[452] An exclusive power of the EC is therefore affirmed, if the Community is given the power to completely regulate a subject matter.[453] In this respect it is necessary to have regard to the position adopted by the Commission, who argues that all areas related to the establishment of the internal market, fall into the exclusive power of the EC. These areas are listed in Article 14(1) EC.[454] Thus, the policies regarding the freedom of free movement of goods, the freedom of establishment, the freedom of services and free movement of workers, as well as some other supporting policies, would fall into the exclusive responsibility of the Community. That should also include parts of the common transport policy.[455] The Commission's position is highly controversial.[456]

---

451 Bogdandy/Nettesheim, in: Grabitz/Hilf, Article 3b, *supra* note 444 at para 15.

452 Konow, *supra* note 436 at 408; Calliess, Exclusive Competence, *supra* note 448 at 693.

453 Zuleeg, Legal Unity, *supra* note 401 at 67f., Langguth, in: Lenz/Borchard, Article 5, *supra* note 435 at para 25; Manfred Zuleeg, 'Das Subsidiaritätsprinzip im Europarecht - Inhalt, Justiziabilität, Entwicklung' (The Principle of Subsidiarity in European Law - Content, Judicial Control, Development), in: Gil Carlos Rodríguez Iglesias, Ole Due, Romain Schintgen and Charles Elsen, eds. (1999), *Mélanges en hommage à Fernand Schockweiler (In Memory of Fernand Schockweiler*), (Baden-Baden, Germany: Nomos Verlagsgesellschaft, 1999) 635 at 638 [Zuleeg, Principle of Subsidiarity].

454 European Commission, cited in, Caliess, Exclusive Competence, *supra* note 448 at 693; Langguth, *ibid.* at para 25; Hirsch, *supra* note 435 at 201; Konow, *supra* note 436 at 407; Jörn Pipkorn, 'Das Subsidiaritätsprinzip im Vertrag über die Europäische Union – rechtliche Bedeutung und gerichtliche Überprüfbarkeit' (The Principle of Subsidiarity in the EU Treaty – Legal Meaning and Judicial Control) (1992) *Europäische Zeitschrift für Wirtschaftsrecht* (*Europ. J. of Commercial L.*) 697 at 699 [Pipkorn]; Wernhard Möschel, 'Zum Subsidiaritätsprinzip im Vertrag von Maastricht' (The Principle of Subsidiarity in the Treaty of Maastricht) (1993) *Neue Juristische Wochenschrift* (*New Weekly L. Rev.*) 3025 at 3026 [Möschel, Principle of Subsidiarity].

455 Calliess, *ibid.* at 693.

456 Peter M. Schmidhuber and Gerhard Hitzler, 'Binnenmarkt und Subsidiaritätsprinzip' (Internal Market and the Principle of subsidiarity), (1993) *Europäische Zeitschrift für Wirtschaftsrecht* (*Europ. J. of Commercial L.*) 8 (according to Schmidhuber and Hitzler the

The German Government, for example, has taken a very restrictive position on that issue and denied such an extension of the Community's powers.[457] It is argued that the Commission's position is very vague and, in particular, with respect to the supporting policies, it is always possible to construe an importance for the establishment of the internal market and thereby justify an exclusive power of the EC.[458] A more restrictive school of thought wants to affirm exclusive powers of the EC only, if any further activity by the Member States in the respective field would endanger future activities of the Community.[459] Proponents of a rather restrictive approach therefore acknowledge only a few areas that belong to the exclusive power of the European Community. The common agricultural and fishery policies, the European Monetary Union, external trade regulations, common tariffs regulations according to Article 26 EC, the control of state aids and the organization of the EC administration are sectors that are exclusively assigned to the Community.[460] Additionally, Zuleeg and Calliess also attribute cross-border transportation according to Articles 71(1)(a), (b) EC to the exclusive responsibility of the EC.[461] According to Vedder the regulation of all common policies should fall into the exclusive responsibility of the European Community and the latter one should be exclusively competent for all internal and external matters.[462] On the other hand, the EC is denied any exclusive powers solely for the purpose to ensure the establishment of a level playing field in the internal market.[463] An exclusive power of the EC can be also approved, as seen above, with respect to external relations pursuant to the ERTA judgement and Opinion 1/76.[464] Some authors even propose to limit exclusive powers of the Community to the external area only and argue that provisions empowering to adopt

---

areas included in the program to establish the internal market belong to the concurrent power of the European Community, at 9) [Schmidhuber and Hitzler, Internal Market]; Konow, *supra* note 436 at 408.

457 German Government, cited in: Calliess, Exclusive Competence, *supra* note 448 at 694.

458 *Ibid.* at 695.

459 Zuleeg, Legal Unity, *supra* note 401 at 67f., Zuleeg, Principle of Subsidiarity, *supra* note 453 at 638.

460 Zuleeg (1993), *Legal Unity, ibid.*; Zuleeg, *Principle of Subsidiarity, ibid.*; Langguth,in: Lenz/Borchardt, Article 5, *supra* note 435 at para 1; Hirsch, *supra* note 435 at 201; Calliess, Exclusive Competence, *supra* note 448 at 694, 695; Hans-Jürgen Lambers, 'Subsidiarität in Europa – Allheilmittel oder juristische Leerformel?' (Subsidiarity in Europe – Omnipotent Instrument or Insignificant Legal Formula?) (1993) *Europarecht* (*Europ. L. Rev.*) 229 at 235 [Lambers]; Peter M. Schmidhuber, 'Das Subsidiaritätsprinzip im Vertrag von Maastricht' (The Principle of Subsidiarity in the Treaty of Maastricht) (1993) *Deutsche Verwaltungsblätter* (*German J. of Admin. L.*) 417 at 418 [Schmidhuber].

461 Zuleeg, Legal Unity, *supra* note 401 at 67f., Calliess, Exclusive Competence, *supra* note 453 at 696.

462 Vedder, Treaty-making power, *supra* note 69 at 139.

463 Bogdandy/Nettesheim, in: Grabitz/Hilf, Article 3b, *supra* note 444 at para 29; *contra* Schwartz, cited in: Calliess, Exclusive Competence, *supra* note 448 at 694.

464 See above at 24ff., whereby it is also my opinion that in the meantime the ECJ has given up its theory of implied powers pursuant to Opinion 1/76.

internal measures are all of a concurrent nature.[465] As set out above, it is my opinion that the Community originally has concurrent powers in external relations, but that, under certain circumstances, these powers can become exclusive ones pursuant to the ERTA judgement or Opinion 1/76.[466] Contrary to Vedder an automatic exclusive, treaty-making power within the common policies is not acknowledged. However, with an increasing internal legislation, such an exclusivity of the EC will also emerge with respect to external relations.

The previous overview has shown that numerous opinions exist, on how it shall determined whether a power belongs exclusively to the Community. While most scholars seem to agree that regulations regarding sea and air transportation pursuant to Article 80(2) EC would not fall into the exclusive competence of the Community, the proponents of an extensive school of thought, predominantly the Commission, could, by invoking Article 14(1) EC, come to the opposite conclusion. Hence it is necessary, to have further regard to this dispute. Furthermore, whether a power belongs exclusively to the EC, shall be decided by having regard to the rules of legal interpretation. In construing a provision of the EC Treaty, attention must be given to its wording, structure, historical background and objective. If one follows the Commission's position that all areas listed in Article 14(1) EC fall into the exclusive power of the Community, this could also include sea and air transport due to the reference that is made in Article 14(1) EC to Article 80 EC.[467] On the other hand, Article 80(2) EC contains a wide discretion for the Council, whether it wants to adopt regulations for these two industries or not. This could contravene the idea of an exclusive power for the EC in these two areas. An exclusive power of the Community has the effect that any corresponding activity of the Member States is precluded. If, at the same time, the Council has a wide discretion and is thus not obliged to act under Article 80(2) EC, it could lead to a complete halt in regulating sea and air transport, a situation that the Member States can hardly have approved, when they adopted the EC Treaty. Additionally, the historical background also contravenes the idea of an exclusive power of the EC in matters of sea and air transport. The exclusion of sea and air transport from the general rules on a common transport policy pursuant to Articles 80(1), (2) EC is due to the low significance these two industries and means of transportation had for the EC, when the EC Treaty was originally signed in 1957. At that time, the Community consisted of only six members that were located very close to each other, so that sea and air transport between these original Member States was not very common.[468] For this reason, a special regulation was included in Article 80(2) EC, excluding sea and air transport from the general rules on transportation. It could be argued now that, with the extension of the EC, the situation has changed and the significance of, in particular, air transportation has increased, thus making it necessary to acknowledge an exclusive power of

465 Stein, cited in: Calliess, *Exclusive Competence*, *supra* note 448 at 695.

466 See above at 24ff.; with respect to Opinion 1/76 see *supra* note 464.

467 Peter Mückenhausen, in: Carl-Otto Lenz and Klaus Borchardt eds. (2003), *EU- und EG-Vertrag* (*EU and EC Treaty*), 3rd edn., (Cologne: Bundesanzeiger), Article 14 at para 4 [Mückenhausen, in: Lenz/Borchardt, Article 14].

468 Frohnmeyer, in: Grabitz/Hilf, Article 80, *supra* note 56 at para 1.

the EC in these two areas. On the other hand, solely the development in the EC cannot be sufficient to attributed a different scope to Article 80(2) EC contrary to its wording, historical background and intention of the signatory states. If the Member States would have wanted to change the scope of Article 80(2) EC, they could have amended this provision in the meantime. But despite the various enlargements of the Community and the introduction of a common market for air transportation services, the Member States have denied to change this provision in all the treaty revisions that took place over the last 15 years. Furthermore, the inclusion of Article 80 EC in Article 14(1) EC, seems to be a legislative lapse. While all other provisions that are listed in Article 14(1) EC and to which it is referred impose a duty to act on the institutions of the Community. Article 80(2) and Article 49(2) EC are the only two provisions that provide for a discretion of the Council, whether it wants to become active. As far as Article 80(2) and Article 49(2) EC are concerned, there is an additional significant difference between these two provisions. Having regard to the regulation contained in the first paragraph of each provision, Article 80(2) EC refers to the regulation of a different subject matter, namely sea and air transportation, compared to transport by rail, road and inland waterway listed in Article 80(1) EC. Article 49(2) EC, on the other hand, does not contain a provision on the regulation of a different subject matter, but extends the regulation contained in Article 49(1) EC to a greater number of people. The subject matter remains the same. Under these circumstances and taking into account the wide discretion granted to the Council, the historical background and the intention of the Member States, it is suggested that the reference to Article 80 EC contained in Article 14 EC should be limited to the first paragraph of Article 80 EC only. Thus, even if one agrees with the extensive approach of the Commission that the EC must have the exclusive power for all areas included in Article 14 EC, this has to be restricted in the field of transportation to the sectors contained in Article 80(1) EC and does not extend to sea and air transportation. Therefore, it is submitted that the European Community can have recourse to Article 80(2) EC to adopt regulations on the distribution of air traffic rights to be undertaken by the Commission. This power is a concurrent one and thus subject to the principle of subsidiarity.[469]

Necessity and scope of an activity of the community

Where a concurrent power of the Community exists, the Community shall only act pursuant to the principle of subsidiarity, if the objective of the proposed action cannot be sufficiently achieved by the Member States and, by reason of scale or efficiency, can be better obtained, if the Community takes the necessary action. Thereby these two conditions are cumulative.[470] Firstly, this can be derived from the wording of the regulation and the use of the term 'and can therefore' between the two conditions

469 See also Frohnmeyer, *ibid.* at para 21.

470 Schreiber, *supra* note 401 at 60; Langguth, in: Lenz/Borchardt, Article 5, *supra* note 435 at para 19; Schmidhuber and Hitzler, Principle of Subsidiarity, *supra* note 435 at 722; Markus Kenntner, 'Das Subsidiaritätsprinzip des Amsterdamer Vertrages – Anmerkungen zum Begrenzungscharakter des gemeinschaftlichen Subsidiaritätsprinzips' (The Principle of Subsidiarity in the Treaty of Amsterdam – Remarks on the Ability to Restrict of the Principle

set out in Article 5(2) EC.[471] Secondly, the legislative history of Article 5(2) EC also leads to this conclusion.[472] While the Commission wanted to establish the principle of subsidiarity solely on the additional-benefit-criterion included in the second part of the provision, the Member States, in particular the member states of Germany, insisted on the sufficiency criterion set out in the first part of Article 5(2) EC. As a compromise, both criteria were combined. But since the sufficiency-criterion is more restrictive, the additional-benefit criterion regularly will not cause any difficulties, once the first condition is fulfilled. Thus, the inclusion of the Commission's favoured criterion was unnecessary.[473] However, some scholars argue that Article 5(2) EC allows the Community to adopt measures, even if the Member States could achieve the respective objective also to a satisfying extent. Contrary to its objective, the principle of subsidiarity is held to encourage the EC to accumulate more and more competences instead of limiting its activities.[474] In this respect reference is also made to the protocol on the application of the principle of subsidiarity, arguing that the 'dynamic concept' of the principle of subsidiarity could be applied in a manner that results in an extension of the powers of the EC. It is shown that the Commission has always ignored the final version of the principle of subsidiarity and has continued to apply the additional-benefit-criterion only.[475] Similarly, Stewing defines the term 'sufficient' included in the first condition as 'being better'. Since 'being better' is also included in the second assessment of Article 5(2) EC, the first condition of that regulation becomes completely obsolete. From his perspective the EC shall be excluded from undertaking a measure only, if it can be better achieved by the Member States. This understanding of Article 5(2) EC is not supported by the proceedings that lead to the adoption of this regulation, the reasons for its inclusion in the EC Treaty as well as its wording. It is, however considered as possible that Article 5(2) EC might not have been applied strictly enough in the past. Still, this does not justify to consider the principle of subsidiarity as a tool to extend the EC's competences.

The greatest difficulty, when applying the principle of subsidiarity, is to determine the criteria upon which it is assessed, whether an activity undertaken by the Member States is insufficient and when the Community can better achieve the objectives of a proposed action.[476] Firstly, the pursued objectives must be objectives of the EC that are laid down in the EC Treaty and, in particular, in the Preamble and in Articles 2,

---

of Subsidiarity) (1998) *Neue Juristische Wochenschrift* (*New Weekly L. Rev.*) 2871 at 2873 [Kenntner]; see also Protocol 30, sec. 5 to the Treaty of Amsterdam.

471 Goppel, *supra* note 436 at 367.

472 *Ibid.*

473 See also Kenntner, *supra* note 470 at 2872f.

474 Achim Schnabel, 'Das Subsidiaritätsprinzip im Vertrag über die Europäische Union', ('The Principle of Subsidiarity in the EU Treaty'), (1992) *Bayerische Verwaltungsblätter* (*Bavarian J. of Admin. L.*) 393 [Schnabel]; similarly Stewing, *supra* note 448 at 1518.

475 Kenntner, *supra* note 470 at 2871, 2874f.; similarly Konow, *supra* note 436 at 409.

476 Zuleeg, Legal Unity, *supra* note 401 at 79; Zuleeg, Principle of Subsidiarity, *supra* note 453 at 639; Hirsch, *supra* note 435 at 210; Lambers, *supra* note 460 at 236.

3 and 4 EC.[477] Thereby, the objectives of establishing an internal market, fostering economic and social integration and creation of a level playing field have to be taken particularly into account.[478] On the other hand, pragmatic considerations, like a higher degree of efficiency that can be obtained, if the EC takes the necessary actions, are usually not sufficient to satisfy the conditions of the principle of subsidiarity.[479] Secondly, it is not necessary that the Member States approach a subject matter in the same manner. The principle of subsidiarity does not require that the Member States act equally.[480] They can approach a situation in different ways, as long as they achieve a satisfying result. Thirdly, the Member States do not have to achieve the best result possible, but only a satisfying one.[481] In the Protocol on the application of the principle of subsidiarity the Member States have laid down that every measure undertaken by the EC must be justified by qualitative and, if possible, quantitative criteria. It should be also taken into account, whether the respective subject has transnational links and therefore cannot be sufficiently regulated by the Member States alone, whether the anticipated measures, if undertaken solely by the Member States, could contravene objectives of the EC Treaty and whether measures, if undertaken by the EC, could, due to their scope or their impact, bring additional benefits over a similar activity undertaken solely by the Member States.[482] Accordingly, based on these criteria and the distinction made in Article 5(2) EC, a two-step approach has been included to determine, whether the EC should be entitled to make use of a concurrent power in accordance with the principle of subsidiarity.[483] Firstly, under the negative-determination test it is hypothetically assessed, whether the Member States have the abilities and possibilities to sufficiently achieve an objective of the EC.[484] Only if this is denied, it is analysed under the positive-determination-test, whether additional benefits can be obtained, if the Community, instead of the Member States, is in charge of the respective task.

---

477 Insofar not very clear Pieper, *supra* note 439 (on the one hand Pieper mentions that 'objectives' in Article 5 (2) EC does not have the same meaning as 'objectives of the EC Treaty' in Article 308 EC, on the other hand Pieper concludes that the objectives of the secondary laws must be in accordance with the objectives of the EC Treaty, at 709); Zuleeg, Principle of Subsidiarity, *ibid.*

478 Schmidhuber and Hitzler, Principle of Subsidiarity, *supra* note 435 at 722f.; Zuleeg, Legal Unity, *supra* note 401 at 80.

479 Schreiber, *supra* note 401 at 60; Langguth, in: Lenz/Borchardt, Article 5, *supra* note 435 at para 35; Schmidhuber and Hitzler, *ibid.* at 723; Zuleeg, Principle of Subsidiarity, *supra* note 453 at 639.

480 Lambers, *supra* note 460 at 236.

481 Konow, *supra* note 436 at 409.

482 Protocol No. 30 to the Treaty of Amsterdam; Langguth, in: Lenz/Borchardt, Article 5, *supra* note 435 at para 20; Schmidhuber and Hitzler, *Principle of Subsidiarity*, *supra* note 435 at 722; Schmidhuber and Hitzler, *Internal Market*, *supra* note 456 at 418f.

483 Pipkorn, *supra* note 454 at 699; Schmidhuber and Hitzler, *Internal Market*, *ibid.* at 419.

484 Schmidhuber and Hitzler, *Principle of Subsidiarity*, *supra* note 435 at 723; Zuleeg, Principle of Subsidiarity, *supra* note 453 at 639.

Negative determination test

Under the negative-determination test it is necessary to assess first, whether the Member States have the ability and capability to organize the distribution of air traffic rights. Generally, the assumption can be made that, due to their experience in distributing air traffic rights in the past and in executing EC laws, Member States will be also able to allocate air traffic rights when the EC is involved in this area. In this respect some Member States have already gained particular experience. For example, Germany and Great Britain have a multitude of air carriers that operate on international routes and both countries have established procedures, how to allocate air traffic rights, if the demand of the air carriers exceeds the number of available air traffic rights.[485] On the other hand, there are also some Member States that have either only one major internationally operating airline or maybe even no such national air carrier at all. These countries have not been able to gain any experience in distributing air traffic rights. Still, the fact that some Member States could not engage in any air traffic rights allocation procedures, does not make these countries inappropriate for this task. As mentioned above, the assumption is that, due to their experience in executing laws and by having the necessary administrative structures in place, the Member States are, at first, the appropriate institutions to distribute air traffic rights. A different conclusion could be drawn however, if particular circumstances exist that could render the Member States as inappropriate for executing this task effectively. Such an extraordinary situation could occur, as described above, due to the new Community nationality clause that will be included in future air service agreements and when multilateral air service agreements are concluded. There are several different kinds of problems that can occur, if Member States continue to distribute air traffic rights in such a situation. Firstly, Member States may favour their own national air carriers. This problem should not occur, since Member States are already precluded by Article 12 EC from discriminating on the basis of nationality. If a Member State favours its national airline, it will be subject to a procedure pursuant to Articles 226, 227 EC. Thus, the possibility of discrimination based on nationality is not a danger specifically connected with the distribution of air traffic rights. It is a situation that can always occur, if Member States are responsible for the execution of EC law. It can be precluded by provisions and procedures currently already laid down in the EC Treaty. Secondly, if several allocation procedures take place independently from each other, it could happen that one air carrier is given air traffic rights in each procedure, which, in the end, may result in a dominant position of this carrier in a specific market. This problem seems to be more difficult to resolve than the first one. It is indeed possible that air carriers are granted air traffic rights, which, on a case by case basis, do not give rise to any concerns, but if all allocations are taken into account, the competitive situation in a specific market may be destabilized. Such a situation could be avoided, if one authority would have control over all distribution procedures. On the other hand, the competitive scenario in a particular market is not an issue that should be dealt with in the context of regulating air transportation, but should be left to the application of the special competition laws. According to

485 See below at 210ff. and 222ff.

Article 82 EC the mere creation of a dominant position in a market is not illegal, but only its abuse is. Whether such an abuse occurs must be examined and cannot be pre-maturely assumed, due to a certain position or strength of a market player. Therefore, it is not necessary to concentrate all distribution procedures of air traffic rights within one institution, only to prevent negative impacts on competition. Any necessary measures can be still undertaken *ex post* once a competition threatening situation occurs. Thirdly, problems may also result from the fact that air transportation services occur in a transnational setting. This fact by itself is, however, not sufficient to take away the competence to distribute air traffic rights from the Member States and assign this task to the EC. When the common air market was established in the European Community and all Community air carriers were give the right to operate in this market, the Member States retained the power to decide upon the licensing of air carriers pursuant to a common regulation.[486] Similarly it can be imagined that Member States distribute air traffic rights pursuant to an EC regulation, which prescribes the applicable procedure. Still, the distribution of air traffic rights differs from the granting of operating licenses significantly with respect to one aspect. It is the fact that the number of available air traffic rights and the conditions of its use can vary greatly, depending on the air service agreement in which they are granted. Therefore, apart from the assessment on the abilities and capabilities of the Member States to organize the allocation procedures for air traffic rights in general, it is also necessary to examine, whether, due to the various constellations that can results from the different air service agreements, it could be infeasible for the Member States to distribute air traffic rights in an EC setting. The distribution of air traffic rights depends on the degree of restrictions that are laid down in air service agreements. Significant limitation can be imposed in respect of the number of permitted designations and in respect of air traffic rights. As defined above, air traffic rights are determined by granted capacities and frequencies. Thus, restrictions of air traffic rights can occur in respect of these two aspects. An air service agreement can be very liberal and not impose any restrictions in respect of designation and air traffic rights or it can be also very restrictive and foresee limitations to one or all of these aspects. The less restrictive the agreement is, the easier is the distribution of air traffic rights. To answer the question, whether Member States can distribute air traffic rights, it shall be distinguished between three different scenarios and kinds of air service agreements:

i) liberal air service agreements;
ii) air service agreements restricting air traffic rights, and
iii) air service agreements restricting number of designation possibilities.

Liberal air service agreements

If an air service agreement is very liberal and does not foresee any restrictions on designation and air traffic rights, the distribution of air traffic rights does not cause any difficulties. In such a situation, every Member State shall be entitled to

486 See Regulation 2407/92, *supra* note 49.

distribute as many air traffic rights as air carriers are asking for. There is no necessity for control on market access as, according to the understanding of the contracting parties to a liberal air service agreement, state interference should be minimized, and the optimum solution and best services shall be determined by market forces and the market players in a competitive surrounding. The only duty a state has in such a situation, is to ensure that no unfair competition occurs. Thus, if a liberal air service agreement is concluded, the Member States can accomplish the distribution of air traffic rights.

Air service agreements restricting air traffic rights

If in an air service agreement, the number of available air traffic rights is limited and the distribution of these rights is left to the Member States, a cooperation and an exchange of information between the Member States would become necessary. Otherwise it could happen that the maximum number of air traffic rights is exceeded, which could result in an infringement of the respective air service agreement. Such a situation must be avoided by any means as it could prove harmful for the international acceptance of the European Community, if international agreements, to which the Community is a party, are breached due to the internal structure of the EC. As described above, air traffic rights could be granted in a multilateral setting by non-EC states to the European Community and the Member States. Thereby the Community and the Member States would be considered as one jointly acting party to the respective air service agreement. Consequently, the EC and the Member States would be jointly responsible to ensure a distribution of air traffic rights that is in accordance with the respective air service agreement. Such a situation, in which the EC and the Member States act together and are jointly granted rights in an international agreement, which they need to distribute among their citizens is not new. To find an appropriate method for the distribution of air traffic rights, it could be helpful to take a look to such other sectors. Two areas, where limited rights exist that are made available to EC citizens are the common fishery policy and the common market for bananas. Although with respect to the latter sector no limited rights are granted to the EC by third countries in international treaties, but the EC itself limits the amount of bananas to be imported from third countries, it might be still helpful to analyse, how these self-imposed quotas are distributed among the Member States and the single importers of the EC.

Within the common fishery policy two different scenarios are foreseen. Firstly, it is possible that a non-EC country grants fishing rights to the Community, which it distributes among its Member States.[487] Secondly, the European Community can set a maximum amount of permitted catches for specific fish or groups of fish in EC waters, the Total Allowable Catch (TAC), and then determines quotas for the Member States within the TAC.[488] Since the quotas and TACs always refer to a specific kind

487 Robin Rolf Churchill (1987), *EEC Fisheries Law* (Dordrecht, The Netherlands: Martinus-Nijhoff Publishers) 183 [Churchill].

488 *Ibid.* at 115; EC, Council Regulation No 2371/2002 of 20 December 2002 on the conservation and sustainable exploitation of fisheries resources under the Common Fisheries Policy [2002] O.J. L 358/59, Article 20 No. 1 [Regulation 2371/2002].

of fish or group of fish, numerous quotas and TACs need to be set. In both cases, each Member State subsequently allocates the quotas it has been assigned to its fishermen according to a method determined by the respective Member State.[489] Although the two scenarios differ slightly from each other, from a procedural perspective they have in common that a limited number of rights is, first, set for the whole European Community, then distributed among the Member States and finally allocated to EC citizens. With respect to the analysis of the distribution procedure it does not matter, whether the EC sets a TAC or limited fishing opportunities are granted to the Community by a third country. For the following analysis it is therefore not necessary to distinguish between the two situations. The most difficult task for the EC is to determine the criteria, according to which the quotas of the Member States are fixed. The regulation dealing with this matter is very broad and prescribes that 'each Member State's relative stability of fishing activities for each of the stocks considered' shall be assured[490] and that 'the interests of each Member State are taken into account'.[491] In finally fixing the quotas three aspects are considered:

- the past fishing performance;
- the needs of regions particularly dependent on fishing from an economic perspective; and
- the losses suffered by Member States as a result of the extension of fishing limits to 200 nautical miles by non-EC states.[492]

Furthermore, Member States can exchange their quotas, after prior notice is given to the Commission.[493] This possibility was introduced to prevent a waste of quotas, if Member States are unable to fully use their allocated shares. It shall be stressed that quotas can be only exchanged, but not sold.[494] Thus, a Member State can obtain additional quotas, if it expects that its fishermen exceed the limits assigned to it for a certain kind of fish. But in doing this, it has to give up quotas for another kind of fish. This procedure ensures that the TACs or granted fishing rights are never exceeded. Objections against this distribution scheme were raised with respect to the non-discriminatory rule of the EC Treaty. The European Court of Justice decided that discrimination only occurs, when similar situations are subject to dissimilar treatment or when dissimilar situations are treated in a similar manner. Under these conditions it would have been discriminatory, if all Member States were given the same amount of fishing quotas as their fishing capabilities can differ significantly. This becomes quite obvious, if one only compares the fishing industry of a country like Luxembourg with the ones of Spain or Greece. Therefore, the ECJ ruled that it

489 Regulation 2371/2002, *ibid.*, Article 20, No 3.

490 *Ibid.*, Article 20, No 1 (this term was already included in the first Council Regulation (EEC) No 170/83 of 25 January 1983 on the establishing of a Community system for the conservation and management of fishery resources and was not changed since that time).

491 *Ibid.*, Article 20, No 2.

492 Churchill, *supra* note 487 at 115.

493 Regulation 2371/2002, *supra* note 488, Article 20, No 5.

494 Churchill, *supra* note 487 at 117.

was permissible to treat Member States differently, if this was done on an objective basis.[495] With respect to quotas, the Court of Justice also decided that it was legal to determine quotas on past fishing performance and not on past consumption within a Member State, which could lead to the result that a Member State was given a higher quota for a certain kind of fish than its citizens actually consumed. This approach was justified by showing that regional specialization would thus be encouraged and least efficient fishermen would be precluded from gaining advantages from the consumption level of the Member State's citizens, where they were registered.[496]

In structuring the common market for bananas a different approach than in the common fishery policy was undertaken. The controversial question, whether this system is in accordance with international law and in particular with the WTO regime shall not be further scrutinized here.[497] The following analysis will concentrate solely on the intra-Community aspects of this system. In distributing quotas for the import of bananas, a distinction is first made between the different kinds of bananas based on their origin. The applicable EC regulation differs between bananas from the Member States, from specific African and Pacific states, with whom the EC has concluded association agreements pursuant to Article 182 EC and bananas from other countries ('third countries').[498] In the following, attention will be given to the regulation of quotas for bananas from third countries, as this situation resembles most that of international aviation. In the respective EC regulation a current limit of 2.2 million tons of banana is foreseen that can be imported from third countries by EC importers.[499] This total amount is divided among all EC banana importers. For this purpose, every importer has to report to its Member State, the amount of bananas that it imported over the past three years. Based on these figures, the respective Member State fixes a yearly average amount, the reference amount, which is then reported to the Commission. Having regard to the previous imports and the maximum current limit, the Commission can decide to equally reduce the reference amounts for all participants. Subsequently, the Member States are informed about any reduction and the corresponding reduction percentage. In applying the reduction coefficient, the Member States allocate the new reference quota, which is valid for one year, to each importer.[500] However, the importers are not free to use their assigned quotas as they wish. It can be prescribed that, for the first three-quarters of a year, importers use

---

495 *Ibid.*

496 John A. Usher, (2001), *EC Agricultural Law*, (Oxford/United Kingdom: Oxford University Press) at 106f. [Usher].

497 See with respect to this question Thomas Jürgensen, 'Gemeinschaftrechtliche Grenzen der Bananenmarktordnung' (Limitation of the Common Market for Bananas derived from Community Law) (1998) *Europäisches Wirtschafts – und Steuerrecht* (*Europ. J. of Commercial and Tax L.*) 357 [Jürgensen].

498 EC, Council regulation No. 2587/2001 of 19 December 2001 on the common market for bananas [2001] O.J. L 345/13, Article 18 [Regulation 2587/2001].

499 *Ibid.*, Article 18(1).

500 EC, Commission Regulation No 756/99 of 12 April 1999 on the implementation of Council Regulation 404/93 regarding the common market for bananas in the EC [1999] O.J. L 98/10, Articles 5, 6 [Regulation 756/1999].

their contingents only to a limited extent.[501] Additionally, the amounts of bananas to be imported from third countries are also allocated with respect to the third countries. To keep track of the imported bananas, the quotas can be further divided between each quarter of the year.[502] In summary, first every importer is granted a yearly quota for bananas to be imported from third countries. At the same time, a yearly quota is fixed for the third countries, from which the bananas are imported and these latter quotas are additionally divided for each quarter of the year. In a second allocation procedure, every importer has to apply for a permission to the respective Member State, for the amount of bananas it intends to import in a quarter of a year from a specific third country. These applications are again forwarded by the Member States to the Commission.[503] The Commission, after having received all applications for the envisaged imports in a quarter, decides whether the applications can be approved or whether they need to be reduced, to not exceed the limits established for each of the third countries for a quarter. If an excess is expected, the Commission again sets a percentage by which all applications for import licenses for bananas from a specific third country must be reduced.[504] This reduction coefficient is transmitted to the Member States, who then can issue the adjusted import licenses to the importers, which are valid only for the respective quarter.[505]

Unlike in the common fishery sector, the real decisions about the allocation of quotas are undertaken by the EC and only transmitted by the Member States to the applicants. The Member States do not have a real decision making power and act only as an assistant to the Commission. The Commission decides, whether the yearly reference amounts and the quarterly imports need to be reduced and fixes the corresponding reduction coefficients. Although it must appear to the importers that all decisions are taken by the Member States, as they are the ones that they have contact with, the real execution of the regulation on the common market for bananas is undertaken by the EC. Furthermore, unlike in the common fishery policy, there are no quotas fixed for the Member States that they can freely distribute among interested importers, but these quotas are directly determined with respect to each importer based on the information that are filed in.

When developing a scheme for the distribution of air traffic rights, one can have recourse to the two aforementioned solutions. A distribution of air traffic rights on a quarterly basis similar to the allocation of import licenses for bananas must be refused. Airlines plan the use of their aircrafts over a longer period of time. This is necessary to be able to schedule maintenance checks, which can take quite long time and which have significant effects on the operations of air carriers. In order to ensure the most efficient use of an aircraft, airlines must know well in advance on which routes they will be able to operate. Still, some features of the approach chosen with respect to the distribution of import licenses for bananas, could be also applied with respect to the distribution of air traffic rights as will be scrutinized in

501 *Ibid.*, Article 14(2).
502 *Ibid.*, Article 14(1).
503 *Ibid.*, Articles 15, 16.
504 *Ibid.*, Article 17.
505 *Ibid.*, Article 19.

the following. If an air service agreement with a non-EC country will be signed that could substitute existing air service agreements signed by the Member States with that respective country, all Community air carriers could report their air traffic rights they hold for operations with respect to this third country to their respective Member State. The Member State could then report this information to the Commission and based on the total air traffic rights available under the new air service agreement the latter one could determine a percentage by which the present air traffic rights held by the airlines could be increased or reduced. The Member States could then adjust the existing air traffic rights of their air carriers accordingly. In case additional air traffic rights available, the Member States could distribute these rights among existing air carriers or assign them to newcomers. The disadvantage of this approach is that the Member States are not really given an autonomous decision power. They function only as a transmitter of the decision of the EC. Their task would be limited to calculating the new number of air traffic rights based on the reduction/increase coefficient determined by the Community. Only if additional air traffic rights become available, the Member States could also decide how these rights should be allocated. This, however, is a fairly limited decision power. From the perspective of the principle of subsidiarity, the activity of the Community seems too extensive. According to the preamble of the EU Treaty, the principle of subsidiarity requires that 'decisions are taken as closely as possible to the citizens'. This cannot be limited to the establishment of a fiction that the Member States have an own decisional power, while in fact major decisions are taken by the EC. The principle of subsidiarity requires that Member States are given a real opportunity to take own decisions. As can be seen from the scheme established in the common fishery policy, a distribution of internationally granted rights can be also achieved by giving the Member States more decision power. Therefore, the latter approach is more favourable compared to the allocation system established in the common market for bananas. Having regard to the solution chosen in the common fishery policy, it will be necessary to distribute air traffic rights among the Member States. For this purpose, the criteria, according to which air traffic rights are assigned to the Member States, must be determined. The following aspects could be taken into considerations:

Population

Firstly, it could be considered to refer to the population of a country. The assumption is that the higher the population , the higher the probability that people will travel and a need for air transportation services exists. The disadvantage of such a quantitative approach is that it does not sufficiently take into account the structure of a society, its level of industrialization and the development of a country's economy. A Member State might have a high population, but if the majority of its citizens are engaged in activities that do not require the means of air transportations, for example agriculture, fishery, and so on, then there is less demand for air transportation services than in a Member State, where its citizens are engaged in a lot of international transactions. Furthermore, attention must be given to the destinations most frequently visited by the citizens of a country. If citizens of one Member State preferably travel to for example South Africa, it does not mean that there is an equal need for travels to other regions of the world as well.[506] An indiscriminate distribution of air traffic rights among the Member States based on their population is therefore not appropriate and might even be discriminatory pursuant to the above-mentioned judgement of the ECJ.

Previous air traffic rights
Another aspect that could be taken into account, is the number of air traffic rights previously granted to a Member State by a non-EC state. It could be argued that a third country may agree to grant a higher number of air traffic rights to the EC as a whole, because it has granted a high portion of air traffic rights to a Member State before. In its negotiations with third countries the EC could refer to the *status quo* and, as a consequence, may receive as many air traffic rights as were previously granted to all Member States together. From this perspective, Member States with a high number of air traffic rights give leverage to the EC for future negotiations. Under these circumstances it is only fair, if such Member States are then given a higher portion of the air traffic rights that are received from the non-EC state. Additionally, it can be argued that, if an air carrier has developed a route between a Member State and a third country, it is justified to grant the necessary number of air traffic rights to that Member State, from whose territory such a route was developed. If strong demand was created on a route, it would by contradictory to undermine this demand and thus the business opportunities for the airlines by reducing the air traffic rights and thus the available flights on this route.

Special interests
Another aspect that could be taken into considerations is, whether Member States and third countries have a special interest in a route. This might be due to a particular relation between these two countries that can have developed for various reasons, like for example a common history or common cultural roots. Here, one may think of the special relationship that can exist between former colonial powers and their former colonies. Many citizens from former colonies live now in the former colonial states and travel frequently between these two countries. Special interests can be also based on economic reasons and the trade level established between a third country and a Member States, which can require that an increased amount of air transportation capacities is made available on a particular route. In general, it might be necessary to take into account the transportation needs of the industry in a Member State. A Member State that hosts many companies, which are involved in international trade businesses may have an increased need for additional air traffic rights.

Past traffic
Finally, past traffic originating in a Member State to particular destinations, could also serve as an allocation criterion. This criterion has the advantage that it implicitly takes into account any special bonds a Member State may have with other states, the necessity for flight services in a specific market and, if applicable, also the need for an increased number of air traffic rights, if a higher contingent was previously granted

---

506 See also below with respect to special interests.

to a Member State by a third country, due to high demand on a route. Under these circumstances, the amount of traffic originating in a Member State, is considered as the most appropriate criterion, when it comes to assign air traffic rights to a Member State. The only problem that could occur under these circumstances is, if a Member State had no previous air services with a non-EC state, with which an air service agreement is concluded. To prevent that such a Member State is not allocated any air traffic rights at all, it is suggested that a certain number of air traffic rights is reserved for those kind of Member States. The distribution of this contingent of air traffic rights could be based on the needs of the industries of the Member States, on any other special interest the concerned Member States might have in these rights and finally on the population of the respective Member States.

Concluding, it is submitted, that, if an air service agreement is concluded that foresees limitations on air traffic rights, these rights, jointly granted to the EC and the Member States, shall be allocated to the Member States based on the past traffic originating from the Member States to the third country. Thereby, the Member States shall be free to distribute their allocated quota of air traffic rights to eligible air carriers and in accordance with EC law.

Air service agreements restricting designation possibilities

While in the previously described sort of air service agreements the number of air traffic rights is limited, it could also happen that the number of possible designations is restricted. Due to the situation in the EC with a high number of internationally operating air carriers, future air service agreements concluded with third countries should contain an unlimited multiple designation clause. Third countries also do not need to be afraid to accept such a clause. If they want to limit competition and traffic from the EC, they can do this by restricting the number of air traffic rights. Still, since the negotiations of air service agreements are dominated by various interests, it is possible that third countries will insist on a limited multiple designation clause or may even opt for a single designation clause. Thus, it may occur that air service agreements provide for a high number of air traffic rights, but allow only for a limited designation. In the most restrictive scenario, air traffic rights and designations possibilities could be limited, resulting in a highly difficult and very complex situation for the EC, when it comes to allocating air traffic rights. It could happen that all Member States are allocated a certain quota of air traffic rights, but that the number of permitted designations is so low that not all Member States would be able to designate their preferred air carrier. For example, one can think of an air service agreement, where a maximum number of seven designations is foreseen. If one Member State allocates its available air traffic rights to five different air carriers and a second Member State to another three different carriers, not only would any other Member State have to distribute its air traffic rights to these previously designated air carriers, but one air carrier would have to give up its air traffic rights, to avoid a violation of the respective air service agreement. Of course, every designated air carrier would be very reluctant to do this and it can be expected that legal disputes may arise. In order not to preclude any decision of a Member State, the distribution of air traffic rights would have to be either coordinated (and maybe even conducted at the same time) or the designated air carriers should obtain their air traffic rights irrevocably only for a limited period of time, to allow other Member States to allocate their quotas to other air carriers. But since very often air carriers must first invest into a route, until it is sufficiently developed and becomes profitable, the possibility of losing a route again, may reduce the airline's willingness to develop and to invest into a route. In the end, this could have a negative impact on the economical development of the EC. Hence, the latter solution must be refused and the only feasible solution is a coordinated allocation of air traffic rights. This could be achieved by conducting the allocation of air traffic rights in all Member States at the same time. In this respected it is partially suggested that, when applying the negative-determination test, it should be also considered, whether a cooperation of the Member States outside the framework of the EC Treaty, if successful, could be considered as ensuring a sufficient achievement of the EC's objectives.[507] A coordination between the Member States outside the framework of the EC Treaty is, according to my opinion, not a valid instrument to be considered under the principle of subsidiarity. The principle of subsidiarity is part of the EC Treaty and thus all measures undertaken within the scope of the EC Treaty by the Member States or the EC must be also based on this treaty and all the principles laid down therein.[508] If a coordination should become necessary, it is an indication that the Member States cannot achieve an objective of EC law by themselves and require the assistance of the Community. In the here presented scenario, a scheme could be established similar to the one in the common market for bananas. Unlike in the above described situation, where only air traffic rights are assumed to be limited, further coordination is necessary, when the number of possible designations is also limited. A solution could be that, after the Member States have chosen the airlines to whom they want to allocate air traffic rights, they report their choice to the EC and the latter one, here most likely the Commission as acting body of the EC, could gather all information and ensure that the total number of possible designations does not exceed the limits set out in the respective air service agreement. If, apart from restrictions on the number of possible designations, air traffic rights are also limited under an air service agreement, these rights must be first assigned pursuant to the procedure set out in the previous chapter by the EC to the Member States. Unfortunately, the solution presented so far bears one disadvantage, It does not take into account one significant difference between the allocation of importation quotas for bananas and the distribution of air traffic rights. Unlike in the common market for bananas, the here envisaged air service agreements contain a limitation on the maximum number of air carriers to receive air traffic rights. This is a condition that neither exists in the common fishery policy nor in the common market for bananas. Contrary to the distribution of banana importation licenses, it is not possible to simply equally reduce the number of designated airlines, if limits are exceeded, but some airlines could not be designated at all, to guarantee compliance with the respective air service agreement. The Member States, who would be denied the possibility to designate certain air carriers, will very likely oppose to this decision and eventually

507 Lambers, *supra* note 460 at 236; *contra* Schmidhuber and Hitzler, Internal Market, *supra* note 456 at 419.

508 See also above at 128ff. with respect to the principle of limited attribution.

take the Commission to court. In addition, the Community would also have to set up another procedure to determine, which air carriers, previously selected by a Member State, would have finally to be denied the designation. Hence, the task to find a procedure to distribute air traffic rights would be aggravated due to the necessity to find a procedure how air traffic rights, foreseen by the Member States to particular air carriers, could be revoked again. The difficulties caused by a restriction on the possible number of designations show that, under these circumstances, the Member States are not sufficiently capable to distribute air traffic rights and that this should be rather accomplished by the European Community.

Positive-determination test

Under the positive-determination test it needs to be shown that the EC can achieve the objective in a more satisfying manner than the Member States. As seen above, the Member States would have a difficulty to distribute air traffic rights, granted in a restrictive air service agreement that foresees a limitation on the number of possible designations. In such a situation, a centralized approach is required and the distribution of air traffic rights would be best accomplished by the Community. Even a coordination and a cooperation that is undertaken by the Member States and the EC in other fields and as it is proposed, if only air traffic rights are limited, but where the number of possible designation is unrestricted, would not solve the problem. In this setting the scale and the effects of the distribution of air traffic rights are such that only the EC can ensure an effective distribution of these rights.

PRINCIPLE OF PROPORTIONALITY

Another principle to be taken into regard is the principle of proportionality set out in Article 5(3) EC. It is applicable to all acts of the EC, including regulations with respect to the administration of EC law. Hence, it must be also assessed, whether a distribution of air traffic rights directly by the Community is in accordance with this principle. Difficulties exist with respect to the distinction between the principle of subsidiarity and the principle of proportionality. According to Schreiber and Langguth the principle of subsidiarity deals with the question of whether the EC can become active, while the principle of proportionality covers the aspect how the Community may act.[509] Von Bogdandy and Nettesheim, on the other hand, argue that these two aspects cannot be separated and must be equally considered, already when applying the principle of subsidiarity.[510] Zuleeg sets out that under the principle of proportionality it must be evaluated, whether an activity of the EC is appropriate, necessary and proportionate. Under the necessity-criterion it is examined, whether other means of activity are available, which could have a less impairing effect on the competences of the Member States. Under the proportionality, it is evaluated whether the chosen measures and the implications resulting therefrom are appropriate

509 Schreiber, *supra* note 401 at 65; Langguth, in: Lenz/Borchardt, Article 5, *supra* note 435 at para 9.

510 Bogdandy/Nettesheim, in: Grabitz/Hilf, Article 3b, *supra* note 444 at para 27.

compared to the objectives attained by the respective measure.[511] It is very difficult to find a clear distinction between both principles set out in Articles 5(2), (3) EC, because, as von Bogdandy and Nettesheim argue, the question of whether and how the Community can act are interrelated and cannot be separated. This is in particular true for concurrent competences, where the principle of subsidiarity is applied. Here, it must be also decided about the scope of a measure of the EC and not only about the question, whether the Community can act at all. Since, in such a case, an activity of the EC is already reduced to the extent necessary, to respect the identity of the Member States and to leave as much competences as possible vested with the Member States, it can be assumed that, if an EC regulation or activity is in accordance with the principle of subsidiarity, it is also appropriate, necessary and proportionate under the principle of proportionality. Particularly the aspect of necessity of an activity of the Community is already evaluated within the application of the principle of subsidiarity. In this respect the wordings of Articles 5(2) and (3) EC are also very similar. While Article 5(2) EC requires that the Community shall take action 'in so far' the objectives of a measure cannot be sufficiently achieved by the Member States, Article 5(3) EC prescribes that the Community's actions shall 'not go beyond' what is necessary. Both provision limit the activities of the EC in the same way and to the same extent. Thus, it is hard to imagine a situation, where an action of the EC based on a concurrent competence is in accordance with the principle of subsidiarity, but infringes the principle of proportionality *vis-à-vis* the Member States.[512] But even if the principle of proportionality is applied with respect to the distribution of air traffic rights by the Community, it will not lead to a different result. In applying the criteria set out by Zuleeg, the direct distribution of air traffic rights by the EC is appropriate. There are no indications that the EC could not successfully distribute these rights. As was set out above, there is also no other solution possible than that the EC is responsible for the distribution of air traffic rights, if an air service agreement provides for a limited designation clause. A cooperation of the EC and the Member States that would preclude a total exclusion of the Member States from the distribution procedure is not feasible in this case either. Therefore, an action by the EC is also necessary. Finally, there is no indication that an allocation of air traffic rights by the EC would be disproportionate. Especially the political, cultural and economical significance of international air transportation requires an effective and fast allocation of air traffic rights. This can be only guaranteed, if it is accomplished by the Community. Therefore, a distribution of air traffic rights by the EC is appropriate, necessary and proportionate and thus in accordance with the principle of proportionality.

CONCLUSION

The previous examination has revealed a very complex situation, when it comes to the allocation of air traffic rights granted to the EC and the Member States by a third country. Although based on the application of the principle of limited attribution

511 Zuleeg, Legal Unity, *supra* note 401 at 56, 84.

512 Similarly Zuleeg, in: Groeben/Schwarze, EU Preamble, *supra* note 259 (according to Zuleeg the principle of proportionality is a part of the principle of subsidiarity, at para 16).

and the principle of subsidiarity, the implementation of EC law is generally undertaken by the Member States, due to the particularities of international public air law this situation must be amended for the distribution of air traffic rights. In one case, if a liberal air service agreement is concluded that provides for unlimited air traffic rights and multiple designation, or if an air service agreement is concluded between the EC and solely one Member State on one side and a non-EC state on the other side, a distribution of air traffic rights can be sufficiently accomplished by the Member States alone. If a multilateral air service agreement is concluded, in which air traffic rights are limited, the distribution of these rights would require a two-step approach and a cooperation between the Member States and the EC. Finally, if air traffic rights are granted in an air service agreement, which foresees a maximum number of designations, an allocation by the EC is suggested, whereby the Member States are precluded from the distribution procedure. Although presently it is foreseen in Article 5 of Regulation 847/2004 that the distribution of air traffic rights is accomplished by the Member States alone, this can be only valid for a limited period of time. Once the European Community becomes involved in concluding comprehensive air service agreements, the necessity will arise to set up a comprehensive distribution system for air traffic rights that should also foresee a participation of the European Community.

Chapter 5

Distribution Methods

When it comes to the distribution of air traffic rights, under present EC law Article 5 of Regulation 847/2004 prescribes that the Member States shall set up a procedure, which ensures a transparent and non-discriminatory allocation. In this chapter it shall be assessed, which procedure is the most suitable one to fulfil these conditions. Thereby, the particularities of international air transportation will be taken into account. Additional attention shall be given to the situation *de lege ferenda*, when as was concluded in the previous chapter, the European Community will become also involved in the allocation of air traffic rights. For this reason, regard will be also given to the EU and EC Treaty examining, whether these regimes contain any principles and provisions that may have to be obeyed, when distributing air traffic rights.

THE DISTRIBUTION OF LIMITED COMMERCIAL PUBLIC RIGHTS – A SECTORAL COMPARISON

States distribute a variety of commercial public rights and have developed different allocation methods.[513] Public rights refer to commercial goods or rights that are considered as the property of the general public. To exploit these rights, usually a permission issued by the respective state is necessary. Commercial goods or rights can be found either as natural resources or as artificially created limited rights. Examples for permissions to exploit these goods or rights are taxi concessions, frequency licences for telecommunication services, contracts for the construction of public buildings or facilities, agricultural quota (for example fishing rights) and also air traffic rights. In order to find the appropriate procedure for the distribution of air traffic rights, it might be helpful to have a look at other sectors, where states distribute limited rights for commercial activities. One such area, in which governments distributed plenty of commercial rights over the last few years, is the telecommunication sector. To show the different methods of allocating limited public rights, the distribution of frequency licenses in the telecommunication sector will be used as an example. The experience that was gained in these procedures might be helpful in assessing, which approach could be the most appropriate one for the allocation of air traffic rights.

---

513 The terms allocation, distribution and designation (of air carriers) are attributed the same meaning and will be used interchangeably.

THE ALLOCATION OF FREQUENCIES IN THE TELECOMMUNICATION SECTOR

Over the past years, various different allocation methods have been used to distribute frequency licenses to commercial entities in the telecommunication sector. Thereby, a shift from the public-trusteeship model to a marketplace approach took place. Under the public-trusteeship model, the electromagnetic spectrum is perceived as a natural scarce resource belonging to the public and entrusted to a company to offer services that are in the public interest. The marketplace approach, on the other hand, allocates scarce resources only under an economic perspective to those entities that are willing to pay the most for the exploitation of the respective resource.[514] In the case of telecommunication the corresponding resources are frequency bands. In the following different methods of allocation that can be classified under one of the two models will be presented. Furthermore, the experiences made with the different approaches in the USA are also discussed. Although reference is only made to the development in the USA, it can be assumed that similar experiences would be made in other countries.

First-come/First-served

The first-come/first-served procedure was the initial distribution method of frequency licenses. The advantages of this approach were low administration costs and a fast allocation.[515] The main problem was that licenses were granted to quick applicants instead of well-capitalized entities, which might have been in a better position to offer the best services, but which failed to file their applications in time. Thus, depending on the technological know-how and the financial situation of the successful applicant(s), an effective use of the spectrum was not guaranteed, the establishment of new services could have been delayed and the offered services might have been less sophisticated than what was technically achievable at that time.[516] These disadvantages finally led to the abolishment of the first-come/first-served procedure.

Beauty contests

Beauty contests were widely used in the distribution of frequencies during the first stages of the privatization of the telecommunication sector. They can be organized as comparative hearings, in which the applicants personally present their applications before an administrative law judge (a procedure sometimes used in the US) or a

514 Todd Bonder, 'A "Better" Marketplace Approach to Broadcast Regulation', (1984) 36 *Fed. Comm. L. J.* 2731 [Bonder]; Andrea M. Settanni, 'Competitive Bidding for the Airwaves: Meeting the Budget and Maintaining Policy Goals in a Wireless World' (1994) 2 *Comm. Law Conspectus*, 117–118 [Settani].

515 William Kummel, 'Spectrum Bids, Bets, and Budgets: Seeking an Optimal Allocation and Assignment Process for Domestic Commercial Electromagnetic Spectrum Products, Services, and Technology', (1996) 48 *Fed. Comm. L.J.*, 511 at 524 [Kummel].

516 *Ibid.*

committee, or as documentary hearings, where the selection is based merely on the applications filed. The most significant advantage of a beauty contest is that, after a thorough examination, the licences are granted to the most suitable applicant(s).[517] But this benefit comes at a high price. First, the procedure is very lengthy and accumulates high costs.[518] It starts with the allocation process itself, in which the applications are verified and the most suitable applicant(s) are selected. But this is not necessarily the end of the procedure. A competitor can disagree with the decision and may ask for a judicial review. Until a final decision is taken, years can go by during which the use of the licence might be barred.[519] It is also possible that incumbent operators abuse the comparative hearings to delay the entry into the market by new competitors.[520] The length of the beauty contest causes considerable costs, which occur on the side of the government for the administrative process, as well as on the part of the applicant(s) for the preparation of the applications and for any possibly following legal proceedings.

The second problem with beauty contests lies in the determination of the applicable criteria, upon which the selection is based and the weight attributed to each of them.[521] This circumstance makes the beauty contest more susceptible to subjective preferences of the persons in charge of the allocation decision.[522] Usually, the criteria applied for the selection are not revealed, which renders the whole process nebulous and thus even more susceptible to litigation.[523] When, on the other hand, the criteria are disclosed, the companies try to comply with them to the highest degree possible, which normally should be considered as a positive effect. However, the downside is that this can finally result in a situation, where all applicants look alike and are equally qualified to obtain the respective licence(s).[524] Therefore, doubts remain, whether a beauty contest is an appropriate method for an objective distribution of limited public rights.

---

517 *Ibid.* at 525.

518 Kurt A. Wimmer and Lee J. Tiedrich, 'Competitive Bidding and Personal Communications Services: A New Paradigm for FCC Licensing', (1994) 3 *Comm Law Conspectus,* 17 at 18 [Wimmer and Tiedrich]; Mark W. Munson, 'A legacy of lost opportunity: Designated entities and the Federal Communications Commission's broadband PCS spectrum auction' (2000/2001) 7 *Mich. Telecomm. & Tech. L. Rev.*, 217 at 220 [Munson].

519 Kwerel and Felker, 'Using Auctions to Select FCC Licensees', in: Thomas G. Krattenmacher, ed. (1994), *Telecommunications Law and Policy* (Durham, NC: Carolina Academia Press) 121 at 122, 125 [Kwerel and Felker]; Jonathan Blake, 'FCC Licensing: From Comparative Hearings to Auctions' (1994) 47 *Fed. Comm. L.J.* 179 at 180 [Blake].

520 Ted Stevens, 'Regulation and Licensing of Low-Earth-Orbit Satellites', (1994) 10 *Santa Clara Computer and High Tech L.J.*, 401 at 410 [Stevens]; George W. Douglas and James C. Miller III (1974), *Economic Regulation of Domestic Air Transport: Theory and Policy* (Washington: Brookings Institution) at 117 [Douglas and Miller].

521 Kwerel and Felker, *supra* note 519 at 122; Stevens, *ibid.*

522 Stephen Breyer, 'Analyzing Regulatory Failure: Mismatches. Less Restrictive Alternatives, and Reform', (1978–1979), 92 *Harv. L. Rev.*, 547 at 566f. [Breyer].

523 Settanni, *supra* note 514 at 119.

524 Blake, *supra* note 519 at 180.

Lotteries

Lotteries were introduced as a reaction to the lengthy and costly procedure of the comparative hearings. They have been referred to as 'methods of pure procedural justice' to ensure that, when no objective criteria were available upon which a choice could be based, then at least a fair procedure should be established, so the outcome would be regarded as fair and correct.[525] Unlike the first-come/first-served procedure, lotteries offered a greater equality by giving the same chances to everybody and not only to the fastest applicants. All participants in a lottery are treated equally, regardless of the differences that may exist in respect of their size, financial capability or technological know-how. The main problem of lotteries is the participation for speculative purposes. Unqualified entities apply for licences only to sell them later.[526] Since there is nothing to lose and everything to gain, it is understandable that everyone tries to obtain a licence first and decides afterwards what to do with it. Thus, it was not uncommon in the past that lotteries generated hundreds of applications.[527] This high number of participants rendered the lottery into a slower procedure than one might have previously expected, as all applications must have been logged, filed and pre-screened, before the lottery itself could take place.[528] The implementation of new services, for which the frequency licences were allocated, was further delayed, when speculators organized their own resale procedures of the licences they had won in the lottery.[529] In addition to that it was not uncommon that the losing applicants filed in petitions to deny the winners the grant of a licence arguing they would not be sufficiently qualified. These petitions were then reviewed in an administrative hearing, a procedure very similar to the comparative hearing. Under these circumstances, it sometimes happened that a lottery turned out to be a lengthier and more expensive procedure than the one, it was supposed to replace.[530]

User fees

User fees were introduced to lower the amount of applications and to shift parts of the administrative process costs onto the applicants. Due to the up-front expenses it was expected that only those entities would apply for frequencies, which could indeed make use of them. It was anticipated that any speculation with the obtained licences would not occur and therefore new services and systems could be implemented faster.[531] The downside of this method was the lack of any secondary market for frequency licences, which could have served as a guide, when it came to fixing the amount of the fees. Without any orientation mark, user fees had to be freely determined by the administration. This was deemed as too complicated and it took

525 Stevens, *supra* note 520 at 413f.
526 Wimmer and Tiedrich, *supra* note 518 at 19; Munson, *supra* note 518 at 221.
527 Kwerel and Felker, *supra* note 519 at 122.
528 *Ibid*. at 125.
529 Munson, *supra* note 518 at 221.
530 Wimmer and Tiedrich, *supra* note 518 at 19; Blake, *supra* note 519 at 181.
531 Kummel, *supra* note 515 at 526.

too much time to fix an appropriate price.[532] Therefore, the user fee approach was given up quite soon.

Auctions

Auctions, also known as competitive biddings, were introduced as a reaction to the development of private auctions, taking place after companies were granted frequencies in a lottery, and to overcome the problem encountered with the user fee approach of not being able to fix an appropriate price without a secondary market. The governments realized that they could auction the frequencies themselves and collect the revenues, instead of permitting a windfall to private entities.[533] The additional funds obtained through an auction may be used for public purposes, thus benefiting the whole society and not only a small group of speculators. This reasoning was further supported by the conviction that the electromagnetic spectrum was a public good and therefore its usage should not be permitted free of charge.[534] Under the Clinton Administration the USA deliberately resorted to competitive biddings to lower their budget deficit.[535] So far, several auctions have taken place worldwide, generating billions of dollars of additional revenues for the respective countries. The USA, being the pioneer in the field of auctioning telecommunication licenses, were able to add a considerable amount of funds to their treasury. As of 30 September 1997 the US have conducted 14 auctions, awarding more than 4,300 licences for a total sum of $23 billion.[536] Contrary to lotteries, auctions require that applicants conduct various analyses on the feasibility of the proposed services, the market demand for these services, the magnitude of revenues that can be obtained and the access to capital by the bidders.[537] These efforts, to be made before an auction, are necessary for the bidders to determine the value of a frequency licence and consequently the maximum price a company shall pay for it. The amount and costs of previous studies should deter any inexperienced, casual or financially weak applicants from participating.[538] A smaller number of participants helps reducing the administration costs and shortens the time of the procedure. Hence, despite the fact that the winner of an auction must pay for its licence, auctions are still considered as the least expensive procedure for allocating limited public rights due to their

---

532 Bonder, *supra* note 514 at 49f.

533 Kummel, *supra* note 515 at 527; Munson, *supra* note 518 at 222; Blake, *supra* note 519 at 179, 181f.

534 Stevens, *supra* note 520 at 419; Blake, *ibid.* at 179; Bonder, *supra* note 514 at 44, 48; Settanni, *supra* note 514 at 127.

535 Janice Obuchowski, 'The Unfinished Task of Spectrum Policy Reforms' (1994) 47 *Fed. Comm. L.J.*, 325 at 326f. [Obuchowski]; Ellen P. Goodman, 'Digital Television and the Allure of Auctions: The Birth and Stillbirth of DTV Legislation' (1997) 49 *Fed. Comm. L.J.*, 517 at 533; Stevens, *ibid.*

536 Munson, *supra* note 518 at 226; Settanni, *supra* note 514 (when auctions were introduced as an allocation method for frequencies it was expected to raise $10.2 billion until 1998, at 118).

537 Wimmer and Tiedrich, *supra* note 518 at 30.

538 *Ibid.*; Munson, *supra* note 518 at 222.

simplicity and short proceeding.[539] It is further believed that auctions ensure the most efficient use of the spectrum, since the companies that can make the most use of a licence value it highest and will offer the biggest bid.[540] Since the winning bidders will have an interest to implement services as fast as possible to recover their large up-front investments, the general public can as well benefit from the introduction of new services within a short period of time. In addition, one can also expect the holders of frequency licences to invest in new technologies, to expand the capacities of their frequencies and to be able to use them in the most efficient manner. On the other hand, as long as the electromagnetic spectrum can be used for free, there is no incentive to invest in research and development and to enhance efficiency.[541] If the use of the spectrum is subject to some costs and if there is a market value attributed to frequency licenses, companies will be deterred from stockpiling frequencies and an incentive exists to invest in new technologies to increase the value of a licence.[542] More research and development of new technologies does not only benefit a country's economy and increase the standard of living, but the knowledge gained through the research and development also enhances the global competitiveness of the companies pursuing the research and development.[543] Considering all the given advantages, it can be said that, compared to the other approaches, auctions are more favourable regarding costs, administrative burdens and delays.[544] Furthermore, they are a very objective, non-discriminatory and transparent allocation method. The main disadvantages of auctions are:

(i) the danger of overestimating the value of the good or right to be auctioned;
(ii) escalation and default;
(iii) delays, and
(iv) collusion among the participants.[545]

Some, but not all of these phenomena, can be avoided by properly structuring the bidding procedure.[546] Defaulting, for example, might result in the necessity to re-auction the concerned licences and thus offsets the time advantage of an auction.[547] Another inherent danger of auctions is the behaviour of the auctioning government. The true objective of auctions as a means for allocating limited public rights might be disregarded in favour of the revenue generating component of an auction. Thereby, the long-term effects on an industry and the impact on the whole economy

539 Compare Kwerel and Felker, *supra* note 519 at 126ff.
540 Obuchowski, *supra* note 535 at 325; Wimmer and Tiedrich, *supra* note 518 at 17.
541 Kummel, *supra* note 515 at 527f.; Munson, *supra* note 518 at 222.
542 Bonder, *supra* note 514 at 32f.
543 Settanni, *supra* note 514 at 123.
544 Munson, *supra* note 518 at 217, 226; Stevens, *supra* note 520 at 420.
545 Kummel, *supra* note 515 at 530; Blake, *supra* note 519 at 183.
546 See e.g. Kummel, *ibid.* at 534.
547 Munson, *supra* note 518 at 248; Martin L. Stern and David Rice, 'Fallout from the C-Block Debacle Threatens to Cost Small Businesses their Best Chance to Participate in Spectrum-Based Telecom Services' *Legal Times* (13 October 1997) S36 [Stern and Rice].

of a country, might be traded off for the short-term gains.[548] Finally, it cannot be denied that auctions favour established companies with a solid financial basis that can afford to spend money for any licence they want to acquire.[549] Burdening them with additional up-front costs, even before they start operating can hamper the establishment of new and innovative start-up companies.[550] Generally, it is one of the biggest constraints for new companies to obtain enough funds to start their businesses. This task becomes even more difficult, if they have to pay a considerable amount of money at the beginning of their activities for necessary licences. Another disadvantage of auctions is that the funds spent for obtaining a licence are not available any more for investments in new equipment and the promotion of new services.[551] Hence, auctions can be also a constraint to competition and reduce the benefits for the customers.

Result

The experiences gained in the telecommunication sector could favour the use of auctions as an allocation method of public limited rights. The first-come/first-served method and lotteries have proven to be inappropriate, as they either benefit the wrong companies or give great opportunities for speculators leading to delays in the implementation of new services. User fees, introduced to exclude speculations in the distribution process, created new problems. It became obvious that governments and their regulatory agencies were not the appropriate bodies to fix prices for the respective licences. Beauty contests are satisfying as far as the results are concerned, but too cumbersome and expensive, if it comes to the procedure. First of all, the length and high costs occurring to the administration and the applicants are one major downside. In addition it is possible that subjective preferences of the selecting persons influence the outcome of the distribution procedure. Since all previous allocation methods have proven to be somehow flawed, governments introduced auctions to distribute frequency licenses. Today this method is acknowledged to be superior in respect of costs, delays, fairness of the selection process and appropriateness of the result compared with all other procedures previously used. The only major downside of this approach seems to be that the establishment of new companies could be hampered.

COMPARISON BETWEEN THE AIR TRANSPORTATION AND THE TELECOMMUNICATION INDUSTRIES

Before every single allocation method is analysed, as to whether it could be used for the distribution of air traffic rights, the airline and the telecommunication industry shall be compared first. It could be helpful to know the similarities between both

548 Compare Settani, *supra* note 514 at 126, 127f.

549 Bruce R. Kraus, 'Pyrrhic Victory in Spectrum Auction', *The National Law Journal* (25 August 1997), B15; Blake, *supra* note 519 at 182; Bonder, *supra* note 514 at 39f.

550 Settanni, *supra* note 514 at 126f.

551 Bonder, *supra* note 514 at 56f.; Stern and Rice, *supra* note 547 at S36.

areas to better assess the appropriateness of the different allocation procedures. A first similarity between both industries is the scarcity of resources they have to deal with. Airlines are burdened with limited slots and gates at airports, as well as restrictions on air traffic rights for international flights. Telecommunication companies face the same challenge in terms of frequencies for telecommunication through the use of satellites or for mobile telecommunication services. Another aspect that the two industries have in common is the fact that both are technologically highly sophisticated and very costly. The necessary assets to operate an airline (aircrafts) or a telecommunication company (satellites and corresponding terrestrial facilities) are very complex objects that are not easily obtained. In particular the financing of the respective assets requires a lot of funds and maybe participation of private investors. Like aviation, telecommunication was referred in most countries, for a long time, as a public service and was either offered by state-owned companies or, like in the US, by private entities that were regulated and operated like a public utility. It was believed that electric, telephone, water and partially public transportation companies were part of the infrastructure of a state and constituted a natural monopoly. It was agreed that competition, if permitted within these industries, would lead to higher costs and prices and should be therefore avoided.[552] This thinking has changed and a privatization and deregulation process in the telecommunication and the aviation sector was initiated at about the same time.[553] Today, these two branches are also recognized as industries of their own, offering commercial products.[554] However, the telecommunication as well as the air transport sector remain part of the infrastructure of a country and have an overwhelming importance for the development of the whole economy of a state.[555] Aviation and telecommunication are also very important for the national security of a country. Aircrafts, owned and operated by airlines, offer a great transportation capacity that can be used in cases of emergency.[556] The most recent use of aircrafts for transportation purposes in a case of emergency could be

552 Robert M. Kane and Allan D. Vose (1975), *Air Transportation*, 5th edn., (Dubuque: Kendall/Hunt), 180 [Kane and Vose]; Vera, A., *Foster Rollo* (2000), *Aviation Law – An Introduction*, 5th edn., (Lanham, MD: Maryland Historical Press), 84 [Rollo].

553 Francis Lyall, '*Expanding Global Communications Services*' in: *Proceedings of the Workshop on Space Law in the Twenty-first Century* (2000), (New York: Office of Outer Space Affairs United Nations Office at Vienna), 63 at 65 [Lyall Communications].

554 Ram S. Jakhu, Virginia R. Serrano, 'International Regulation of Radio Frequencies for Space Services', in: *Proceedings of the Project 2001 Workshop on Legal Framework for Commercial Satellite Telecommunications* (2000), (Cologne: Institute of Air and Space Law at the University of Cologne/Deutsches Zentrum für Luft- und Weltraumrecht) 41 at 108 [Jakhu and Serrano]; Francis Lyall, 'On the Privatization of INTELSAT' (2000) 28 *J. Space L.*, 101–110 [Lyall Intelsat].

555 Jonathan Ira Ezor, 'Costs Overhead: Tonga's Claiming of Sixteen Geostationary Orbital Sites and the Implications for US Space Policy', (1993) 24 *Law & Pol'y Int'l Bus.*, 915 at 919 [Ezor]; Stéphan Le Goueff, 'Licensing Global Mobile Personal Communications by Satellite: The Quest for the Holy Grail?' (1997) 22 *Ann. Air & Sp. L.*, 417 at 431 [Le Goueff]; Lyall Communications, *supra* note 553 at 63; see also above at 87f.

556 John H. Frederick (1961), *Commercial Air Transportation*, 5th edn., (Homewood, IL: Richard D. Irwin), at 274 [Frederick].

witnessed after a tsunami stroke several countries in the pacific area in South-East Asia in 2005. Many airlines from various countries immediately started offering air services to these regions and flew in medical equipment and staff as well as food, water and other necessary goods. At the same time air carriers helped evacuating tourists from these regions back to their home countries. Usually, in cases of national defence, carriers are obligated to make their aircrafts available to their governments on request. Especially in times of war, countries make use of the aircrafts of their national airlines.[557] Telecommunication has an even greater importance for the security of a state as it is used for the transfer of sensitive information. Therefore, governments must be able to rely on the stability, security and trustworthiness of the telecommunication systems they use. This dependence increases in times of a crisis and especially during wars. It does not happen by accident that, in armed conflicts, telecommunication and broadcasting facilities are one of the first targets.[558] Another similarity between aviation and telecommunication is the irrational and very emotional behaviour of governments, if it comes to both industries. Airlines are regarded as representatives of their countries, which by flying all over the world show the country's flag around.[559] By now almost a custom has evolved that every state requires a national air carrier to underline its sovereignty.[560] A very similar understanding is associated with telecommunication. The ability to have access to a worldwide information network and to communicate with other people through a national and self- owned telecommunication system, creates or reinforces the feeling of being connected to the rest of the world.[561] Taking into account the great deal of similarities between the airline and the telecommunication industry, one might draw the conclusion that auctions could be also the most suitable procedure for the allocation of air traffic rights, since this method has already proven to be very effective in the distribution of telecommunication frequencies. This assumption will have to be further examined in the following.

ANALYSIS OF THE VARIOUS DISTRIBUTION METHODS FOR THE ALLOCATION OF AIR TRAFFIC RIGHTS

In the following, the various methods for the allocation of limited public rights will be examined with respect to air traffic rights. Particular attention shall be given to auctions due to the results found in the previous chapter.

First-come/First-served

The biggest disadvantage of the first-come/first-served method could be the uncontrolled and random distribution of air traffic rights. Such an approach might not be beneficial to the customers. In the telecommunication industry, frequency

557 See above at 87f.
558 Compare Lyall Intelsat, *supra* note 554 at 116.
559 Burkhardt, *supra* note 24 at 136, 138; see above at 88f.
560 Compare Burkhardt, *ibid*. at 18.
561 Ezor, *supra* note 555 at 919; Le Goueff, *supra* note 555 at 431.

licences were mostly granted for the introduction of new services. Therefore, giving away frequencies on a random basis did not interfere with existing structures and incumbent operators. The air transportation sector with its hub-and-spoke system and airline alliances is an established, integrated, interrelated and complex structure. Neglecting this aspect and distributing air traffic rights on a random basis could be a waste of valuable resources. Depending on which airline would obtain the new or additional air traffic rights, the number of customers benefiting from it, could vary significantly. If an airline with an existing network of flights is allowed to add additional flights to its system, it is not only beneficial for the air carrier as it can increase its load factor and operate more efficiently, but will also benefit the customers, allowing them to reach more destinations, to choose from more flights and thereby to be more flexible in planning their travels. If, on the other hand, air traffic rights are given to a new entrant, who has no established network, the only customers benefiting from it are the ones living close by the origin and destination airports. This does not mean that air traffic rights shall be only granted to incumbents and not to new entrants. However, a random and uncontrolled selection procedure does not take into account sufficiently the existing structures according to the significance they have for the travelling community. Furthermore, it has to be taken into account that the air transport sector is, as set out above, a technologically highly sophisticated and very costly industry. It could have fatal consequences, if valuable air traffic rights are granted to companies that cannot make the best use of it. Airlines invest vast amounts of money into the establishment of their businesses. They expand worldwide, build facilities for the maintenance of their aircrafts or the training of their crews, employ huge numbers of people and have significant overspill effects on other branches of the economy. In this respect it is necessary to accept that there are two different kinds, not to say classes, of airlines. On the one hand, there are the long established trunk carriers that can be spotted in almost every country and which are usually referred to as 'flag carriers'. In most developed countries these airlines were usually founded a long time ago and grew and expanded, while they were protected and shielded from competition by their registration states. On the other hand, there is a second category compromising smaller commuter and low-cost airlines. In particular the low-cost carriers were, with some exceptions, established only a few years ago and the investments, they have made since, were, in general, smaller than the ones of the 'flag carriers'. An uncontrolled application of the first-come/first-served method could maybe result in the allocation of air traffic rights to commuter and low-cost carriers. While this is not negative *per se*, it can have serious consequences for the air transportation industry, the whole economy of a state and the customers. Since airlines of the second category usually are not members of any of the major airline networks, they often will not be in a position to make the most extensive use of an air traffic right. Thus, less people could benefit from the offered services. Additionally, depriving the flag carriers of necessary air traffic rights could cause them financial problems, possibly even leading to bankruptcies. As a consequence, high amounts of invested capital would be simply lost and a large number of persons could find themselves unemployed. Furthermore, since the air transport sector is still part of a country's infrastructure and very important for the development of a state's economy, the collapse of a major airline could have widespread effects. Although this shall not be an advocacy for the established trunk airlines or an attempt to shield them from competition, it is still necessary to stress that there is a lot to lose, if air traffic rights are given away indiscriminately, unsupervised and uncontrolled. A first-come/first-served method, neglecting these realities and distributing air traffic rights randomly not to the most suitable airline, but to the one, which can file its application fastest, cannot be regarded as an appropriate allocation method.

User fees[562]

The downside of imposing user fees for granting public rights is the determination of the appropriate price. The fixing of an appropriate user fee for an air traffic right would require the government to investigate a lot of different aspects, thereby taking uncertainties like market expectations, political risks or future trends into account. While an extensive market analysis is not necessarily a government's primary task, for which it might not even have the necessary resources, such an examination would also render the allocation process lengthier and more expensive. Therefore, user fees do not seem appropriate for the allocation of air traffic rights.

Lotteries

The Us Airline Deregulation Act of 1978 contained a provision, according to which, air traffic rights, which had been previously granted to incumbent airlines and not sufficiently used by them, had to be re-allocated among those competent carriers that applied first. Hence, the responsible regulatory authority intended to use the first-come/first-served approach to distribute these 'dormant permissions'. However, it switched over to lotteries, when airline representatives queued up in front of its office a week before the Airline Deregulation Act was to come into force.[563] Thus, lotteries were not only used for the distribution of frequency licences in the telecommunication sector, but also already for the allocation of air traffic rights. This, however, happened only in a particular case, where additional and previously unused routes were distributed. The biggest disadvantages of lotteries are the participation for speculative purposes, the possibly lengthy procedure and the fact that not the most qualified applicants, but the luckiest ones obtain the traffic rights. The experienced difficulties with speculations could be easily prevented, if the transfer of air traffic rights is forbidden. Control of such a prohibition could be easily exercised, as governments are responsible for the designation of airlines for international scheduled air services to other states. Such a designation could be denied, if the holder of an air traffic right did not obtain it directly through a lottery, thereby rendering the air traffic right worthless. Under these circumstances no airline would be willing to obtain air traffic rights on the secondary market. Thus,

562 Compare Terry A. Ferrar, 'Route Assignment and the C.A.B.', (1973) 5 *Transp. L. J.*, 215 (author proposes to sell routes in profitable high-density markets and suggests that the collected money should be spent for subsidizing socially desirable but economically unprofitable markets, at 216) [Ferrar].

563 Christopher T. Tourtellot, 'Competitive Policy in Airline Deregulation', (1978–79) 28 *Am. U. L. Rev.*, 537 at 561 [Tourtellot].

the danger of speculation is not pre-eminent. The other problem that, due to the high number of applications and possible subsequent petitions to denial by losing applicants, lotteries turned out to be as lengthy and costly as beauty contests, does not exist to the same extent in the air transportation industry. Airlines require an operating licence and to obtain it, they must fulfil some very stringent conditions.[564] This ensures that not everybody, but only highly qualified and serious companies are allowed to offer air transportation services. The number of airlines is thus limited from the beginning. The danger that hundreds or thousands of airlines participate in a lottery for air traffic rights, which would slow down the whole process significantly, can therefore not occur to the same extent as it happens in the telecommunication sector. Presently there are about 130 air carriers in the EC.[565] However, the possibility that a subsequent appeal for judicial review of the allocation may bar the exercise of an air traffic right cannot be excluded. Still, this is a problem that can occur any time, regardless of which allocation method is used. But, as is the case with the first-come/first-served method, the major disadvantage of lotteries is the possibility that air traffic rights are granted to not necessarily the most qualified applicant. Although the procedure ensures greater equality than the first-come/first-served procedure, the problem remains the same. It is the unsuitable randomness of the allocation procedure. The same reasons that disqualify the first-come/first-served method for the distribution of air traffic rights can be brought up here as well. Hence, lotteries are not recommended for the allocation of air traffic rights.

Auctions[566]

As described above, auctions have been attributed several advantages like additional revenues for the treasury, fast and less expensive allocation procedure, efficient use of the allocated frequencies, high degree of objectivity, transparency and non-discrimination in the allocation process and the promotion of technological developments. So far, the only perceived disadvantage is the possible negative impact of auctions on competition. It will be evaluated, whether and how the advantages and disadvantages could also occur, if air traffic rights were auctioned. Furthermore, it shall be assessed, how the additional costs, that would burden the airline industry, as well as the international nature of air transportation, can affect the choice of auctions as an allocation method. Finally it will be examined, whether any legal aspects could prevent the use of auctions for the distribution of air traffic rights.

564 See Regulation 2407/92, *supra* note 49, Articles 5, 6.

565 See above at 121.

566 Ferrar, *supra* note 562 (it is not clear whether Ferrar wants the Civil Aeronautics Board, the former US regulatory agency for air transport, to sell or to auction air traffic rights since from his description both solutions would be possible; however, Ferrar discusses only the allocation of domestic routes within the US and not the special case of international air traffic rights, at 216).

Revenue-generating capacity

By auctioning air traffic rights, governments could collect additional revenues for their treasuries. If one considers air traffic rights as public property that is mainly to be used to increase a state's budget, then auctions should be certainly selected as the applicable allocation method. However, such an approach ignores the infrastructural element and neglects the value and significance of the air transportation industry for the whole economy of a state. The double function of air transportation, on one hand as an own industry, which needs to be regulated, and on the other hand as a part of a country's infrastructure, should be always kept in mind. The short-term gains for the treasury could be more then offset by the negative long-term effects on the whole economy due to a weakened air transportation industry. One may argue that the telecommunication sector is also part of the infrastructure of a country and that this does not prevent governments from auctioning frequency licenses. The difference is that the auctioned frequency licenses are used for offering additional and enhanced services, like for example mobile telecommunication services. The basic supply of fixed telecommunication services is, on the other hand, usually in place and its continuity is guaranteed. International air transportation services are not so much enhanced additional services, but rather comparable with the basic supply of a fixed phone line. Under these circumstances it is my opinion that the revenue generating capacity of auctions should not prevail in selecting the appropriate allocation method for air traffic rights.

Efficiency of the allocation process

As the winning bidders have to pay for their air traffic rights, only those companies will take part in the auction that have the financial capabilities and that are truly willing to make use of the granted routes. Thus, it can be expected that auctions will reduce the number of applicants, helping to reduce the costs and the amount of time that is necessary for the allocation procedure.[567] This will directly benefit the airlines through lower transaction costs as less personnel must be assigned to the application process, especially in comparison with beauty contests.[568] Hence, the general advantages of auctions, as far as speed and expenses are concerned, can be expected to occur with respect to the distribution of air traffic rights as well.

Fast implementation of services

While other procedures might be very formal and time consuming, auctions are easy to accomplish and will enable the air carriers to start operations quickly.[569] This will not only benefit the air carriers by granting them a faster return on their investment, but also the general public. Not only are new services available sooner

567 Kwerel and Felker, *supra* note 519 at 121, 125.

568 *Ibid.* at 125.

569 See also Lawrence D. Roberts, 'A Lost Connection: Geostationary Satellite Networks and the International Telecommunication Union' (2000) 15 *Berkeley Tech. L. J.*, 1095 at 1115f. [Roberts].

to the customers, but any additional service that is offered can increase competition, which, in turn, usually leads to lower prices.[570]

Objectivity, transparency and non-discrimination

In an auction, the scarce goods or rights are distributed to the highest bidder. The only criterion is the price that the applicants are willing to pay. No other factors influence the decision. Under these circumstances, auctions are a very transparent, objective and non-discriminatory procedure. With respect to EC law auctions would thus also fulfil the conditions set out in Article 5 of Regulation 847/2004.[571]

Promotion of research and development

Another benefit, mentioned with respect to auctions, is the promotion of technological developments, which finally shall lead to a stronger economy and an enhanced competitiveness of the companies operating in the global markets. It is doubtful though, whether this advantage could also be achieved to the same extent in the air transportation industry. While in the telecommunication area, the licence holders might be also developing new technologies, this is not the case in the air transport sector. Airlines are offering air transport services and manufacturers are the ones responsible for the development of new technologies with respect to aircrafts, engines or avionics. Airlines can improve their business operations only to a limited extent with respect to research and development. Their main tool to increase income or lower expenses is a more efficient management. The airlines' possibility to contribute to the development of new technologies is thus only marginal and works indirectly through cooperations with the manufacturers. Hence, the argument that auctions promote the development of new technologies, cannot be brought forward in the air transportation sector.

Impairing competition

As far as competition issues are concerned, it could be argued that the big and financially strong airlines are in a better position to acquire the necessary traffic rights and that, in the long run, the international scheduled air transportation market might end up in the hands of only a few carriers.[572] Indeed, this scenario seems quite possible. Depending on the price a route would cost, it is not unlikely that small carriers will not be in the position to participate in the biddings. On the other hand, it is also possible that the additional expenditures resulting from auctions could reduce the competitive ability of the bidding airlines. This would put the smaller and financially weaker airlines in a better position to compete with the established airlines on domestic routes and, in case of airlines established in the EC, on intra-Community routes, where no additional costs are imposed for obtaining the necessary

570 Kwerel and Felker, *supra* note 519 at 125.

571 See above at 51.

572 See also Jakhu and Serrano, *supra* note 554 (the authors describe the same problem for the telecommunication industry, at 93).

route permissions. From this perspective it is hard to evaluate, whether auctions will impair competition or, to the contrary, might even foster it.

Internalization of costs

Another advantage of auctioning air traffic rights, instead of granting them for free, would be the internalization of the costs caused by aviation. By using the airspace free of charge, the air carriers burden the society with pollution and noise, while the income from the operations accrues only to the air carriers.[573] Charging the airlines for the use of the airspace would, of course, increase their expenses, which they would cover through higher fares. But the question is, whether these higher fares could be justified as they would represent the true costs of travelling by air.[574] Similarly, railway companies must purchase or lease the land on which the railway track is built and include these costs in their fares. Airlines, on the other hand, do not need to pay for the medium they use for their operations. Making the airlines pay for air traffic rights would increase the awareness for the use of the airspace and could promote a more efficient use of these rights.[575] It would also allow for a more equal treatment between the different industries providing transportation services.[576] From this point of view it would make sense, to impose additional charges for the use of the airspace and to distribute international routes through auctions.

International aspect of air transportation

When trying to transfer the successful application of auctions in the telecommunication sector to air transportation, despite the many similarities existing between both areas, one substantial difference between both sectors must be also taken into account. The distribution of telecommunication frequency licences occurs within the sovereignty and territory of a single state. Any subsequent activities in exploiting the granted licences are also limited to the territory of the distributing country. The winners of an auction compete with each other on a level playing field, as each of them has to bear the additional expenses for obtaining the necessary licences. Airlines engaged in international air transportation are, on the other hand, not competing with each other in a closed market within one single state. International civil aviation is based on bilateral relationships and the competitors, facing each other in one market, usually come from different states. While the bilateral air service agreements aim at creating a level playing field for the airlines from both countries, they do not cover all possible

---

573 See also Roberts, *supra* note 569 at 1115f.

574 See also Richard Janda, 'Auctioning Airport Slots: Airline Oligopoly, Hubs and Spokes, and Traffic Congestion' (1993) 18(1) *Ann. Air & Sp. L.*, 153 at 191.

575 See also Jakhu and Serrano, *supra* note 554 (authors have the same expectations for the telecommunication industry, at 93).

576 See also EC, Commission, *Communication on Air Transport and the Environment – Towards meeting the Challenge of Sustainable Development*, COM (1999) 640 final of 1 December 1999, online: Eur-Lex <http://europa.eu.int/eur-lex/lex/LexUriServ/site/en/com/1999/com1999_0640en01.pdf> (date accessed: 26 October 2005) at 11 [Commission, Sustainable Development].

aspects of commercial aviation. The question of how each state distributes the air traffic rights it receives from the other state party, is always left out in air service agreements. Thus, insofar states are not precluded from auctioning the air traffic rights they are granted by another country. However, from an economic perspective this could be a harmful approach. A carrier, which has to pay for its air traffic rights, will need to amortize these additional costs through increased fares. If the airline of the other state, operating on the same route, receives the air traffic rights without any charges, it might, due to lower costs, be able to offer cheaper fares and attract more passengers. In the long run, this competitive disadvantage could cause financial problems to the air carrier, which had to pay for its traffic rights and force it to leave this market or, in the worst case, even lead to its bankruptcy. The aforementioned positive effect of internalizing the cost of the use of the airspace would be spoilt. It seems therefore not feasible, if only one of several air carriers is burdened with additional charges. The idea of auctioning air traffic rights becomes completely infeasible, when routes are granted in a very restricted air service agreement, which might include predetermined regulations on fares, frequencies and capacities. In such a case, the carriers, obtaining their traffic rights through an auction, could not even compensate their costs through an effective management, as any necessary flexibility is precluded. The fact that international air transport services are offered by companies from different countries, is, not from a legal, but from an economic perspective, a serious obstacle to unilaterally auctioning air traffic rights.[577]

In this respect it is also necessary to evaluate the legal status of air traffic rights, if they are to be auctioned. The main question is, if an air traffic right can be regarded as property of an airline. In respect of telecommunication frequency licenses, some US courts have already denied to regard them as property of the holding companies.[578] According to another school of thought, property rights must be acknowledged, if frequency licences are auctioned. It would be contradictory to sell a right, without also transferring property.[579] The latter position indeed seems to be more consistent. I am not aware of any legal system in a market economy country, where the trade of rights is not combined with the transfer of property in these rights. A system, based on the free tradability of goods and rights, requires also that legally protected property rights are guaranteed. Hence, the question arises, whether property status can be also maintained for air traffic rights. Attributing a value and legal protection to traffic rights would have the additional benefit that these rights could become real assets of the airlines and might be used as securities in obtaining capital, thereby improving the access of the air carriers to the financial markets. However, the fact that air traffic rights are exchanged in an international context causes difficulties, when it comes to attributing property status to these rights. The allocation of telecommunication licenses occurs within a single country and the entities, acquiring these licences, can always rely on the law established in that country. A certain degree of protection

577 See also similarly for the imposition of a kerosene tax: Commission, Sustainable Development, *ibid.* at 13.

578 Cited in Pantelis Michalopoulos, 'Money for Nothing?' *Legal Times* (23 September 1996) 20.

579 *Ibid.*

and predictability can be guaranteed. Due to its international nature, a similar level of reliability and judicial protection cannot be ensured, when air traffic rights are distributed. As explained above, a country for the use of its airspace grants air traffic rights. The state, receiving these permissions, distributes these rights among its national airlines. The problem is that the receiving country does not own the traffic rights itself and the granting country can terminate an air service agreement and revoke the air traffic rights at any time, leaving the other state without any possibility to prevent this. Therefore, a country cannot guarantee the continuity of air traffic rights to the airlines, it receives from other states. There is no legal instrument or provision, which would grant a protection to the owner of an air traffic right comparable to property rights.[580] Thus, if a state cannot grant any legal protection, it should be also denied the possibility to auction international air traffic rights.

Legality of charges

When assessing the viability of auctions as a distribution method for air traffic rights, it is necessary to examine, whether the selling of air traffic rights is permitted under the existing laws. Thereby due regard shall be given to the Chicago Convention and in particular to Article 15 of this Convention. This provision could forbid the allocation of air traffic rights through auctions. Article 15 says:

> (1) Every airport in a contracting State which is open to public use by its national aircraft shall likewise, subject to the provision of Article 68, be open under uniform conditions to the aircraft of all the other contracting States. The like uniform conditions shall apply to the use, by aircraft of every contracting State, of all air navigation facilities, including radio and meteorological services, which may be provided for public use for the safety and expedition of air navigation.
>
> (2) Any charges that may be imposed or permitted to be imposed by a contracting State for the use of such airports and air navigation facilities by the aircraft of any other contracting State shall not be higher
>
> (a) As to aircraft not engaged in scheduled international air services, than those that would be paid by its national aircraft of the same class engaged in similar operations, and
>
> (b) As to aircraft engaged in scheduled international air services, than those that would be paid by its national aircraft engaged in similar international air services.

580 See Martin Staniland (2003), *Government of Birds – Air Transport and the State in Western Europe*, (Lanham, MD: Rowman & Littlefield) (the author refers to domestic routes as property rights, at 213) [Staniland]; but see Alfred E. Kahn and Michael Roach, 'A Paean to Legal Creativity' (1979) 31 *Admin. L. Rev.*, 91 (the authors deny to attribute property status and constitutional protection to domestic routes issued by the CAB, at 100) [Kahn & Roach]; Civil Aviation Authority (1984), *Airline competition policy*, (Cheltenham/UK: Civil Aviation Authority) (the CAA also clearly sets out that 'route licences are not property', at 16) [CAA Competition].

> (3) All such charges shall be published and communicated to the International Civil Aviation Institutionization: provided that, upon representation by an interested contracting State, the charges imposed for the use of airports and other facilities shall be subject to review by the council, which shall report and make recommendations thereon for the consideration of the State or States concerned. *No fees, dues or other charges shall be imposed by any contracting State in respect solely of the right of transit over or entry into or exit from its territory of any aircraft of a contracting State or persons or property thereon.*[581]

According to the legislative history of Article 15 and to its objectives, the main purpose of this provision is to oblige the contracting states to refrain from any discriminatory activities. The paramount concern of the state parties to the Chicago Convention was the possible unequal treatment of their airlines by other nations. Article 15 of the Chicago Convention was included to prevent this.[582] Pursuant to Articles 15(1) and 15(2) the examination of discrimination is assessed by looking at the treatment that is granted to an airline by its registration or 'home' state.[583] Pursuant to Article 15(2) a contracting state must not impose higher charges on foreign airlines than it does with respect to its own national airlines. As far as the auctioning of air traffic rights is concerned, one could argue that no such forbidden discrimination of foreign airlines occurs, if a state charges its national carriers a fee for allocating air traffic rights, as long as foreign carriers are not charged higher fees or, as would be the case here, if they are completely exempted from these charges. Thus, under these circumstances, a state would not be prevented from auctioning air traffic rights among the applicant air carriers.

On the other hand, such an understanding might be incompatible with the wording of Article 15(3) of the Chicago Convention. This provision stipulates that any kind of fees, charges or 'tolls' collected solely for the use of the airspace is forbidden.[584] The reference in Article 15(3) to the foregoing provision ('All such charges ….') and the explicit prohibition that 'no other charges shall be imposed …' Indicate that the only permitted fees are set out in Article 15(2), which are imposed for the use of either airports or air navigation facilities. Such an understanding of Articles 15(2) and (3) is also confirmed through the subsequent state practice, manifested in their bilateral air service agreements. Any regulation contained in these agreements, dealing with fees

581 Emphasis added by author.

582 Edwin O. Bailey, 'Article 15 of the Chicago Convention and the Duty of States to Avoid Discriminatory User Charges: The US–United Kingdom Of Great Britain And Northern Ireland London Heathrow Airport User Charges Arbitration', (1994) 19(2) *Ann. Air & Sp. L.*, 81, 89, 99 [Bailey]; Rod D. Margo and William T. MacCary, 'The Chicago Convention and the Resolution of Landing Fee Disputes' (1994) 19(2) *Ann. Air & Sp. L.*, 501 at 505; L. Welch Pogue and Calvin Davison, 'User Charges in International Aviation' (1979) 73 *Am. J. Int'l L.*, 42 at 47 [Pogue and Davison].

583 Bailey, *ibid.* at 99.

584 Bailey, *ibid.* at 81; Pogue and Davison, *supra* note 582 at 47; Gunnar Finnsson, 'Airports and Route Facilities: International Cost Recovery Policies and their Applicability in the Framework of new Forms of Infrastructure Provision' (1994) 19(2) *Ann. Air & Sp. L.*, 283 at 286.

or (user) charges, refers to airports or other (navigation) facilities.[585] The subsequent state practice is of particular importance as it can be referred to, to determine the intention of the signatory states of how a specific provision of international law was intended to be understood and applied.[586] Further clarification with respect to Article 15(3) Chicago Convention is required for the terms 'entry into,' 'exit from' and 'transit over' the territory of a state. While 'transit' can be defined as 'the act of passing through or over,'[587] the terms 'entry into' and 'exit from' the territory are less clear. Firstly, Article 2 defines 'territory' as 'the land areas and territorial waters adjacent thereto'. Considering this definition, the term 'entry into' a territory would need to be understood as landing, while 'exit from' a territory would have the meaning of departure. Although the definition of 'territory' strongly suggests such an understanding, a systematical interpretation of other provisions of the Chicago Convention leads to a different result. It does not seem that the state parties to the Chicago Convention, by using the term 'territory' in Article 15(3), meant to refer to the landing and departure of an aircraft. When relating to landings or departures, these terms were explicitly used, as can be seen in various provisions throughout the Chicago Convention.[588] Furthermore, Article 10 uses the terms 'entering the territory' and 'landing' simultaneously, but, as can be inferred from the context, attributes a different meaning to both expressions. Additionally, Part I, Chapter II of the Chicago Convention, of which Article 15 is a part, deals with flights over the territory of contracting states, but not with landings or departures. Finally, it would contradict Article 15(2), if 'exit from' and 'entry into' the territory of a state were understood as landing and departure, since this provision explicitly allows the imposition of charges for the use of airports, while Article 15(3), on the other hand, would forbid any charges for landings and departures. Therefore, 'entry into' and 'exit from' the territory of a state must be understood as entry into and exit from the airspace of a state. Under these circumstances, the construction of Article 15(3) Chicago Convention is considered as a legislative lapse. Thus, it is concluded that Article 15(3) contains a prohibition to impose charges for the entry into and exit from the airspace of a country. As was defined above, the allocation of air traffic rights and the subsequent designation can be understood as the permission to enter a market and to offer air transportation services.[589] Additionally, the allocation of air traffic rights and the act of designation must be attributed a second meaning and also include the right to leave the designating country, when flying into other, foreign states, and the permission to re-enter the designating country on the way

585 See e.g. Article III(1) of the US–UK Bermuda I Agreement, reprinted in Bailey, *ibid.* at 87; Article 1(j) of the US Model Agreement, reprinted in Richard W. Bogosian, 'Aviation Negotiations and the US Model Agreement' (1981) 46 *J. Air L. & Com.* 1007 at 1023.

586 Michael Akehurst (1992), *A Modern Introduction to International Law*, 6th edn., (London: Routledge), 204; Anthony Aust (2000), *Modern Treaty Law and Practice*, (Cambridge, United Kingdom: Cambridge University Press), 194; Bleckmann, European Law, *supra* note 75 at para 561; Article 31, Sec. 3 (b) Vienna Convention.

587 *The Penguin English Dictionary*, 2nd edn., s.v. 'transit'.

588 See also Articles 5, 10, 16 Chicago Convention.

589 See above at 8.

back from other states. It would make no sense to designate an air carrier for international air transportation services and not to grant the mandatory exit and re-entry permissions. Therefore, the allocation of air traffic rights and the subsequent mandatory designation must also implicitly include the right to exit and re-enter the airspace of the designating state.[590] In conclusion, it is submitted that Article 15(3) of the Chicago Convention contains a prohibition to impose charges for the entry into and exit from a state's airspace. Since the allocation of air traffic rights and the act of designation include the permission to enter and exit the airspace of a country, it would result in a violation of Article 15(3) Chicago Convention, if a state imposes charges for the distribution of air traffic rights.

The only difficulty that remains to be discussed is the fact that international law usually is not directly applicable to a state's own nationals.[591] Therefore, one could argue that the allocation of air traffic rights among the air carriers of a state is a purely national procedure, to which international lawn, and in particular Article 15(3) Chicago Convention, is not applicable. However, while Article 15(2) forbids the imposition of higher charges on airlines of another contracting party, this distinction is missing in Article 15(3) Here reference is made to any aircraft of a contracting state. Thus, the latter provision refers not only to the aircraft of another contracting state, but also to all aircrafts operating in the airspace of a country. Article 15(3) Chicago Convention prohibits to a state to impose charges for the entry into or the exit from its airspace even on those aircrafts that are registered in the respective state. Hence, contrary to the general principle of international law, this provision also extends to purely national situations.

Result

When assessing auctions as a method of distributing air traffic rights, it is concluded that auctions offer the greatest benefits as far as efficiency of the allocation process, fast implementation of services and objectivity and transparency of the procedure are concerned. Furthermore, imposing charges on the airlines would help reflecting the true costs of travelling by air. It is unresolved, whether auctions might hamper competition in the air transportation sector. However, due to the international nature of air transportation and because of Article 15(3) Chicago Convention, auctions are inapplicable for the distribution of air traffic rights.[592]

---

590 Compare e.g. § 92(1) *Luftverkehrszulassungsordnung* (German Air Traffic Licensing Regulations) according to which aircrafts registered in Germany and permitted for international scheduled flights do not require a separate permission to exit the German airspace. Here the prior designation includes even a *de jure* exit permission. According to §§ 1c Nr. 1, 2(7) *Luftverkehrsgesetz* (German Aeronautics Act) aircrafts registered in Germany do not need any special permissions to enter the German airspace.

591 Ipsen, Intl Law, *supra* note 28 at 95, para 1.

592 14. C.F.R., § 399.120 (1986), (it is interesting to note that the US Department of Transport asked air carriers for comments about its future route assignment policy; thereby many airlines opposed to the use of auctions or lotteries as allocation methods, and not even one single company supported either of these procedures).

Beauty contest

The last allocation method to be evaluated, is the beauty contest. Due to the thorough examination of the applications filed, beauty contests have proven to be an appropriate tool in choosing the most suitable applicant(s). The major problems are the lengthy and costly procedure, as well as the ambiguity of the decision making. It will be analysed below, how these negative aspects can influence the allocation of air traffic rights and whether they can be avoided.

Procedure

The main disadvantage of a beauty contest is the amount of time it takes, until a final decision is made. Experience gathered in the allocation of telecommunication licences shows that solely the administrative process can take up to 18 months.[593] Subsequent litigation can extend the delay, until use of a licence is possible, up to a couple of years. While the danger of judicial proceedings can be hardly excluded, regardless of which distribution method is used, it is possible to influence the length of the administrative process. Beauty contests in the telecommunication sector were partly conducted as comparative hearings, in which each applicant was allowed to present its case personally in front of a committee or a judge.[594] This rendered the procedure to be very time consuming. It can be questioned though, whether a comparative hearing is always necessary. A personal appearance might be useful, if a technically highly sophisticated area is concerned, to give the applicants the opportunity to explain their projects and to ensure that the deciding persons have all the information they need to ensure a fair decision and allocation of the limited rights. However, this does not seem mandatory in less complicated areas, which do not require a high degree of expertise. Since the telecommunication sector is a highly complicated and dynamic area, where new technologies are developed continuously, it is useful to have the applicants appearing in person to explain their endeavours. In reaching a decision on which project is of the highest public interest and requires the allocation of a frequency licence, it can be necessary, to hear the different opinions of the parties involved. Air transportation services are, on the other hand, from a technological perspective less complicated. The number of different business possibilities is limited and it does not require an extensive presentation to explain them. Therefore, comparative hearings do not seem necessary to properly allocate air traffic rights. A comparison of the applications filed, without a hearing and a personal appearance of the airlines' representatives, should be sufficient. Conducting the beauty contest as such a documentary hearing should already reduce the length of the administrative process significantly. Another factor, influencing the duration of a beauty contest, is the number of applicants. As described above, one can expect the total number of airlines applying for air traffic rights to be much smaller, than the number of companies applying for frequency licenses in the telecommunication

593 Evan Kwerel and Alex Felker, 'A Free Market solution – FCC Should Auction Licenses for Radio', *Legal Times*, (22 September 1986).

594 See above at 159.

sector.[595] Hence, the danger of having an enormous number of applications is not pre-eminent in the air transportation sector. Under these circumstances, the length of an allocation procedure should remain within a reasonable time period, thereby keeping the administration costs fairly low.

Ambiguity of the decision making

The other main point of criticism with respect to beauty contests is the ambiguity of the decision making, if the selection criteria are not disclosed. When, on the other hand, the criteria are made public, it results in that all applicants, in their effort to comply with them to the highest degree possible, end up to be all equally qualified. With respect to a distribution of air traffic rights in the EC, Article 5 of Regulation 847/2004 requires the allocation to be transparent. Therefore, the criteria, upon which the allocation decision will be based, must be disclosed. The danger is the emergence of a situation, in which all companies will look alike in the end. To assess the possibility of the rise of such a situation, it is first necessary to determine, which criteria shall be applied. Subsequently, it can be evaluated, how likely the danger is that all airlines will appear to be equally qualified. However, the choice of the selection criteria seems to be the only problem related to beauty contests with respect to the allocation of air traffic rights. This problem maybe can be even solved.

RESULT

The foregoing analysis has revealed that auctions, as the primary method of allocating frequency licences in the telecommunication sector, might not be an appropriate instrument to distribute air traffic rights. The legal and economic circumstances of international air transportation do not allow to apply this method. User fees, firstly, have proven to be infeasible and secondly, as it is with auctions, must face economic and legal obstacles. Lotteries and the first-come/first-served method are, due to their randomness, inappropriate to allocate air traffic rights. Unlike in the telecommunication sector, it seems that beauty contests, in the form of documentary hearings, could be the most appropriate method to allocate air traffic rights. The only question that needed to be answered, is the determination of the appropriate criteria, upon which the selection should be based. In the following an examination of the methods of distributing air traffic rights in various countries will be undertaken. The objective will be to, firstly, examine, whether the conclusion reached in this chapter, that is that air traffic rights should be allocated through beauty contests, is valid and confirmed by practice. Secondly, one may find criteria that could be applicable in the beauty contest.

595 See also above at 121 (in the EC there are only about 130 air carriers).

THE DISTRIBUTION OF AIR TRAFFIC RIGHTS – A STATE-BASED COMPARISON

After the general analysis of different available allocation methods in the previous chapter, it will be shown now how various states have solved the problem of distributing air traffic rights among their national air carriers. To fully assess their approaches, an overview of the corresponding legal and political circumstances shall also be given.

USA

The first country to be examined, are the USA. Here, it will be necessary to distinguish between the period before and after deregulation.

Pre-deregulation period

Background

In the Civil Aeronautics Act (1938) the Civil Aeronautics Board (CAB), a regulatory body responsible for aviation matters, was established. One of the CAB's most important duties was the assignment of air traffic rights. To operate on a route an air carrier had to prove, first, that it was fit, willing and able to offer air transportation services and, second, needed to obtain the Certificate of Public Convenience and Necessity (certificate).[596] While the first pre-requisite was a formality that had to be fulfilled mainly for safety reasons, the certificate was the core element of the route assignments. When adopting the Civil Aeronautics Act in 1938, Congress decided not to opt for competition in the air transportation industry.[597] The experiences made with railroad companies in the USA in the second half of the nineteenth century and the losses that were generated in that industry due to destructive competition, led to the conviction that a similar development needed to be prevented in the air transportation industry.[598] As destructive competition might not only have an impact on consumers, but could also deter financial investors from investments and thus mostly affected the respective players, an unleashed competition was in nobody's

596 Burkhardt, *supra* note 24 at 17.

597 Paula Brown Sinclair, 'Case Notes – Administrative Law', (1978–1979) 44 *J. Air L. & Com.*, 849 at 854f. [Sinclair]; 'Airline Regulation by the Civil Aeronautics Board – Summary of the Report of the Senate Subcommittee on Administrative Practices and Procedures', reprinted in (1975) 41 *J. Air L. & Com.*, 607 (according to the legislative history of the Civil Aeronautics Act, new companies should be able to enter the industry as the market for air travel expanded and it was not believed that the application of the Civil Aeronautics Act would prevent new entries; however, this assumption proved to have been wrong and the Civil Aeronautics Act lead to the opposite result, at 614) [CAB Report].

598 A.J. Thomas Jr (1951), *Economic Regulation of Scheduled Air Transport*, (Buffalo: Dennis & Co.) at 53, [Thomas]; Russel A. Klingman, 'Predatory Pricing and other Exclusionary Conduct in the Airline Industry: Is Antitrust Law the Solution?', (1991–92) 4 *DePaul Bus. L.J.*, 281 at 285 [Klingman]; Sinclair, *ibid.* at 859.

interest.[599] The certificate was introduced as a tool to prevent the development of such a cut-throat competition and to protect and nurture the infant air transportation industry.[600] Air transportation was not only perceived to be part of a state's infrastructure, upon which the whole economy was dependent, but also considered as a natural monopoly, where competition was only possible to a limited extent.[601] A natural monopoly describes a market, where necessary investments are so high that the demand can be satisfied at the lowest price, if only one company offers the respective service. Competition in such markets is inefficient. It was assumed that too many operating airlines could lead to a wasteful duplication of services. Therefore, competition should be limited to promote a better use of resources. To prevent the airlines from abusing their monopolistic or *quasi*-monopolistic positions, the CAB had to regulate prices to protect consumers from unjustified high rates.[602] However, soon it was understood that a controlled monopoly was also not in the public interest and regulated competition was needed instead.[603]

Criteria

Domestic routes

Although the following outline will concentrate on the distribution of domestic routes, the approach taken by the CAB and the experiences gained thereby were influential for the distribution of international routes. When issuing the certificate, the CAB had to consider public necessity and convenience. Under the Civil Aeronautics Act (1938) the CAB had to take into account the development of an air transportation industry that was properly adapted to the present and future needs of foreign and domestic commerce, postal services and national defence of the USA. Competition should have been also fostered, but only to the extent necessary to promote the desired development of the air transportation industry.[604] It was not a goal by itself, but rather a tool to achieve other objectives.

In the beginning, the CAB did not only decide which air carrier was granted a route licence, but also which route had to be established and which communities were to be served.[605] This two-fold approach was due to the fact that most air transportation services were subsidized through air mail contracts sponsored by the state-owned postal service.[606] This wide discretion of the CAB was held necessary as available

599 Thomas, *ibid.* at 54.

600 *Ibid.* at 53; Sinclair, *supra* note 597 at 852, n. 27, 858.

601 John H. Shenefield, 'Regulation and Deregulation – Where Do We Stand?', (1976) 45 *Antitrust L.J.*, 244 at 245.

602 Klingman, *supra* note 598 at 285.

603 Thomas, *supra* note 598 at 229, p.241.

604 *Ibid.* at 225.

605 Williams K. Jones (1962), *Licensing of Domestic Air Transportation by the Civil Aeronautics Board* [Unpublished, Archived at Columbia University School of Law] at 12 [Jones].

606 Thomas, *supra* note 598 at 241; Rollo, *supra* note 552 at 93; Klingman, *supra* note 598 at 285.

subsidies were limited and to ensure that the taxpayer's money was used in the most efficient manner. In reaching a decision, the Board examined four questions:[607]

- Which useful public purpose according to public needs would the new route serve?
- Could existing carriers fulfil the new service?
- Would a new air carrier impair the operations of existing carriers contrary to the public interest?
- Would the benefits from the proposed service to the public outweigh the cost to the government?

To be able to answer the first question, an intensive study of different kind of information like for example population of the communities to be connected, living standards, wealth, past traffic between both destinations or growth potential was required.[608] The evidence brought forward by the applying air carriers was always very voluminous and detailed and its analysis was very time consuming.[609] At the next stage the CAB had to evaluate, whether the existing air carriers could sufficiently provide the proposed services and whether a new airline could impair the operations of the incumbent operators. The answer to this question required the decision, whether present airlines were shielded from further competition or whether additional competition was allowed in the market. The decision of this conflict was interrelated with the last question and whether the benefits of a new service would outweigh the costs occurring to the public. The government intended to promote the establishment of a network of flights across the USA by granting subsidies through mail contracts.[610] The admission of a new competitor on a route could have distracted traffic from another carrier, reducing its profit and making it more dependent on subsidies.[611] In the worst case, an airline that was previously built up with governmental mail contracts could have gone bankrupt and the invested capital would have been lost. For this reason, the Board had also vigorously denied non-scheduled airlines the access to the domestic air transportation system. It was argued that non-scheduled airlines would concentrate on the more profitable markets and might divert traffic from the scheduled air carriers, which the latter ones needed to cross-subsidize the less profitable routes.[612] Admitting new entrants, on the other hand, could have increased competition, led to lower fares and thus attracted more people to use air transportation services. An increase use of air transportation by the general public would have resulted in more profit for the airlines and a reduced need for subsidization. Thus, it was a dilemma to strike the balance between these two contradicting possibilities. In the end, the CAB decided to protect the incumbents at

607 Thomas, *ibid.* at 226.
608 Sinclair, *supra* note 597 at 853.
609 Thomas, *supra* note 598 at 74.
610 *Ibid.* at 95.
611 Richard E. Caves (1968), *Air Transport and its Regulators – an Industry Study*, (Cambridge, MA: Harvard University Press), 169, 173 [Caves]; Jones, *supra* note 605 at 16.
612 Caves, *ibid.* at 173.

the expense of competition.[613] Additional competitors were only allowed on a route, if enough traffic was available to keep more than one air carrier operating profitably and permissions were strictly denied, if competition might have had a negative impact on the incumbent.[614] New entrants were also allowed to a market next to an existing airline, if they were small and weak and the expected diversion of income for the incumbent was negligible.[615] Sometimes the CAB set up two weak carriers to operate in the same market in order to strengthen them both.[616] Furthermore, if an existing carrier developed a route well and was offering satisfying services, the Board was reluctant to allow additional competition. If, on the other hand, the services offered were poor and contrary to the CAB's expectations, the Board was more inclined to allow the entrance of a new airline.[617] Once the Board had decided, whether a route should be established and if an existing carrier or a new entrant should serve it, the choice had to be made as to which air carrier should be given the route permission. This was maybe the most difficult part of the route assignment procedure and many aspects had to be taken into account. The CAB could have preferred a weak carrier to make it stronger and to increase its competitiveness, rendering it less dependent on subsidies. Sometimes small, commuter carriers were permitted to operate on the more profitable long routes, in order to use the income from these routes to cross-subsidize the economically less profitable, but socially much desired short routes.[618] Another possibility was to promote the extension of the air transportation network and to grant routes to those air carriers that offered the most connections and carried the most traffic and thus exploited the routes most efficiently. This approach favoured the bigger and more advanced airlines. Thereby, competition was subordinated to route integration and passenger convenience. For a limited time, the CAB pursued this single-carrier approach and preferred carriers with the most integrated network of flights. But soon it was realized that this solution extensively favoured the largest carriers with the most developed route structures and thereby extensively impeded competition. Therefore, the single-carrier approach was abolished and sacrificed for a more stable and regulated development of a healthy and competitive air transportation system, which foresaw a multitude of air carriers operating in one

613 Tourtellot, *supra* note 563 at 560; Sinclair, *supra* note 597 (no new trunk carriers have been certified since 1938 when the CAA was adopted, at 857); CAB Report, *supra* note 597 (the CAB has never allowed an air carrier to go bankrupt, at 614); John T. McGoldrick Jr., 'The Impact of the Civil Aeronautics Board's Regulations on Price Structure in the Airline Industry' (1975-76) 10 *Ga. L. Rev.* 619 at 629 [McGoldrick]; Paul Stephen Dempsey, William E. Thoms and Sonja Clapp, 'Canadian Transport Liberalization: Planes, Trains, Trucks & Buses Rolling Across the Great White North' (1990–91) 19 *Transp. L. J.* 113 at 119 [Dempsey, Thoms and Clapp].

614 Douglas and Miller, *supra* note 520 at 115; Rollo, *supra* note 552 at 85; McGoldrick, *ibid.* at 630.

615 Caves, *supra* note 611 at 199.

616 Rollo, *supra* note 552 at 208f.

617 Jones, *supra* note 605 at 15f.

618 Douglas and Miller, *supra* note 520 at 117; Sinclair, *supra* note 597 at 857, n. 55; Rollo, *supra* note 552 at 85; Thomas, *supra* note 598 at 82f.

market.[619] Another aspect that was considered when routes were assigned was the historic interest of an airline in a route and its permanent presence in a market.[620] Due to the fact that an air carrier previously had taken the risk to develop and invest into a market, it was held to be unfair, if it was not also granted a route in this market.[621] Furthermore, a carrier, that has established itself in a market, might have already undertaken investments in airport facilities. These additional expenses, that would have to be made by the newcomers as well, could have been avoided by allocating routes to the incumbent(s). But, unlike in the case when routes are 'grandfathered,' the historic-interest approach did not guarantee the allocation of a route. It was only one aspect that was considered and it almost never prevailed, when it was balanced with the objective of strengthening competition through admitting new entrants.[622] It is interesting to note that, during this period, fares were never a factor in the selection process. The CAB believed that competition could lead to enhanced services, but not to lower prices.[623] Although the cost structure of an airline was sometimes taken into account, it never prevailed over other aspects. The Board members were afraid that choosing a carrier with the lowest cost structure would have made the big airlines even bigger and might have endangered a beneficial development of the whole air transportation industry. Therefore, the cost structure as a criterion was soon completely abandoned and more emphasis was put on strengthening all air carriers equally instead.[624]

Once the aviation industry became stronger, the trunk carriers could generate enough income from their ordinary operations and went of subsidy. The CAB ceased to strictly apply its four-question-approach and started to concentrate more on the aspect of competition.[625] Despite the fact that this might have again increased the need for subsidies, the Board stopped putting too much emphasis on the possible impairment of the existing operations of incumbent air carriers.[626] Enhanced competition and the strengthening of weaker airlines were perceived as being superior in the public interest than possible higher governmental expenses.[627] This was even more valid, if air carriers were willing to start operations on a route and thereby risked their own funds and not the taxpayers' money. In such a case there was no reason for the CAB to supervise the airlines and to decide, where and how they should operate.[628] Another reason for the change of the Board's approach was that the political surrounding also changed. Congress started to be less concerned with public subsidies and wanted instead to promote competition in the air transport

619 Caves, *supra* note 611 at 200.

620 Rollo, *supra* note 552 at 222.

621 Jones, *supra* note 605 at 18.

622 Rollo, *supra* note 552 at 223.

623 Douglas and Miller, *supra* note 520 at 116, n. 17.

624 Caves, *supra* note 611 at 200f.

625 Thomas, *supra* note 598 at 240; Rollo, *supra* note 552 at 213.

626 Thomas, *ibid.* at 241; Caves, *supra* note 611 at 199; Kahn and Roach, *supra* note 580 at 105.

627 Thomas, *ibid.* at 242; Caves, *ibid.* at 204; Rollo, *supra* note 552 at 226; Sinclair, *supra* note 597 at 856.

628 Kahn and Roach, *supra* note 580 at 109.

sector. For this reason, it increased political pressure on the CAB to give up its protectionism.[629] To achieve the objective of more competition in the air transport sector, Congress adopted the Airline Deregulation Act in 1978. But even before this act came into force, the CAB had already started to apply the existing regulations in a very pro-competitive manner, thereby establishing a *de facto*, administrative deregulation. For this purpose the CAB, when granting route licences, considered low fares as a factor in carrier selection, allowed for a multitude of air carriers to operate on new routes and granted a limited pricing freedom to air carriers.[630] In this context it is interesting to note that, in justifying its new approach, the CAB also addressed the issue of destructive competition, which was feared in 1938 to also arise in the air transportation industry. After scrutinizing the four possible types of destructive competition, the Board concluded that three variations occurred only under very unusual and particular circumstances and the last one, predatory behaviour, could be controlled and avoided by the government.[631]

International routes

When it came to allocating international air traffic rights, the CAB was more in favour of competition.[632] Since the advent of international air transportation, Pan Am was the only US carrier operating on international routes. In order to gradually introduce competition, the Board divided Europe, the Middle East and Asia into different regions. In addition to Pan Am, which operated into all of these areas, the CAB assigned routes to each of the regions to a different air carrier. Only two destinations, London and Lisbon, were served by more than one airline, as these markets were expected to provide high traffic and thus allowed for increased competition. On all other routes only one carrier was initially allowed to operate.[633] To be assigned an international route, the applicants needed to have a well-established institutionalized corporate structure, sufficient experience, a solid financial basis and outstanding managerial abilities. Since US carriers would compete with airlines from other nations, only the most advanced air carriers were allowed to operate in international markets.[634] Like in the allocation of domestic air traffic rights, the Board also considered the air carriers' ability to integrate an international route into their existing domestic networks, thereby ensuring the most efficient use of an air traffic right.[635] A significant difference between the allocation of domestic and international routes was that the latter ones were formally assigned by the US President and the CAB could only suggest a carrier to the President for designation. It occurred several times that the President did not follow the proposal of the Board and designated a different air carrier than the suggested one.[636] The reason for this ultimate presidential power is vested in the US Constitution, which gives the President the responsibility

629 Caves, *supra* note 611 at 192.
630 Dempsey, Thoms and Clapp, *supra* note 613 at 117.
631 Sinclair, *supra* note 597 at 860.
632 Frederik, *supra* note 556 at 271.
633 Thomas, *supra* note 598 at 236.
634 *Ibid.* at 242.
635 *Ibid.*
636 Frederik, *supra* note 556 at 298f.

for foreign relations. It is understood that the question which US carrier operates between the US and other countries is part of foreign policy.[637] Indeed, it is possible that a country may oppose a specific foreign carrier to fly into its territory. This is particularly true, if a carrier is already offering services to another state, with whom the first country is in conflict.[638] Here, the representative function of air carriers becomes obvious.

Procedure

To complete the US method of route allocation, a short summary of the procedure shall be also given. Usually, an air carrier applied to the CAB for permission to start operations on a specific route, chosen by the airline. The Board published a summary of the application in its weekly newsletter and scheduled a hearing before an administrative law judge.[639] Any person or entity having property or financial interests that might be affected by the new route, could have been permitted to join the proceedings as a party. Typical parties were other air carriers, civic groups, cities included in the route and industrial or other professional groups.[640] Airlines often joined a procedure for the allocation of air traffic rights, not only to oppose the granting of a route to an applying opponent, but also to request the granting of the route in question to themselves. After a pre-hearing, in which procedural matters regarding the trial were decided, and before the hearing itself took place, the applicant had to file in an exhibit, showing the public need for the new route.[641] To substantiate the application, considerable amounts of evidence and data were presented regarding the expected amount of traffic, the wealth level of the communities to be served, the expected income for the air carrier, the amount of incurred costs, the degree of route integration, and so on. The exhibit was distributed to the other parties, to give them the possibility to support or contradict the applicant.[642] Apart from the applicant, the other parties also presented evidence during the following hearing, which usually resulted in vast amounts of detailed exhibits and lengthy testimonies.[643] The airlines involved intended not only to show that they were competent to serve the suggested route, but also that they were superior to all other air carriers. Evidence was presented to support their own application and to discredit the other airlines' proposals.[644] At

637 *Ibid.* at 298f.

638 A case came to my knowledge, where the Dutch air carrier KLM was denied air traffic rights to fly to China by the Chinese Government, because the airline was already offering air transportation services to Taiwan. In another case the Israeli Government, when granting air traffic rights to Germany, insisted that a certain quota of these rights was allocated to German air carrier Aero Lloyd. The Israeli Government wanted to reward the airline for having previously continued air services to Israel despite a crisis in the Middle East. Unlike Aero Lloyd, many other airlines had ceased operations into this region during the crisis.

639 Kane and Vose, *supra* note 552 at 187.

640 *Ibid.*

641 *Ibid.* at 188.

642 *Ibid.* at 189.

643 Sinclair, *supra* note 597 at 854.

644 Kahn and Roach, *supra* note 580 at 104.

the end of the hearing the judge presented his opinion, whether and to whom a route should be granted. Thereafter the parties were given time to prepare and to file their briefs. After the briefs were received, the judge made his decision, which was, at this stage not binding yet, but only a recommendation to the CAB.[645] If no party or the CAB opposed, the decision became effective 30 days later. Each party could file a petition for discretionary review and the CAB could order such a review itself.[646] If a review was granted, the objecting party had to submit a brief on the disputed matters and an oral hearing before the Board could have been requested.[647] After the oral argument before the CAB, which was the only possibility for the parties involved to speak directly to the Board members during the entire proceedings, a final order was issued. However, the losing party could have attacked this order again with a petition for reconsideration. It was necessary to show that the CAB committed a legal error or did not take certain evidence sufficiently into account. If this petition was denied, the losing party could appeal the Board's decision in courts. This appeal was limited to legal errors, but did not extend to errors in facts.[648] As can be seen, the procedure was very formal and cumbersome. More than two-thirds of all applications for route assignments were pending for more than two years and more than one-quarter were pending for more than five years.[649] The average time for allocating a route from the pre-hearing conference to the final decision by the CAB was about three and a half years.[650]

Post-deregulation era

Background

In the 1970s politicians took the position that regulation might have been necessary to shield an upcoming and still fragile air transportation industry at its early stage, but that this industry has also become mature and could survive without any governmental protection. Furthermore, regulation resulted in higher fares, less services and fewer competitive alternatives.[651] A subcommittee of the Congress examined the CAB's work and reached four conclusions:[652]

---

645 Kane and Vose, *supra* note 552 at 190.

646 *Ibid.* at 191.

(The petitions for review could have been filed only for certain reasons:
(a) a material fact was in error
(b) the legal conclusion was without any governing precedent or was a departure from, or it was contrary to law, CAB rules or a precedent
(c) an important question of law, policy or discretion was involved
(d) a prejudicial error has occurred.)

647 *Ibid.*

648 *Ibid.* at 192.

649 CAB Report, *supra* note 597 at 615.

650 Jones, *supra* note 605 at 109.

651 Senator Edward M. Kennedy, cited in: Rollo, *supra* note 552 at 93.

652 CAB Report, *supra* note 597 at 608.

- The Board's practices were efficient in promoting industry growth, technological improvement and reasonable industry profits, but failed to maintain low prices.
- Some procedures applied by the CAB lacked the openness, intelligibility and impartiality that were required by the notion of procedural fairness.
- The Board should concentrate on making air transportation available to a broader public instead of promoting the aviation industry's growth. More price competition, a more liberal entry policy and less antitrust immunity were required.
- To achieve the above-mentioned objectives and to avoid the undesired effects, it was suggested to limit the CAB's powers.

It is remarkable that the subcommittee, when addressing the problem of destructive competition, was of the opinion that there was no substantial historical, empirical or logical reason for believing that competition might lead to predatory pricing, destructive competition or risk of monopolization.[653] With this conclusion the basis for the Civil Aeronautics Act (1938) was withdrawn. Based on the subcommittee's report, the Airline Deregulation Act was adopted in 1978. It introduced competition as the final goal for domestic commercial aviation and eliminated the possibility for the CAB to introduce a protective policy again.[654] The certificate was abandoned for the allocation of domestic routes and airlines only had to prove that they were fit, willing and able to provide the suggested services.[655] The CAB's role was limited to ensure compliance with antitrust laws, consumer protection and avoidance of predatory pricing.[656] As far as international air transportation was concerned, it remained necessary to obtain the certificate.[657] This is valid until today. The objectives for international air transportation, set out in the new Airline Deregulation Act, were fostering sound economic conditions and service at reasonable prices.[658] The Airline Deregulation Act further introduced expedited administrative procedures and required the CAB to issue final decisions at latest 180 days after the initial consideration of the application.[659] To minimize political influence the rights of the US President with respect to the designation of air carriers were also limited. The CAB's decision could be only disapproved for reasons of foreign relations or national defence, but not based on economic or other grounds.[660] Finally, the Airline Deregulation Act also foresaw the elimination of the CAB in 1985 and the transfer

653 *Ibid.* at 612.

654 Rollo, *supra* note 552 at 104; Leonard R. Stein, 'The Airline Deregulation Act of 1978' (1979) 20 *Harv. Int.* at 385, 389 [Stein]; Tourtellot, *supra* note 563 at 558, 561; Dempsey, Thoms & Clapp, *supra* note 613 at 118.

655 Rollo, *supra* note 552 at 93f., 103.

656 *Ibid.* at 102; CAB Report, *supra* note 597 at 635.

657 Stein, *supra* note 654 at 385f.; Jeffrey J. Ward, 'Transportation Law – Agency Interpretation of Regulations for Awarding International Air Carrier Routes' (1995–96) 64 *Geo. Wash. L. Rev.*, 1450–1452 [Ward].

658 Stein, *ibid.* at 386.

659 *Ibid.*

660 *Ibid.* at 387.

of its powers to the Departments of Transport and Justice. Therefore, since 1985, the Department of Transport is, among others, responsible for the certification of domestic and international routes.[661]

Criteria[662]

The Department of Transport (DOT), which took over the duty to allocate air traffic rights from the CAB, has issued a policy statement on 1 December 1986 on the distribution of international routes.[663] The DOT first addressed the issue, whether international routes should be granted permanently or only temporarily. The Airline Deregulation Act included the possibility to grant limited routes as 'experimental certificates'.[664] The CAB had made use of this possibility and had issued experimental certificates, limited to a period of five years. The DOT requested comments, whether this practice should be continued or whether permanent licenses should be issued. Proponents of the latter solution argued that indefinite certificates provided for security and encouraged carriers to invest in the development of their routes. Limited permissions, on the other hand, had the disadvantage that the renewal proceedings would burden the airlines with additional time and money that needed to be spent, even if a carrier would have previously performed satisfactorily and a renewal is approved in the end. This would put the US carriers at a disadvantage compared to foreign airlines, if the latter ones possessed unlimited air traffic rights for the same route. Proponents of a temporary experimental certificate argued that competition would be fostered, if incumbents had to fear to lose their route licences. Indeed, issuing temporary route permissions creates a situation that is comparable to the one that underlying the Theory of Contestable Markets. According to this theory, market players, if put in a surrounding, where free access to a market is guaranteed, behave as if they were in real competition although no competitors may be present, to deter any new entrants from the start.[665] Similarly, the threat of possibly losing a route licence in a renewal proceeding would encourage airlines to keep their fares low and their service standards high. Additionally, the renewal process would regularly allow other airlines to contest the incumbents and to substitute them in a market, if their services have been unsatisfying. Thus, competition could be simulated in a market, which is usually occupied only by a limited number of market players and where free competition is very often hindered by bilateral constraints. It was further disagreed that temporary certificates might discourage airlines from investing in their routes. To the contrary, temporary certificates provided for an even greater incentive to pursue the development of a route. Incumbent air carriers would have only a limited time period, during which they could prove that they were qualified

661 *Ibid.* at 388.

662 See 14 C.F.R. § 399.120 (1986).

663 Officially the DOT refers to this procedure as the distribution of certificates for limited-entry markets.

664 These 'experimental certificates' can be still found today in the Federal Aviation Act, see Federal Aviation Act, 49 USC § 41,102 (2003).

665 Paul S. Dempsey and Laurence E. Gesell (2004), *Air Commerce and the Law*, (Chandler, AZ: Coast Aire Publications), 57, n. 51 [Dempsey and Gesell].

to operate on that route and to offer satisfying services. As far as the time and costs of the renewal proceedings were concerned, it was argued that the procedure could be organized as a simplified, non-oral, show-cause procedure.[666] The DOT has adopted the second position and decided that it would continue to issue temporary experimental certificates limited to a period of five years. This has been deemed to be the most appropriate solution. It would give the incumbents enough time to develop a route and, at the same time, would also guarantee that the current holders of air traffic rights act responsively according to the needs of the markets. Additionally, other airlines are provided with a real opportunity to enter the market and the DOT is given enough flexibility to respond, if circumstances should change. The DOT describes this approach as 'a carrot and a stick'. It encourages air carriers to fulfil their proposals with respect to fares and services and it discourages them from abusing their monopoly. Should they, however, act contrary to their route proposals, when they applied for air traffic rights, they can be punished by losing their route licences. Another advantage of the issuance of temporary licences is that the performance of the air carriers must be really reviewed by the DOT on a regular basis. As far as the allocation process is concerned, the DOT decided to adopt the criteria used by the CAB and has also declared to continue the practice of varying the weight of the criteria on a case-by-case basis.[667] The following criteria are considered, when international air traffic rights are distributed:

Market structure

Under the market-structure-criterion the DOT examines the effects a route award might have on the overall level of competition in a particular market. For this purpose the DOT has regard to potential competition between the airports. It examines which competitive effect the proposed flights may have on existing flights offered to the same destination from different airports. It is also taken into account, to which degree the applying air carrier can increase competition by making more intensive use of its existing services. Hub dominance of an airline, as a separate criterion, is rejected. The DOT holds the opinion that this is mainly a matter to be discussed in connection with domestic market issues. The single decision with respect to an international route is not considered as sufficiently significant to have an impact on the dominance of an air carrier on a specific airport. Nevertheless, it would be possible to take hub domination into consideration within the market-structure criterion, if the market power of an airline at a particular airport is extremely high.

Route integration

Within the route-integration criterion the DOT analyses the air carriers' ability to fly traffic beyond the primary route to and from destinations behind the chosen US gateway, as well as the possibility to fly traffic beyond the foreign gateway. This criterion is considered as very significant, since it shows the economic viability of

666 See 14 C.F.R. § 399.120 (1986).

667 14 C.F.R. § 399.120 (1986); DOT Order 96-4-23, Docket OST-95-422, online: Department of Transport <http://dms.dot.gov/search/searchResultsSimple.cfm> (date accessed: 1 August 2004) [DOT Order 96-4-23]; DOT Order 93-12-27, Docket 49052 at 28 [DOT Order 93-12-27].

an air carrier's proposal and the benefits it brings to passengers outside the primary route.

Fare and service proposals

In connection with passengers' benefits the DOT also examines the fares and services a carrier intends to offer on a route. Preference is given to low or innovative fares and high frequencies or capacities.

Incumbency

In addition to the criteria formerly applied by the CAB, the DOT has included a rebuttable presumption in favour of the incumbent, when a route is to be renewed, if the airline performed well in the past. It is presumed that the incumbent is the best applicant to continue providing services, unless another air carrier can show that its services are substantially superior. The incumbent does not enjoy any positive presumption, if it has deviated from its suggested fare and service proposal without a reasonable justification.

Quick entrance

Finally, the DOT also considers the applicant's ability to enter a market quickly.

The DOT added that this enumeration was not exclusive and other criteria might be added, should this become necessary. Taking some of the cases decided by the DOT into account, one can discover that the Department did not always adhere to its own criteria. While it has not included in its policy statement the ability of US carriers to compete effectively with foreign carriers and to divert traffic from foreign airlines, it explicitly made use of this criterion in the past.[668] The general competitive situation between US air carriers was also taken into account.[669] Any prior assignments with respect to other routes never had any influence on the allocation in question.[670] In some cases, the DOT based its decision on an evaluation as to which community had the highest economic interest in a route, an aspect, which was previously regarded as not relevant.[671] Therefore, in addition to the above mentioned criteria, the DOT also takes the following aspects into consideration:[672]

- impact on domestic inter-carrier competition;
- ability to compete effectively with foreign airlines;
- economic interest of communities in a route.

---

668 Ward, *supra* note 657 at 1451; DOT Order 93-12-27, *ibid.* at 43, 46; DOT Order 96-4-23, *ibid.*

669 Ward, *ibid.* at 1459.

670 DOT Order 96-4-23, *supra* note 667.

671 DOT Order 93-12-27, *supra* note 667 at 4.

672 14 C.F.R. § 399.120 (1986); DOT Order 93-12-27, *supra* note 667 at 27ff.

As far as the procedure is concerned, under existing law the DOT has the possibility to use the expedited show-cause-procedure to allocate air traffic rights, which does not require an oral hearing any more.[673] This approach is usually applicable, unless:

(i) it prejudices a party;
(ii) material issues of fact need to be resolved by an oral hearing; or
(iii) the public interest requires an oral hearing.[674]

Result

The main problem with the US method is that the two conditions under the certificate public convenience and necessity are very vague, obscure and indefinite.[675] Convenience is affirmed, when something is 'suitable and fitting to the public need' and necessity is described as an improvement, which is highly important to the public convenience and desirable to the public welfare.[676] This broad definition allows it to subsume almost every interest under these two terms.[677] It is for this reason that the CAB was very inconsistent in its decision making and never succeeded to correlate and reconcile the principles it had applied in previous decisions. At one point the

---

673 Ward, *supra* note 657 at 1450, n. 7.

674 *Ibid.* at 1454; but see DOT Order 96-4-23, *supra* note 667 (despite the changes in the procedure, the allocation of a highly requested route can still take up to eight months, even if it is conducted without any hearings; the problem is that a lot of documents are filed and generated; in the here cited case (DOT Order 96-4-23), 216 different documents including orders, affidavits, responses, motions, various correspondence, exhibits, briefs and objections were filed or issued).

675 Thomas, *supra* note 598 at 69.

676 *Ibid.* at 69f.

677 See also Thomas, *ibid.* at 71f.; (the declaration of policy to the Civil Aeronautics Act (1938) contained objectives that should be taken into account, when evaluating the public interest; these objectives were weighed equally and without any priority to one or the other objective:

(a) The encouragement and development of an air-transport system adapted to the needs of foreign and domestic commerce, of the Postal service and of national defence;

(b) The regulation of air transportation in such a manner as to recognize and preserve the inherent advantages of, assure the highest degree of safety in, and foster sound economic conditions in, such transportation and to improve the relations between, and coordinate transportation by, air carriers;

(c) The promotion of adequate, economical, and efficient service by air carriers at reasonable charges without unjust discriminations, undue preferences or advantages, or unfair or destructive competitive practices;

(d) Competition to the extent necessary to assure the sound development of an air-transportation system properly adapted to the needs of the foreign and domestic commerce of the United States, of the Postal Service, and of the national defence;

(e) The regulation of air commerce in such a manner as to best promote its development and safety, and

(f) The encouragement and development of civil aeronautics.

Reprinted in Sinclair, *supra* note 597 at 852, n. 29, 30).

CAB used 22 different criteria in the allocation of air traffic rights.[678] The CAB consistently changed the criteria, upon which it based its decisions. Once it preferred big incumbent air carriers to avoid the diversion of traffic and of income, to reduce the airlines' needs for subsidies. Then the Board wanted to promote route integration and pursued the single-carrier-approach. After that small and weak airlines were supported to strengthen them and to enhance competition. Other criteria were the provision of services required by the travelling public, the needs of local communities, the air carrier's equipment or the historical interest of a carrier in a route.[679] The CAB changed its criteria, either, because it was not satisfied with the development in the air transportation sector and did not achieve its objectives or due to political pressure.[680] The Board never previously defined which weight would be attributed to each criterion and instead decisions were taken on a case by case basis, varying the significance of the applicable criteria in each case.[681] This inconsistency, especially when similar cases were decided differently, left the impression that each selection was rather a random choice. It was also speculated that air traffic rights were allocated according to political considerations and preferences and justified with a hollow and meaningless decisional formula.[682] The inconsistency and unpredictability combined with limited judicial control did not only frustrate the parties involved in the allocation process, but in the end also raised suspicion of corruption.[683] The subcommittee, that examined the CAB's practices, concluded that some of the Board's practices were infringing elementary notions of procedural fairness.[684] Even after the Airline Deregulation Act of 1978 has been enacted, many of these problems have not vanished. The DOT continues to apply the ambiguous manner of allocating air traffic rights by relying on a multitude of criteria, whose weight is determined on a case-by-case basis. Very often the outcome of a selection procedure depends on the view of the DOT's officers and the weight they attribute to a criterion. Hence, the subcommittee's demand to establish a simple and objective procedure for the distribution of air traffic rights is still unfulfilled. The DOT's decisions remain to be inconsistent and unpredictable, leaving irritated applicants, who regularly contest the DOT's findings, arguing that their services are superior and that the Department should attribute a different weight to its criteria. In addition to that, politicians still tend to interfere with the selection process and try to influence the DOT. Members of Congress support route assignments that bring the most benefits to their electoral

---

678 Robert George Sedding (1971), *Regulatory Police Making in the Civil Aeronautics Board*, (Ph. D. thesis; Political Science, Princeton University) [unpublished].

679 Breyer, *supra* note 522 at 568; Sinclair, *supra* note 597 at 854.

680 Caves, *supra* note 611 at 203.

681 Sinclair, *supra* note 597 at 854.

682 Breyer, *supra* note 522 at 568; CAB Report, *supra* note 597 (it is persuading, that the CAB's own staff described the selections of the Board as 'random', at 616); Kahn & Roach, *supra* note 580 (Kahn himself admits that the carrier selections by the CAB have often proven to be seriously mistaken, at 101, 106).

683 Breyer, *ibid.*; Kahn & Roach, *ibid.* at 99; CAB Report, *ibid.* at 616, 618.

684 CAB Report, *ibid.* at 608, 633.

districts.[685] Thus, with respect to the distribution of international air traffic rights, the deficiencies that already existed before deregulation occurred in 1978, are still present today.

AUSTRALIA

Background

Australia's aviation history began in 1936, when the privately founded Australia National Airways (ANA) started offering air transportation services. The airline performed very well and was about to dominate the Australian aviation market. To prevent the emergence of a private monopoly, the Australian Government decided to nationalize ANA in 1944. This endeavour failed for legal reasons.[686] Instead, the government established a second, state-owned airline, Trans Australian Airlines (TAA; later to be renamed Australian Airlines), in 1946. From the start the government intended to use TAA to drive ANA out of the market.[687] But, despite a strong support of TAA with governmental contracts, air mail subsidies and discrimination of ANA with higher airport and navigation charges, the governmental air carrier had difficulties to sustain competition from its private counterpart and incurred heavy losses.[688] However, after a while the governmental support started to bring the desired effects and ANA was facing bankruptcy. It was only due to the change of the government and a new aviation policy that ANA, for the time being, survived.[689] The new government decided to expose TAA to real competition with ANA and to cut any governmental subsidies unilaterally supporting TAA. This was the beginning of the Two-Airlines Policy. According to this policy, the government intended to have both airlines operating and competing on the domestic market.[690] The Australian Government believed that too much as well as too little competition in the airline industry was economically and socially undesirable.[691] Having two airlines operating was considered as the perfect compromise. Like in the USA, aviation was still an infant, upcoming industry going through a critical development period and needed governmental protection until it would be able to stand on its own feet.[692] Over supply and wasteful competition in a high-cost and high-technology

685 Ward, *supra* note 657 at 1453, 1460; DOT Order 96-4-23, *supra* note 667; DOT Order 93-12-27, *supra* note 667 (in this case some Members of Congress even argued, that a route should be granted to a city in Utah, because the Church of Jesus Christ of Latter Day Saints had its headquarters there, at 25).

686 D. Campbell, 'Competition Policy in Domestic Aviation or The Rise and Fall of the Two Airline Policy', in: S.G. Corones, ed. (1992), *Competition Policy in Telecommunication and Aviation*, (Sydney: Federation Press) at 17 at 18 [Campbell].

687 Williams, *supra* note 40 at 98.

688 Campbell, *supra* note 686 at 18.

689 *Ibid.* at 19.

690 *Ibid.*

691 *Ibid.* at 20.

692 *Ibid.*

industry like the airline sector could lead to tremendous and spectacular failures and must be prevented.[693] Taking into account the geographical structure of a country like Australia, which is characterized by only a few metropolitan areas and tourist places with considerable distances in between, aviation is of particular importance for national communication and transportation.[694] The government controlled entrance into this industry by adopting trade regulations that prohibited the import of aircrafts into Australia without the government's permission.[695] Unfortunately, ANA was, due to the prior government's interference, in such bad condition that it could only survive through a take-over by Ansett Transportation Industries in 1957. From this time on, the airline operated under the name Ansett-ANA.[696] It is remarkable that the government maintained its Two-Airlines Policy also with respect to Qantas, the other state-owned carrier and the only Australian airline allowed to offer international flights to and from Australia. Neither the domestic Australian air carriers were allowed to offer international services, nor was Qantas permitted to operate on the domestic market.[697] The Two-Airlines Policy proved to be very efficient and the airline industry could stabilize and prosper during the 1960s and early 1970s. But in the middle of the 1970s criticism towards this policy was raised, in particular because competition had not been fostered. In fact, both airlines were undistinguishable, using the same aircrafts, operating on the same routes at the same times and even charging the same fares.[698] Finally, deregulation and a shift to more competition in 1990 marked the *de jure* end of the Two-Airlines Policy.[699] The government decided to grant free access to domestic routes and even international airlines were allowed to offer domestic flights, if they were not engaged in international air transportation to and from Australia at the same time.[700] The separation between airlines operating on domestic or international routes was initially supposed to remain in place, as concerns arose that, if Qantas was permitted to serve the domestic market, then domestic airlines would have to be granted access

---

693 *Ibid.*

694 *Ibid.* at 21.

695 *Ibid.* at 20.

696 *Ibid.* at 19f.

697 Peter Forsyth, 'The Regulation and Deregulation of Australia's Domestic Airline Industry', in: Kenneth Button, ed. (1991), *Airline Deregulation – International Experiences*, (Washington Square, NY: New York University Press) 48 (in fact, Qantas was allowed to operate on some domestic segments, but only to carry passengers from and to its international routes, at 54) [Forsyth].

698 Campbell, *supra* note 686 at 21; Ross Jones, 'An International Perspective on AntiCompetitive Pricing Practices by Dominant Carriers and Regulatory Rules to Facilitate Competitive Entry – Australia', (2001–2002) 14 *DePaul Bus. L. J.* at 243 [R. Jones]; Williams, *supra* note 40 (it was even detected that both airlines were engaged in collusion, partially encouraged by the government, at 101).

699 Jeffrey Goh (2001), *The Single Aviation Market of Australia and New Zealand* (London, UK: Cavendish Publishing) 76 [Goh].

700 Olivier de Marolles, and Alexis Lenoir, (1993), *The Deregulation of Air Transport in Australia* (Paris/France: Institute of Air Transport,) 39 [Marolles and Lenoir].

to international routes as well.[701] This might, however, not have been possible due to the limited amount of available air traffic rights and the single designation clause, contained in air service agreements Australia had concluded with other countries.[702] In the end, when the Australian Government permitted a merger between Qantas and Australian Airlines in 1992, it also approved Qantas' access to the domestic market and, in return, allowed all Australian Airlines to operate on international routes.[703] The last remarkable event in Australia's aviation history occurred on 5 March 2002, when Ansett-ANA declared bankruptcy and ceased operations. This marked the *de facto* end of the Two-Airlines Policy.

Criteria

Procedure and general overview on applicable criteria

With opening international routes to all Australian Airlines, the government had to solve the problem of distributing its limited air traffic rights. For this purpose the International Air Services Commission Act was adopted in 1992. It foresaw the installation of the International Air Services Commission (IASC).[704] The IASC consists of three members, who are appointed for five years and are responsible for the allocation of international air traffic rights.[705] When discharging its duty, the IASC has to comply with a policy statement issued by the Minister for Transport and containing the criteria to be applied in the allocation of limited air traffic rights.[706] As far as the procedure is concerned, the Commission is supposed to act with as little formality and as quickly as possible. Hearings are not mandatory and it is left to the discretion of the Commission, whether it wants to summon a hearing.[707] The IASC also determines, how much time an air carrier is granted, until the allocated capacity must be fully used. This period is supposed to be as short as is reasonable to

701 Forsyth, *supra* note 697 at 69, 70.

702 *Ibid.* at 72f.; Williams, *supra* note 40 at 101.

703 Williams, *ibid.* at 102; Marolles and Lenoir, *supra* note 700 at 39; Jones, *supra* note 698 at 244.

704 International Air Services Commission Act 1992, Act No. 103 of 1992 as amended, online: International Air Services Commission <http://www.iasc.gov.au/iascleg.aspx#IASCAct> (date accessed: 3 August 2004) [Air Service Commission Act 1992].

705 Air Services Commission Act 1992, *ibid.*, s. 6; the Act does not speak of allocation of air traffic rights or routes, but refers to capacity. This is based on the fact that the Australian Government insists in international negotiations to include a multiple designation clause in its air service agreements. Therefore the restrictions contained in these agreements flow from limited capacities or frequencies. Hence, it is only consequent to talk of distributing capacity and not air traffic rights, although it is possible to allocate a whole capacity to only one carrier, which equals the allocation of all available air traffic rights to one single airline. In the following it will be, nevertheless, referred to the allocation of capacities as distribution of air traffic rights.

706 Goh, *supra* note 699 at 76; Air Services Commission Act 1992, *ibid.*, s. 6(3), s. 7(3), s. 11(2); the Air Service Commission Act 1992 does not speak of limited air traffic rights, but of scarce capacity.

707 Air Services Commission Act 1992, *ibid.*, s. 28, at 30(1).

commence operations and shall not exceed three years.[708] Transfers of route licences should be generally not permitted, if this could encourage speculation.[709] Depending on the situation and the underlying bilateral air service agreement, air traffic rights are granted for a limited period.[710] When allocating air traffic right, the Commission has to maximize the benefits to the public that can be gained from its use.[711] In the allocation procedure, the Commission applies two different sets of criteria. It depends on the situation, which of the sets is applied. The first set contains the general criteria. This test is similar to the 'fit, willing and able' assessment made in the US. The Commission has to evaluate, whether the carriers are:[712]

(i) reasonably capable of obtaining the necessary approval on the route; and
(ii) reasonably capable of implementing their applications.

The second set of criteria, the additional criteria, is more specific and gives more guidance to the IASC, how capacity should be allocated. Thereby the following aspects shall be taken into account:[713]

Competition
To evaluate the impact an allocation might have on competition, the Commission has to consider:

- the air carrier's ability to compete effectively with foreign airlines;the number of air carriers on a route and the distribution of capacity between Australian carriers;
- possibility of lower fares, increased choice and frequency of service and introduction of innovative products;
- the degree to which capacity, granted to Australia by a foreign country, will be used by foreign carriers due to code-sharing agreements;[714] and
- the assessments and decisions made by the Australian Competition and Consumer Commission or the Australian Competition Tribunal.

Tourism
To asses the impact an allocation will have on tourism, the Commission has to study:

---

708 International Air Services Commission Act (1992) Policy Statement, Paragraph 9, online: International Air Services Commission <http://www.iasc.gov.au/pubs/PSV520040519.pdf> (date accessed: 4 August 2004) [Policy Statement].

709 Policy Statement, *ibid.*, paragraph 10.

710 See also below at 204f.; The formal decision issued by the IASC on the allocation of air traffic rights or capacities respectively, is called 'determination'.

711 Policy Statement, *supra* note 708, paragraph 3.2.

712 Policy Statement, *ibid.*, paragraph 4.

713 Policy Statement, *ibid.*, paragraph 5.

714 Policy Statement, *ibid.*, paragraphs 3.3, 3.6 (applications of capacities to be used in code-sharing operations can be approved, but greater benefit will be attributed to services offered by own Australian aircrafts).

- the degree of promotion, market development and investment the applicants intend to make; and
- route service possibilities to and from points beyond the Australian or the foreign gateway (route integration).

Consumers
In determining the benefits to consumers, the following aspects have to be taken into account:

- the possibility to have a choice among several air carriers;
- the possibility of lower tariffs and improved services;
- the promotion of innovation; and
- route service possibilities to and from points beyond the Australian or the foreign gateway (route integration).

Trade
In examining the benefits to trade, the availability of frequent, low-cost and reliable freight movement is taken into account.

Industry
It has to be evaluated, whether and how a certain allocation of air traffic rights might be beneficial for the Australian aviation industry.

Additionally, the Commission can also apply other criteria, which it considers as necessary and relevant. Furthermore, not all criteria must be taken into account, if the Commissioners believe that those criteria, which are predominantly relevant for the allocation, have been met.[715] Particular consideration is given to competitive benefits that can be gained through an allocation. The government desires to promote competition in the international air service transportation sector, which is supposed to come either from real competitors or from the threat that new entrants might be allowed into the market.[716]

Application of the two sets of criteria

As indicated above, the application of the general and additional criteria varies and depends on the situation. One can distinguish between situations, in which only the general criteria are applied, the allocation of capacities in the start-up phase and the renewal of route licences. Only the general criteria are applicable when:

- capacity is not limited under a bilateral agreement;
- capacity is limited, but only one carrier files in an application for it;

715 Policy Statement, *ibid.*, paragraph 5.2.

716 Explanatory Statement to the International Air Services Commission Policy Statement, No. 5 at 3, Online: International Air Services Commission <http://www.iasc.gov.au/pubs/PSV5EM20040519.pdf> (date accessed: 4 August 2004) [Explanatory Statement].

- several air carriers apply for limited capacity, but their consolidated applications do not exceed the available contingent, or
- a carrier requests a variation of an existing air traffic right allocation to be more flexible in its operations, including to use the allocated capacity in code-share operations, and no submission by another party regarding such a request is filed in. If a submission is filed, the Commission may apply the special criteria to determine the request.

Different criteria are applicable during the start-up phase. The start-up phase describes the period, until a new entrant is permitted to operate on a route, on which one or several other Australian air carriers already offer air transportation services.[717] The start-up phase ends, when the new entrant is granted sufficient capacity so the development of efficient and commercially sustainable operations is possible.[718] Once a new carrier has been permitted on a route, the start-up phase for this particular route is terminated and will not be reinstated, even if the new entrant might leave this market again.[719] To promote competition, the criteria to allocate capacity to new entrants are modified in favour of new entrants. The Commission only takes the following aspects into account:[720]

- the level of capacity to be granted must be sufficient to support efficient and commercially sustainable operations by the new entrant and the incumbent Australian carrier;
- the new entrant's tariffs and services must enhance competition on the respective route;
- the approval must not lead to a reduction of tourists visiting Australia;
- benefits for consumers or trade shall not be impaired; and
- the new entrant must be reasonably capable to obtain any other necessary approvals and to commence operations as suggested.

If more than one new air carrier applies to start operations on a particular route, the general and additional criteria are applied to select between the applicants. The IASC can request applications for operations on a route during the start-up phase. If this invitation is answered only by incumbents and no new entrants respond, the selection among these applicants is based on the general and additional criteria, unless one of the cases is at hand, in which only the general criteria are applicable (see above).[721]

---

717 Policy Statement, *supra* note 708, paragraph 2.1 (a new entrant is defined as an Australian carrier that has not been previously allocated any capacities on the respective route).

718 *Ibid.*, paragraph 7.1; Explanatory Statement, *supra* note 716 at 4.

719 International Air Service Commission, – Procedures of Handling Applications for Capacity in the Start-Up Phase, paragraph 4, online: International Air Services Commission <http://www.iasc.gov.au/pubs/instart.pdf> (date accessed: 4 August 2004).

720 Policy Statement, *supra* note 708, paragraph 7.1; Goh, *supra* note 699 at 76.

721 Policy Statement, *ibid.*, paragraph 7.5.

Depending on the respective air service agreement, route allocations are temporally limited. When it comes to their renewal, a distinction is made between different kinds of route allocations.[722] Normal route permissions are issued for five years, if the available capacity is limited, and for 10 years, if capacity and air traffic rights are unrestricted. Route permissions, issued during the start-up phase can, but do not need to, be limited to a period of three years (interim determination).[723] In case a normal route permission is renewed, a rebuttable presumption in favour of the incumbent exists. The renewal of the incumbent's route permissions can be denied only for certain exclusive reasons. Further distinction is made, whether the renewal occurs during or after the start-up phase.

Renewal during the start-up phase:

- the specific start-up phase allocation criteria apply to that portion of the capacity that is reasonably necessary for the development of efficient and commercially sustainable operations; and
- for the exceeding capacity, not necessary for the development of efficient and commercially sustainable operations, it is assessed:
  - whether the air carrier, requesting the renewal, failed to serve the route efficiently; and
  - whether the allocation of this part of the capacity, in whole or in part, to another Australian air carrier, would be more beneficial to the general public, thereby taking into account the general and additional criteria.

After the start-up phase, it is scrutinized:

- whether the carrier, requesting the renewal, failed to serve the route efficiently; and
- whether the allocation of the capacity, in whole or part, to another Australian air carrier, would be more beneficial to the general public, thereby taking into account the general and additional criteria.

If an interim determination is renewed, no rebuttable presumption in favour of the incumbent exists. Distinction is made again, whether the interim determination is renewed during or after the start-up phase. During the start-up phase, the criteria that are originally used to allocate capacity during the start-up phase are applicable. After the start-up phase, the general and additional criteria are fully applicable.

Result

Australia's initial approach to separate domestic and international air transportation services was very similar to the US method. The main difference was that the US, due their size and the number of possible route combinations, had more domestic airlines, which differed very much in size and financial strength. Australia's domestic aviation industry, on the other hand, consisted mainly of two equally

722 Policy Statement, *ibid.*, paragraph 8.

723 Policy Statement, *ibid.*, paragraph 7.6.

strong airlines. It is interesting to note that the reasons for the introduction of the Two-Airlines Policy and for its termination were the same in both cases, namely to ensure effective competition.[724] From today's perspective both manoeuvres seem to have failed. While in a strictly regulated domestic air transportation market the existence of two healthy airlines was maintained, competition was hindered at the same time. The subsequent deregulation allowed for competition for a limited time, until Ansett-ANA, one of the competitors, went bankrupt. With Qantas, as the only remaining considerable airline in Australia, the level of competition in the Australian air transportation market is limited. As far as the distribution of air traffic rights is concerned, Australia's approach is very similar to the US one with respect to the temporal limitation of the route permissions. Also the criteria applied resemble those that are considered by the DOT. However, unlike in the US, an independent Commission and not the Ministry itself is responsible for the allocation. Furthermore, the procedures and applicable criteria are more predefined and fixed. This shall lead to less ambiguity and uncertainty than in the US.

CANADA

Background

Due to Canada's vast area and the distances that lie between major industrial centres, like in Australia air transportation is an important tool when in comes to unifying the country and conducting trade.[725] In 1937, the government founded state-owned Trans-Canada Air Lines, which was renamed Air Canada in 1964. Simultaneously, regulations to control market entry and fares were also introduced. Like in the USA and in Australia, these regulations were intended to eliminate wasteful and destructive competition. As a result Air Canada was shielded from any competitor on domestic routes, until in 1959 a second air carrier, Canadian Pacific Airlines (CP Air), was permitted to offer limited domestic air transportation services.[726] But it was not until 1979 that capacity constraints on domestic routes were completely removed.[727] Like in the US, new airlines needed to prove that their entrance into a market was required by the present and future public convenience and necessity and that their operations would not affect the profitability of the existing carriers.[728]

---

724 Campbell, *supra* note 686 at 24.

725 Tae Oum, William Stanbury and Michael Tretheway (1991), 'Airline Deregulation in Canada', in: Kenneth Button, ed. (1991), *Airline Deregulation – International Experiences* (Washington Square, NY: New York University Press) at 124, 126 [Oum, Stanbury and Tretheway]; George Petsikas (1989), *Airline Deregulation and Competition in the Canadian Air Transport Industry Today, and Prospects for the Future* (LL.M. Thesis, Institute of Air and Space Law, McGill University, [unpublished]) at 23 [Petsikas]; Dempsey, Thoms and Clapp, *supra* note 613 at 114.

726 Oum, Stanbury and Tretheway, *ibid.* at 126; Williams, *supra* note 50 at 94.

727 Oum, Stanbury and Tretheway, *ibid.* at 127; Williams, *ibid.*; Petsikas, *supra* note 725 at 26f.

728 Oum, Stanbury and Tretheway, *ibid.* at 129f.

International routes initially were granted exclusively to Air Canada. In 1948, CP Air was designated as Canada's second flag carrier and was assigned to operate in the pacific region. The remaining international routes continued to be granted mostly to Air Canada.[729] In 1965, CP Air and Air Canada decided to split the world among themselves into several regions, into which each of them would operate as the sole Canadian flag carrier.[730] Influenced by deregulation in the US, the Canadian Government decided to deregulate its airline industry as well.[731] Since most Canadian cities are located close to the US border, many Canadians went south to fly on the cheaper US carriers. Deregulation and enhanced competition was the only possibility for the Canadian Government to safeguard Canadian air carriers from bankruptcy. Over a period of 10 years, regulations were gradually liberalized until full economic freedom was achieved in 1988.[732] With the beginning of deregulation, the Air Transport Committee, the regulatory body responsible for licensing and fare control in air transportation, was advised to put more emphasis on competition, when applying the public convenience and necessity test. Furthermore, air carriers should lose their routes, if they did not use them.[733] In 1987, new rules were introduced, which were very similar to the ones adopted in the US a few years before. The corresponding public convenience and necessity test was abolished and new entrants only needed to prove that they were fit, willing and able to operate.[734] However, this new freedom was locally limited, as the sparsely populated north of Canada was still subject to some regulations.[735] This part of the country, with its remote communities, is heavily dependent on air transportation services, but, due to its low population, does not offer great chances for profit. To ensure that socially desired and needed air transportation services to, from and within this area continued to be offered, the government decided to maintain a certain level of control and protection from competition for this region.[736] This distinction between a liberalized south and a regulated north was given up with the adoption of the Canada Transportation Act in 1996 and equal rules were established for the whole of Canada. Before full deregulation was achieved, CP Air already started to prepare for future competition by acquiring all regional airlines and was able to complete this endeavour by the mid-1980s, before it was overtaken itself by Pacific Western Airlines (PWA) in 1987. The new group was renamed Canadian Airlines International and, after acquiring

729 *Ibid.* at 127; Williams, *supra* note 40 at 94.

730 Oum, Stanbury and Tretheway, *ibid.*; Harold Sandell, (1976) *Government Regulation of Air Carriers in Canada* (LL.M. Thesis, Institute of Air and Space Law, McGill University [unpublished]) at 200f., 210ff.

731 Oum, Stanbury and Tretheway, *ibid.* at 133f.; Dempsey, Thoms and Clapp, *supra* note 613 at 127.

732 Oum, Stanbury and Tretheway, *ibid.* at 124.

733 Dempsey, Thoms and Clapp, *supra* note 613 at 128-30.

734 Oum, Stanbury and Tretheway, *supra* note 613 at 137; Petsikas, *supra* note 725 at 63.

735 Oum, Stanbury and Tretheway, *ibid.* at 124, 135, 137.

736 Dionigi (Dan) M. Fiorita (1995), *Safety and Economic Regulation of Air Transportation in Canada* (LL. M. Thesis, Institute of Air and Space Law, McGill University [unpublished]) at 100f.; Petsikas, *supra* note 725 at 91f.

Wardair, the last independent air carrier in 1985, the Canadian air transportation industry was left, like in Australia, with only two major airlines.[737] Like in Australia, the Canadian duopoly also came to an end in 2000, when Air Canada took over Canadian Airlines.[738] Unlike in the US, deregulation did not cause new entries into the air transportation sector, but primarily led to a consolidation of the existing airlines, finally leaving Canada with only one dominant air carrier. Although, new airlines like Air Transat, WestJet and Jetsgo were also founded, the outcome of their efforts to secure a piece of the Canadian air transportation market remains to be seen. Only recently Jetsgo had to cease operations and declared bankruptcy. Thus, the long-term effects of the Canadian deregulation cannot be fully assessed yet.

Criteria

Licences for scheduled international air transportation services are granted to Canadian air carriers by the Canadian Transportation Agency (Agency) according to the rules stipulated in the Canada Transportation Act (CTA). Pursuant to sec. 69 of the CTA, the Agency issues a licence to operate a scheduled international route, if the applicant was designated to operate on the requested route by the Minister of Transport. Hence, the Agency acts solely as an administrative institution, upon the order of the Minister and without any autonomous decisional power. The Minister of Transport has released a policy statement on 21 May 2002, in which the selection procedure of Canadian air carriers for scheduled international air transportation services is explained. According to this statement, every Canadian air carrier is eligible for designation to operate on all international routes.[739] The designated air carrier is given a one-year period, to fully implement its air transportation services. In the allocation process, the following rules and criteria are applied:[740]

- If the requested routes do not exceed the available contingent available under a bilateral agreement, all applications are approved.
- If all applications cannot be accommodated, the Minister of Transport decides which air carrier is designated according to the following criteria:
  - the quality of service, whereby a hierarchy exists according to which non-stop services are considered as offering the highest quality, followed by one-stop services and code-sharing services;
  - the frequency/capacity (including cargo) offered;
  - the number of Canadian cities served (route integration);
  - the category of service, whereby passenger and passenger/cargo services are preferred over all-cargo services, and

737 Oum, Stanbury and Tretheway, *supra* note 725 at 142f.

738 Williams, *supra* note 40 at 97f.

739 Transport Minister announces new Policy for International Scheduled Air Services, online: Transport Canada <http://www.tc.gc.ca/mediaroom/releases/nat/2002/02_h051e.htm> (date accessed: 6 August 2004) (multiple designation by the Canadian Government was previously permitted only in markets, where more than 300,000 passengers were travelling on one-way scheduled routes per year; this cap was removed in the meantime).

740 *Ibid.*

- the level of competition, thereby taking into account:
  - the frequency and capacity a foreign air carrier offers on the route; and
  - whether air traffic rights should be divided among Canadian carriers to enhance competition between Canadian airlines or whether it is preferable to concentrate all international routes with one air carrier, that would be enabled to offer higher frequency and capacity, to enhance efficiency.

Result

Compared to the US and Australia, Canada has recourse to only a few criteria and, unlike the other two, it partially discloses the weight and significance it attributes to each criterion. Still, if different air carriers file in several applications, it can become very difficult to decide, which proposal offers the highest benefit to the public. For example, how would a case be decided, in which one carrier proposes non-stop services, while another applicant offers one-stop services and code-sharing flights, but also foresees a higher frequency and capacity? Additionally, when it comes to assess the level of competition, the Agency is given considerable discretion. Like in the US, one can expect that decisions, involving this aspect, may lead to many disputes, challenges and appeals. Unfortunately, there are no cases presently at hand, where the application of these rules could be examined under practical conditions.

GREAT BRITAIN

Background

Like in most other European countries, Great Britain's air transportation policy was strongly connected with its main national flag carrier, British Airways (BA). On the other hand, unlike in other EC Member States, significant changes to Great Britain's air transport policy and airline industry occurred before any of the liberalization packages of the European Community entered into force. The restructuring of Great Britain's aviation policy and airline industry started in 1969, when the Edwards Report suggested that air transportation should be regarded as any other commercial activity and should take place in a competitive environment. It was believed that the best way to ensure airline efficiency and adequate service to users was competition. In such a surrounding innovation would be promoted and, in the end, would enable British airlines to compete more efficiently with foreign carriers.[741] For this purpose, more airlines needed to be established in the market and an independent Civil Aviation Authority (CAA) should be created to regulate this industry. The first proposal of the report was implemented three years later, when in 1972 the CAA was founded and given the mandate to enhance competition in the air transportation industry.[742] So far, the airline industry was dominated by the two state-owned airlines. British Overseas Airways Corporation (BOAC) was founded in 1939 and engaged in long-haul flights. The second governmental carrier, British European Airways

741 CAA Competition, *supra* note 580 at 2.

742 Staniland, *supra* note 580 at 192f.

(BEA), was established later and limited to offer services within Europe. For more competition to emerge, the most practical solution was to consolidate the British airline industry, to establish another powerful airline next to BOAC and BEA. For this purpose a merger between Caledonian and British United Airways was envisaged, which finally occurred in 1970. The new airline started offering services under the name British Caledonian (BC). However, solely creating a new big air carrier was not sufficient to enhance competition. It was also necessary to supply the new company with the critical mass of business opportunities, to make it viable. Therefore, air traffic rights, with a value of over £5 million, were transferred from BOAC to BC. But despite this transfer, BC remained too small to effectively compete with BOAC and BEA.[743] After further competitors were allowed into the air transportation industry, the government decided to merge its two airlines in 1972 under the name British Airways (BA).[744] The new company had 58,000 employees and 190 aircrafts.[745] But, contrary to the expectations, the new airline hardly increased efficiency and productivity. Staff cutting and integration of operations occurred only at a very slow pace. The problem was the lack of true competitive pressure for BA to improve its operations. It was still shielded by the government from competition by domestic and foreign airlines. The situation was not better for BC and even worsened when the airline, due to high fuel costs following the oil crisis, faced a loss of £4.3 million in 1974. In order to save its Multi-Airlines Policy, the government reacted by splitting up the markets between the two airlines and thus prevented any direct competition.[746] In 1981, it was decided to privatize BA.[747] On the one hand, the government still wanted to promote competition. For this purpose BA should be put in the same situation as all other private airlines. The airline should have all the possibilities a private entity has, but, at the same time, should also be forced to compete. On the other hand, there was still some residual nationalism left to grant a special treatment to BA, to preserve the positive overspill effects it had on other parts of the economy, especially with respect to the aviation industry.[748] When the decision to privatize the airline was taken, BA found itself, as a result of the oil crisis in 1979, in deep financial trouble.[749] Therefore, it was necessary to improve BA's overall condition and performance, before it could be privatized. The government contributed to the following restructuring by further protecting BA from competition and by preserving its dominance. In general, the government's behaviour was very controversial at this time. On the one hand, the government normally would had to undertake measures that were in the consumers' interests. This would have required it to further promote competition. On the other hand, it was also BA's shareholder

743 Staniland, *ibid.* at 194.

744 Lucien Rapp and François Vellas (1992), *Airline Privatisation in Europe*, 2nd edn., Vol. 28 (Paris, France): Institute of Air Transport), 75 [Rapp and Vellas]; Staniland, *ibid.*, at 195.

745 Staniland, *ibid.*

746 Staniland, *ibid.* at 196.

747 Staniland, *ibid.* at 200.

748 Staniland, *ibid.* at 198ff.

749 Rapp and Vellas, *supra* note 744 at 71; Staniland, *ibid.* (revenues have dropped sharply from £90 million in 1978/79 to a loss of £141 million in 1980/81, at 200).

and had an interest in promoting the business opportunities of the company, in order to achieve the highest possible price in the coming privatization.[750] The government decided for the latter approach. The only guarantee for competition in the air transport sector came, at that time, from the CAA. The Civil Aviation Act mandated the CAA to 'have regard to the advantages that might arise from competition' when issuing licences. This urged the Agency to relax its regulations on entry and price control in the 1980s and turned into an administrative *de facto* deregulation already in 1982. Like the CAB in the US before, the CAA started to grant more licences to independent carriers and enabled them to compete with BA on domestic routes. Fare control was relaxed and limited to the prevention of predatory pricing. The CAA justified its approach by arguing that potential and real competition was in the interest of the users.[751] Still, the Agency did not regard a competitive environment as an end by itself, but only as a tool to achieve other goals in the air transportation policy.[752] Before BA's privatization was to become effective, the other independent and private British air carriers asked for a redistribution of BA's routes, to ensure that competition would be still possible in the thereafter.[753] The government arguing that, due to BA's critical condition, it was not the right time to take away routes from the airline denied this motion.[754] Even the CAA, normally a proponent of competition, had concerns that the hub established at Heathrow by BA could be seriously weakened. Therefore, any transfer of routes from BA to other carriers should be balanced against the impact this might have on BA's existing network. The same was true of the benefits, BA could offer due to its size.[755] The advantages of additional competition had to be balanced against the disadvantages that would incur due to a reduction in services by BA. BA's initial public offering took place in 1987 and generated a revenue of £900 million.[756] Although, from that moment, BA was not given any preferential treatment

750 Staniland, *ibid.* at 199.

751 Button and Swann, *supra* note 35 at 96.

752 CAA Competition, *supra* note 580 at 3.

753 *Ibid.* at 16; Staniland, *supra* note 580 (BC was even willing to pay up to £250 million for BA's routes, but the latter one refused to sell, at 206).

754 Staniland, *ibid.* at 205 (in 1981/82 all inherited and anticipated losses of BA were concentrated in one financial year to clear out all deficits and obligations at once and to allow for a new beginning; this resulted in an accounted loss of £544 million, leading to a negative equity of £257 million. Technically, the airline was bankrupt and survived only, because the government decided to bail it out with an amount of £550 million, at 203).

755 See also Dempsey and Gesell, *supra* note 665 ('economies of scale are realized when increases in total production simultaneously decrease unit costs; long-run average cost decreases as output increases; as the scale of production grows, the enterprise becomes more efficient;' thus economies of scale describe a situation in which the costs created by a production unit decrease, if the total number of production units is increased; hence, with respect to airlines it means that bigger air carriers shall be able to offer service at lower prices, at 363, n. 110).

756 Rapp and Vellas, *supra* note 744 at 76; Staniland, *supra* note 580 (at the time of privatization, BA's debt amounted to £647 million and the debt–equity ratio was 2:1; the government contributed another £86 million to the airline to make it an even more attractive investment, at 209; only four years earlier, in 1983, the debt amounted to £1,053 million resulting in a debt–equity ratio of 89:11, at 200).

from the government any more, it remained the dominant carrier for international routes to and from Great Britain.[757] In the aftermath of its privatization, British Airways expanded significantly, to position itself for the coming single European market and to be able to withstand the competitive pressure from the powerful US airlines. For this purpose it acquired British Caledonia in 1987. This take-over was highly disputed as it threatened Great Britain's Multi-Airlines Policy. In the end it was approved, because of BC's bad financial situation and due to the fact that the airline would have needed a stronger partner to survive anyway.[758] The overall condition of British Airways has improved tremendously, since it was privatized and it has become one of the most profitable airlines of the world.[759] Changes in Great Britain's airline industry were completed to a large extent, when the consolidation of US airlines took place in mid-1980s. Due to its early privatization, it was in a better position and had more flexibility in dealing with the consequences of the first Gulf War in 1990 and with the rising global competition coming particularly from across the Atlantic and from Asian airlines.[760] Today, BA is the dominant air carrier in Great Britain. Nevertheless, the airline also has to face serious competition from other airlines like Virgin Atlantic Airways and British Midland Airways. Hence, so far the Multi-Airlines Policy of the British Government is, unlike Australia's Two-Airlines Policy, still in force.

Criteria

In examining the CAA's approach to distribute air traffic rights, a distinction will be made between the time before BA's privatization and today's situation.

Allocation of international routes before BA's privatization

The CAA's approach at that time was to grant international air traffic rights to every British carrier that was able to compete effectively with foreign airlines. All applying British airlines were designated, if sufficient air traffic rights were available.[761] Still, before admitting an additional airline, the CAA analysed, whether a route generated enough traffic and volume to allow a reasonable prospect of profitable operations by a multitude of British air carriers.[762] The CAA willingness to allow additional air carriers to a route was higher, if the existing services were inadequate, enough traffic

757 Rapp and Vellas, *ibid.* at 81.

758 Staniland, *supra* note 580 (at that time SAS showed strong interest in BC, but had to face strong opposition as many politicians did not want to see a continental carrier overtaking BC and intruding into the British market, at 210; BC's chairman later blamed the government for his airline's failure and complained that BC had not been provided with sufficient resources to become strong and competitive enough. He especially pointed to the denial of granting more routes to BC in 1984 before BA's privatization, at 212).

759 Rapp and Vellas, *supra* note 744 (passenger traffic rose by almost 50 per cent between 1987 and 1990 and airfreight traffic even increased by 66 per cent, at 84).

760 Staniland, *supra* note 580 at 192, 214; Rapp and Vellas, *ibid.* at 82f., 85f.

761 CAA Competition, *supra* note 580 at 23.

762 *Ibid.* at 24.

was available or the additional carrier could offer an improvement in services.[763] In case the available international routes were limited, the CAA favoured that carrier, which offered the greatest user benefits. In assessing user benefits, the following aspects were taken into account:[764]

- choice of services in terms of
  - airports served;
  - seat availability (capacity);
  - product range; or
- greater carrier efficiency reflected in
  - lower tariffs;
  - improved standards of services; or
  - the stimulation of innovation.

When allocating air traffic rights, the CAA always tried to promote a sound development of the airline industry and also had regard to the regional development of Great Britain. If all airlines offered an equal degree of user benefits, environmental factors were taken into account and airlines favoured that used the quietest aircrafts.[765] Similarly to the historic-interest criterion in the US, an air carrier could be given a route, due to its past involvement and development of that market.[766] The CAA retained the right, to revoke a route licence and to substitute an incumbent with a new carrier, if this was in the interests of the users. When deciding upon a possible substitution, the Agency considered the following circumstances:[767]

- benefits of an extension of the users' choice through the substitution;
- improvements in quality of services measured by:
  - capacity and seat availability;
  - frequency and timing;
  - regularity and reliability;[768]
  - in-flight and on-ground services;
  - fares and conditions of the existing and envisaged services;
  - competition;
- the overall effects of the substitution on all operations of the airlines concerned.

When evaluating the quality of the services offered so far by the incumbent air carrier, opinions of user representatives and affected communities were taken into

763 Robert Baldwin (1985), *Regulating the Airlines – Administrative Justice and Agency Discretion* (Oxford, United Kingdom: Clarendon Press), 167 [Baldwin].

764 CAA Competition, *supra* note 580 at 23f.

765 *Ibid.* at 25f.

766 Baldwin, *supra* note 763 at 170.

767 CAA Competition, *supra* note 580 at 24.

768 This criterion was applied to evaluate the incumbent's past services.

account.[769] The CAA also granted protection to air carriers that were engaged on infant routes, to give these airlines a chance to develop a market and to receive a fair return on their up-front investments.[770]

Present allocation of international routes

In the distribution of air traffic rights Great Britain presently follows a very unique approach. Unlike in all other countries, licences are not issued for a specific route, but granted as a general route permission to operate between Great Britain and any place in the world. However, since international routes are mostly limited by bilateral agreements, the route licences are narrowed down as well. In the procedure for obtaining air traffic rights, airlines consequently do not request the granting of a specific route, but ask for the licences of competitors to be limited to such an extent that they can continue operating and are only restricted by the limitations set out in the applicable air service agreement.[771] This process, called scarce bilateral capacity procedure, is initiated by the Secretary of State by notifying the CAA that the capacity available to British carriers under an air service agreement will be insufficient to enable all of them to offer unlimited services to a particular country. The CAA then determines, how the available capacity will be allocated among the existing British air carriers.[772] As far as the procedure is concerned, hearings are not mandatory, but can be requested by the parties involved. According to the CAA, the length of the procedure can vary between at least one month, and up to two-and-half months in certain cases, when a hearing is required.[773] The parties involved can appeal the decision of the CAA to the Secretary of State.[774] In discharging its duties, the CAA is not completely independent as the Secretary of State can issue directions to the CAA.[775] When allocating air traffic rights, the CAA must pursue the objectives listed in Section 4 of the Civil Aviation Act. The first goal is to secure that services, provided by British airlines, satisfy all substantial categories of public demand to the maximum extent and at the lowest rate possible. Secondly, the safety standard is to be maintained at a high level and the airlines should be given the possibility to generate enough profit to ensure a sound development of the airline industry. Thirdly, the CAA has to promote the reasonable interests of the users of air transportation services.[776] Apart from the objectives listed in sec. 4 of the Civil Aviation Act, the CAA, according to sec. 68 of the Civil Aviation Act, has to take into account additional aspects. Firstly, British carriers shall be enabled to compete

---

769 CAA Competition, *supra* note 580 at 24.

770 Baldwin, *supra* note 763 at 167.

771 Civil Aviation Authority, Official Record Series 1, 'Licensing Airlines in the UK', (12 August 2002, revised 31 March 2004) at 13, online: Civil Aviation Authority <http://www.caa.co.uk/docs/213/Series%201-%2031%20March%202004.pdf> (date accessed: 6 August 2004) [CAA Licensing].

772 *Ibid.* at 15.

773 *Ibid.* at 45.

774 Civil Aviation Act, 1982 (U.K.), 1982, c.16, s. 67, ss.5.

775 *Ibid.*, s.6.

776 *Ibid.*, s.4.

as efficiently as possible with other airlines. Secondly, airports in Great Britain are to be used in the most efficient way. Thirdly, the Agency shall have regard to benefits that may arise, if services are offered by two or more airlines. Fourthly, attention must be given to the effects a new service can have on existing services. Finally, any adverse effects on the environment and any disturbances of the public possibly arising from noise, vibration and atmospheric pollution shall be minimized.

Pursuant to sec. 69 of the Civil Aviation Act, the CAA issues a policy statement that contains, among others, the criteria the Agency intends to apply, when it awards route licences. The latest statement was published in December 2003. The CAA takes a very pro-competitive position and declares that the interests of users of air transportation services are best served, if airlines are given the opportunity to compete with each other, limited only by the application of the general competition laws. The CAA aims at making the international aviation markets of Great Britain as contestable as possible and, for this purpose, intends to eliminate all unnecessary restrictions within its licensing process that are in conflict with this goal.[777] However, since in many international markets the number of air traffic rights is limited, the CAA's main objective in allocating these rights must be the maximization of economic efficiency.[778] Additionally, the CAA takes into account the effects its decisions can have on competition in all relevant markets. For this purpose, it will weigh the benefits a certain allocation may bring to the users, against any detrimental impact it may have on competition.[779] In evaluating the economic benefits of every proposal, the CAA has recourse to the budgetary information provided by the airlines that must contain detailed three-year-business plan for the requested route.[780] Apart from purely economical aspects, the Agency has also regard to direct user benefits. For this purpose the proposals are assessed with respect to two aspects: travel costs and passenger convenience. Travel costs are further split into direct fares and indirect time costs. Travel costs are reduced and/or passenger convenience increased through:[781]

- greater flight frequency;
- shorter flight times;
- more direct flights;
- services to more accessible airports;
- higher quality of in-flight service;
- greater choice among different airlines;
- higher capacity and greater seat availability; and
- lower fares.

Parts of the listed benefits are difficult to assess and vary from passenger to passenger. Especially the amount of reduced travel time costs depends on the value,

777 CAA Licensing, *supra* note 771 at 42.
778 *Ibid.*
779 *Ibid.* at 42f.
780 *Ibid.* at 52f.
781 *Ibid.* at 50f.

time has for each passenger. While tourists will attribute a lower significance to their time, business travellers value their time higher. Another consumer benefit that is difficult to assess is the quality of in-flight services.[782] As set out above, the CAA compares direct user benefits with any possible negative impact its choice might have on competition. This assessment is further complicated, if one takes into account that competition itself can have an indirect impact on consumer benefits by forcing competitors to reduce air fares or increase in-flight services.[783] Generally, the CAA holds the opinion that the less competitive a market is, the greater are the benefits for the consumers, if new service is added.[784] According to the Agency, competition among British airlines and between British and foreign air carriers is the best means to ensure that users have the widest possible choice of products, services and airports, that a high standard of quality is maintained and that fares are set at a reasonable level in relation to the incurred costs. Thus, the interests of users are served best, if a multitude of efficient and profitable British airlines exist, which are strong enough to compete with each other and with foreign airlines.[785] For this purpose, the CAA retains the possibility to substitute a carrier with another airline, if competition between both players on a route cannot be established due to bilateral constraints. Such a substitution will be only undertaken, if this will help to promote the objectives of the Civil Aviation Act and is in the interests of the users. Thereby, the duration of the incumbent's involvement on a route and its commitment in a market are taken into account. Furthermore, the quality of services in respect of capacity, seat availability, frequency, timing and fares offered by the incumbent and the newcomer are also compared.[786] As far as fares in general are concerned, the airlines are free to determine the prices they want to charge for their services. Still, the CAA retains the right to intervene, if air carriers should abuse their market power at the expense of the users. As long as the airlines offer an appropriate level of seat availability, freedom to change reservation, commensurate in-flight facilities and charge a fare, which is reasonably related to the costs, including a reasonable return on capital, the CAA will not object to the airline's fare system.[787] Finally and despite the detailed regulations foreseen, the CAA also retains the right to depart from its policy, if special circumstances will require it to do so.[788] An examination of past decisions of the CAA with respect to the allocation of international air traffic rights, shows that the Agency has regularly departed from its own standards. In particular, the following aspects were considered in various decisions:

---

782 *Ibid.* at 54.

783 *Ibid.* at 51.

784 *Ibid.* at 49.

785 See e.g. Civil Aviation Authority, Decision on Air Transport Licences and Route Licences 2/98 (25 August 1998) at 8, online: Civil Aviation Authority <http://www.caa.co.uk/docs/213/ald9802.pdf> (date accessed: 6 August 2004).

786 CAA Licensing, *supra* note 771 at 43.

787 *Ibid.* at 43f.

788 *Ibid.* at 44.

Quality of service

The CAA argued that the quality of a service should be only considered, if compelling evidence was presented. According to the CAA, the best way to secure high quality standards is to enhance competition. The Agency prefers to widen the users' choice than to guess which choice the users would make.[789] Still, the CAA has considered aspects like legroom, in-flight entertainment, pampering in the departure hall, departure and arrival time and the circumstance that connecting flights were located in the same terminal.[790]

Route integration

In one case, route integration, although clearly present and strongly favouring one applicant, was not given any additional value due to the low number of possible connecting passengers. The decision was appealed and in the re-hearing of the case, this notion was not amended, but the Agency acknowledged a benefit for the consumers resulting from the respective air carrier's services, due to the fact that it had an existing network of flights.[791] In another case, the CAA declined to put too much weight on the scope of an airline's network and argued that, if it considered this criterion too much, the big and established carriers would always have an advantage and the objective of creating more competition among British air carriers could not be achieved.[792]

Capacity

In evaluating the most efficient use of an air traffic right, capacity was very often a decisive factor.[793] On the other hand, in one decision the CAA agreed that incumbents might have an incentive to propose and offer a higher capacity when applying for air traffic rights, to protect their position and to preclude new carriers to enter into the market. It conceded that in such a situation it might be appropriate, to sacrifice the short-term benefits of a higher capacity for the long-term benefits of enhanced competition.[794]

---

789 Civil Aviation Authority, Decision on Air Transport Licences and Route Licences 3/99, (25 October 1999) at 7, online: Civil Aviation Authority <http://www.caa.co.uk/docs/213/ald9903.pdf> (date accessed: 6 August 2004).

790 Civil Aviation Authority, Decision on Air Transport Licences and Route Licences 4/98, (26 November 1998) at 10, online: Civil Aviation Authority <http://www.caa.co.uk/docs/213/ald9804.pdf> (date accessed: 6 August 2004) [CAA decision 4/98].

791 *Ibid.* at 11.

792 Civil Aviation Authority, Decision on Air Transport Licences and Route Licences 5/98, (10 December 1998) at 9, online: Civil Aviation Authority <http://www.caa.co.uk/docs/213/ald9805.pdf> (date accessed: 6 August 2004) [CAA decision 5/98].

793 Civil Aviation Authority, Decision on Air Transport Licences and Route Licences 3/98, (25 August 1998), online: Civil Aviation Authority <http://www.caa.co.uk/docs/213/ald9803.pdf> (date accessed: 6 August 2004) [CAA decision 3/98].

794 Civil Aviation Authority, Decision on Air Transport Licences and Route Licences 4/99, (14 December 1999) at 7, online: Civil Aviation Authority <http://www.caa.co.uk/docs/213/ald9904.pdf> (date accessed: 6 August 2004).

Profitability

It is interesting to note that the CAA, contrary to its policy statement, very often did not attribute any significance to the profitability or the yield forecast of an air carrier and has not even regarded it as negative, if a company assumed to incur losses on a route in the initial years of operation.795

Frequency

As far as frequency is concerned, the CAA holds the opinion that 'the higher the frequency enjoyed by an incumbent, the stronger the reasons must be in favour of the incumbent for the CAA not to grant capacity to the new entrant'.796

Competition

The question of competition has become especially relevant in relation to BA and whether the granting of a route to another British air carrier would widen the users' choice. Thereby competition among British airlines could be also increased to such an extent that it might be beneficial to the consumers. In this regard, the existing network of an applicant and its ability to put competitive pressure on BA has some bearing for the final decision.797 Still, the CAA has also declared that it cannot be concluded that smaller airlines would be automatically given preference over BA.798 However, due to the reference that is made in every case to the industry structure in Great Britain, one can also expect the CAA to use the allocation of routes to subsidize air carriers, to be able to continue pursuing its multiple-airline policy. Although it has never been an influential factor so far, the CAA in this respect always also examines, whether an applicant requires the allocation of a certain route to survive.799 It is noteworthy that in none of the recent decisions, the CAA considered the competitive situation of British carriers in vis-à-vis to foreign airlines.800

Fares

Although the CAA discussed the level of the proposed fares in past cases and deliberated upon which applicant was able to offer the lowest rates, this aspect has not been decisive so far. Usually, the applicants also offered very similar and equally competitive fares.

Slots

Slots were considered, but never had a profound influence on the decision of the CAA.

---

795 See e.g. CAA decision 4/98, *supra* note 790.

796 See e.g. Civil Aviation Authority, Decision on Air Transport Licences and Route Licences 1/99 (20 September 1999), online: Civil Aviation Authority <http://www.caa.co.uk/docs/213/ald9901.pdf> (date accessed: 6 August 2004).

797 CAA decision 4/98, *supra* note 790 at 11.

798 CAA decision 3/98, *supra* note 793 at 10.

799 See e.g. CAA decision 5/98, *supra* note 792 at 9.

800 See especially CAA decision 5/98, *supra* note 792.

Historic interest

Contrary to its policy statement, the CAA never considered the duration of the presence of an air carrier in a market in its deliberations. The same is true for any possible lobbying sometimes undertaken by the airlines in foreign countries for additional air traffic rights. With respect to the latter issue, the CAA argues that it is not the airlines' duty to pursue the government of a foreign country for additional air traffic rights and therefore, they cannot be given the merits, if they are successful.[801]

Result

The CAA has attempted to prevent its allocation method from becoming nebulous and inconsistent by establishing a clear hierarchy among the criteria it takes into account and by pre-determining the weight, it attributes to each criterion. When deciding the allocation of air traffic rights, the Agency distinguishes between the benefits a service offers to the users and the impact it might have on competition. Although the CAA's effort is remarkable, it does not lead to consistent results. The services airlines propose, either bring the same user benefits or offer different, but equally valuable advantages. In the end the CAA has to decide, which user benefits it considers higher. At this stage, it must substitute its intended objective evaluation with a subjective judgement. Further subjectivity influences the decision, when it comes to the evaluation of competition on a route. Here again the CAA has to determine, whether and to what extent competition on a certain route is beneficial and desirable.

GERMANY

Background

Like many other European countries, Germany had only one major flag carrier. The major impact on the German air transportation sector did not even come from the German Government, but came from the EC. The only real impetus, given to the airline industry by the German Government, was the privatization of the country's flag carrier, state-owned Lufthansa (LH). Despite the fact that originally the German Government owned 74.5 per cent of LH, the airline was traditionally exposed to very little political influence.[802] This changed in 1982, when the first attempt to privatize the airline was undertaken, after a coalition of the conservative and the liberal parties came into power. This endeavour failed due to strong opposition within the ruling coalition. It was argued that governmental control of the airline was needed for matters of national defence and for matters of foreign policy, as LH could be used to strengthen the ties with other states by establishing air transportation services into those countries. Thereby it was explicitly mentioned, whether such routes might not have been profitable.[803] Finally, governmental control over LH was also necessary to

801 See e.g. CAA decision 3/98, *supra* note 793.

802 Staniland, *supra* note 580 at 232.

803 Here the political role is exemplified that is sometimes attributed to airlines.

ensure that future aircrafts would be purchased from Airbus, to support the domestic industry and in which Germany had a stake.[804] Since 1990 the situation changed significantly for LH for two reasons. The first one was a purely political one and resulted from Germany's reunification and the emerging European Monetary Union. The second reason was a serious crisis, LH found itself in and which threatened its survival.[805] Competition had become fierce, especially on the North-Atlantic route, one of LH's primary markets. The US airlines, freed from governmental regulations and enjoying the benefits of free markets, put the European air carriers under pressure. Delta Airlines, after having purchased a network of routes from the bankrupt Pan Am, even established a hub in Frankfurt, LH's home base, and started offering fifth freedom traffic to Eastern Europe, Russia, the Middle East and South Asia. Additionally, British Airways and KLM started making use of their sixth freedom rights and flew passengers from Germany, via their hubs in London and Amsterdam respectively, to North America.[806] Apart from the competitive pressure, LH had manoeuvred itself into a disastrous financial situation, when the airline was preparing itself for the liberalized common market. To benefit from the deregulation in the EC, the air carrier had strongly invested in a new fleet and expanded extensively.[807] While being in the middle of this expansion process, the first Gulf War started. The final outcome was a debt of almost €3.2 billion in 1992.[808] Unlike British Airways, LH was not blessed with a government that was willing to help the airline out, but had to manage on its own. Apart from an increasing competition and its disastrous financial situation, LH had also a problem with its institutionalized structure and the mentality of its employees, who were not used to thinking in a commercial manner.[809] It was mainly due to a new CEO, Jürgen Weber, who intended to run LH like a true commercial entity and less like a public utility that the airline survived.[810] In the process of saving the airline, the government interfered once, when it came to decide about the future of Lufthansa CityLine, LH's regional division. The executive board intended to sell it, because of its unprofitability, but the Ministry of Transportation precluded this plan.[811]

Another influential factor for Germany's air transportation policy was the reunification of Germany and the preparation for the European Monetary Union. The rebuilding costs for the former German Democratic Republic have caused the German Government serious financial difficulties and, at the same time, the preparation for the European Monetary Union forced the government to lower its deficit to fulfil the

804 Staniland, *supra* note 580 at 232; Rapp and Vellas, *supra* note 744 at 113f.

805 Staniland, *ibid.* at 233.

806 *Ibid.* at 234.

807 *Ibid.* (within three years about €4 billion were spent and the fleet doubled from 151 aircraft to 302 between 1987 and 1992; staffing also expanded by 40 per cent to 50, 283 between 1985 and 1991).

808 *Ibid.*

809 *Ibid.* at 236.

810 *Ibid.* at 236ff. (staff was cut down by almost 20 per cent between June 1992 and January 1994 to 40,000 employees, almost 30 per cent senior management positions were eliminated and the Unions agreed to concessions valued €250 million, at 237f.).

811 *Ibid.* at 238.

criteria and to be able to be part of this Monetary Union.[812] The German Ministry of Finance announced in 1990 that all means for reducing the budget deficit would be considered, including privatizations and that this time LH would not be excluded from this process.[813] Thus, the financial constraints of the government accelerated the privatization of the airline. It was a beneficial situation for the government and for LH. Both needed fresh capital for either financing the rebuilding of East Germany and for reducing the budgetary deficit or for surviving. Restructuring and privatizing LH enabled the airline to have access to capital markets. The carrier was privatized in three steps and full privatization was achieved in 1997.[814]

Criteria[815]

The distribution of route licences in Germany is based on § 21 *Luftverkehrsgesetz* [Aeronautics Act]. Like in most other countries and unlike in Great Britain, the airlines must apply for every route separately.[816] A hearing is not mandatory, but can be arranged, if this is considered as necessary. Route licences are, in general, issued for an unlimited period. Applications for route licences can be denied, if the proposed air transportation service would interfere with public interests. The term public interest is not specified and it is left to the administration to determine, whether and how the public interest could be affected. When defining the public interest, international obligations, undertaken by Germany in air service agreements, are taken into account. For example, the public interest is considered as affected, if Germany would have to issue a route licence to an air carrier, which, in fact, could not be designated under the applicable air service agreement.[817] The public interest is also held as being affected, if additional air carriers would have to be permitted into a market, which is already sufficiently served by the incumbents. However, the latter interpretation aims at shielding LH from competition. Due to the deregulation initiated by the EC, doubts must be raised, whether such an understanding of the public interest exemption is still valid and in accordance with EC law.[818]

On 13 February 2001 the German Ministry for Transportation issued guidelines on the distribution of air traffic rights and frequencies to German air carriers.[819] The guidelines provide that, where unlimited air traffic rights and frequencies are granted to Germany, they will be allocated without any limitations to interested airlines. When distributing air traffic rights, the government pursues the following objectives:

812 *Ibid.* at 239, 241; Rapp and Vellas, *supra* note 744 at 115.

813 Rapp and Vellas, *ibid.* at 112.

814 *Ibid.* at 240, 241ff.

815 Information on the criteria applied in Germany were provided by the German Ministry of Transportation.

816 Hofmann/Grabherr, *supra* note 3 at § 21, notes 3, 6.

817 Schwenk/Giemulla, *supra* note 13 at 697.

818 *Ibid.* at 696.

819 Federal Ministry of Transportation and Construction, Guidelines for the Allocation of Air Traffic Rights and Frequencies to German Air Carriers of 13.2.2001, LS 12/20.00.50; According to the Ministry these guidelines are also applicable to applications filed in by community air carriers.

- to guarantee the highest stimulation of the public transportation interests;
- to promote competition and to allow new entrants into a market, if sufficient demand exists to justify an increase in capacity; and
- to take into account, on an equal basis, the interest of the following groups:
  - consumers;
  - export industry;
  - German air carriers;
  - German airports; and
  - the different regions in Germany.

The possibility to distribute air traffic rights through auctions or a quota system is explicitly rejected and a free distribution based on the discretion of the administration is preferred. The airlines are obliged to make use of their licences within one year after the allocation. Every applicant must also guarantee that it has sufficient route-related capacity available for at least two flight-plan periods. Unlike any other country, Germany applies a grading system, in which the applying airlines receive between one and 10 points depending on how much they comply with the respective criterion. The route is then given to that airline, which scores highest. The Ministry of Transportation applies the following criteria:

- degree of service to the public (consumers and shippers), thereby taking into account:
  - product quality:
    - frequency;
    - duration of flight;
    - route integration;
    - category of service (non-stop, single-stop with or without change of aircraft, code-sharing, interline);
    - type of aircraft used (capacity);
  - tariff level and tariff structure;
  - connections offered to various regions in Germany and in the area of destination;
- competition in the corresponding market:
  - economic and technical potential of the applicant to further develop the route;
  - competition between German and foreign airlines;
  - competition among German airlines;
  - promotion of national aviation interests;
  - ensuring competitiveness of German industry;
- consideration of scarce resources and environment:
  - relieving congestion in the hubs;
  - use of wide-body aircrafts;
  - use of low-noise aircrafts;
  - use of low-polluting aircrafts;
  - propulsion;

- arrival and departure times;
- use of slots at off-peak periods.

These criteria are not exclusive and each application is decided on a case-by-case basis. Air traffic rights that are not used during two seasons are revoked and re-allocated (use-it-or-lose-it principle). Until 2004 no dispute has ever arisen over the allocation decision. However, according to an information by the Ministry of Transportation a rejected (German) air carrier appealed the decision of the Ministry on the allocation of air traffic rights to the administration law courts in 2004. Unfortunately, this appeal was withdrawn again and thus so far no judicial review exists on the allocation of air traffic rights in Germany.

Result

National deregulation in Germany, if one acknowledges LH's privatization as deregulation, was very different from other liberalization processes. Unlike in the US, Canada, Great Britain and Australia, Germany did not deregulate its airline industry to enhance competition, but to raise money for the rebuilding of East Germany and to fulfil the criteria for the European Monetary Union. Deregulation, with the aim to enhance competition, mostly occurred through the liberalization process initiated by the European Community. Unlike other airlines, which entered deregulation from a protected surrounding and usually were in sound financial situation, LH went through a crisis, was in deep financial trouble and struggled to survive. Still, LH's management welcomed the privatization process and used it to change the corporate culture from a government-like bureaucratic thinking of its employees to a service-orientated, competitive attitude.[820] On the other hand, it shall be also stressed that LH did not have to face any serious competition on the domestic market, as it was the only German air carrier offering scheduled air transportation services. Furthermore, it was also Germany's major flag carrier on international routes and did not have to share significant international routes to and from Germany with other German airlines. As can be seen from the examples of BA and LH, the scheme of privatization in the European Community was very similar, yet occurred at different times. LH, and similarly Air France, was privatized against the background of deregulation in the EC, while BA, as also KLM, was privatized by the British Government and before the EC became involved. All airlines, going through this process, faced the same financial and cultural crisis. The only difference was the moment, when it occurred. BA went through its troubles before the Gulf War and the rising competition from US carriers. It was therefore better prepared to face these challenges.[821] LH, on the other hand, had to master the Gulf crisis, rising competition and the restructuring of its business at the same time. This is even more remarkable, considering that the German Government refused to give any support to the airline.

As far as the distribution of air traffic rights is concerned, Germany has mostly adopted the approach applied in the other states. Many criteria established by the

820 Staniland, *supra* note 580 at 239f.

821 *Ibid.* at 192, 214.

German Ministry of Transportation can be found in the US, Canada, Australia and Great Britain. The only difference is the inclusion of environmental aspects in the selection and the application of a grading system. Although a grading system, as applied in Germany could give the impression that a high degree of objectivity exists, it is still not flawless. The greatest difficulty is to decide, how each proposal is graded. For example: How many points are to be given for the use of a seven-year-old aircraft? How many points is a departure time at 08.00 worth? These decisions again require a subjective assessment by the person in charge of the selection process and thus depends on subjective preferences.

CONCLUSION

In the foregoing analysis, different countries and their approaches to the distribution of air traffic rights were shown. The first remark that can be made is that all countries examined used beauty contests to distribute air traffic rights. This fact strengthens the previous conclusion that air traffic rights should be allocated through this method. It is remarkable that all countries used the same allocation method despite the partially quite significant differences of their airline industries. On one side of the spectrum, one can place the USA that liberalized its air transportation sector quite early and which has a multitude of internationally operating private airlines. On the other side, one will find Germany that only reluctantly liberalized its airline industry and which has only one, previously state-owned, major flag carrier. In between these two extremes are Canada, Great Britain and Australia, which have pursued a Multi-Airlines Policy with a public and a private airline competing against each other. Remarkably the approach to establish a system with two major airlines started to become defunct, when the respective governments became actively involved and tried to promote it. In all three cases, the formerly state-owned airline in the end overtook its private counterpart (Great Britain, Canada) or drove it out of the market (Australia). Only slowly, did new competitors emerge and expand to a considerable size in these countries (WestJet in Canada, Virgin Atlantic and British Midland Airways in Great Britain). Apart from the US, all major airlines, in the countries included in this analysis, were originally state-owned and then privatized between the late 1980s and the late 1990s.

Apart from Great Britain, licences are allocated on a route-by-route basis. Hearings are not mandatory any more, but can be summoned in most cases, if the distributing agency wishes to do so. The decision on the distribution of international routes is either taken by an independent committee (Australia, Great Britain) or by the Ministry of Transport (USA, Canada, Germany). As far as the selection is concerned, it is remarkable that, despite the differences among the countries, the states mostly used very similar criteria. Sometimes criteria were given up, modified or resumed again. The following outline provides an overview of the different criteria used:

- competition
  - competition among airports (USA);
  - competition within one airport (USA);

  - ability to compete effectively with foreign airlines (USA, Australia, Canada, Great Britain, Germany);
  - competitive situation among national airlines (USA, Australia, Canada, Great Britain, Germany);
- quality of service
  - route integration (USA, Australia, Canada, Great Britain, Germany);
  - low fares (USA, Australia, Great Britain, Germany);
  - high frequencies (USA, Australia, Canada, Great Britain, Germany);
  - high capacities (USA, Australia, Canada, Great Britain, Germany);
  - improved in-flight and ground services (Australia, Great Britain);
  - promotion of innovation (Australia);
  - higher choice among airlines (Australia, Great Britain);
  - category of flights (non-stop, one-stop with or without change of gauge, code-sharing) (Australia, Canada, Great Britain, Germany);
  - category of services (Passenger, Passenger/Cargo, all-Cargo) (Canada);
  - preferable arrival and departure times (Great Britain);
  - duration of flight (Great Britain, Germany);
  - higher choice among airports served (Great Britain);
- airlines
  - cost structure (USA, Great Britain);
  - financial and technical potential to develop the route (Germany);
- externalities
  - promotion of tourism (Australia);
  - benefits to trade (Australia, Germany);
  - benefits to aviation industry (Australia, Germany);
  - regional development (Great Britain);
  - economic interest of communities in flight services (USA);
- diverse
  - quick entrance into the market (USA, Germany);
  - environmental issues (Great Britain, Germany);
  - protection or support of airlines during the development of a route (Great Britain, Australia);
  - protection of airlines on low traffic routes (Germany, Great Britain);
  - historic interest in a route (USA, Great Britain);
  - slots (Great Britain, Germany);
  - relieving congestions in the hubs (Germany).

All states take the aspect of competition into account and, at the same time, try to select that carrier, which they expect to offer the best service to the customer. In general, the regulatory bodies intend to strike a balance between introducing more competition by distributing air traffic rights to newcomers and enhancing the quality of services by granting additional routes to the incumbents, which usually are in a position to offer more user benefits due the networks they have established and the routes they already possess. The most rational decision would be to grant the routes to that airline, which offers the best service and most benefits to the users of air transportation services. The difficulty is that this would mostly favour the big air carriers, because through their established hub-and-spoke systems they can offer the greatest choice of destinations and frequencies and thus the greatest benefits to travellers. If air traffic rights are always granted to a certain group of airlines, these carriers might grow to such an extent that they could drive all of their remaining competitors out of the market and finally establish an oligopoly or, in the worst case, even a monopoly. Not only is such a situation extremely disadvantageous in the long run, but also the huge amounts of funds that were invested in smaller start-up airlines would be also lost. This decision between short-term gains in form of higher user benefit through enhanced services and long-term effects with respect to increased competition is mostly to be made by those countries, which once had owned airlines that they subsequently privatized. The US, which has never been engaged in the air transportation industry, is less affected by this problem. It has a multitude of big airlines that can compete against each other on the same footing. Canada, Australia, Great Britain and Germany, on the other hand, protected their major airlines for decades from competition and must now ensure that these carriers do not establish a monopoly in their countries or, if they already did, do not abuse their market power. Finding the right balance between the two opposing possibilities, promoting competition or enhancing higher user benefits, causes the main dilemma for states. This task becomes even more complicated, when, especially like in the US, the applicants offer equal user benefits. This is, as mentioned above, also the typical problem that occurs, when beauty contests are used as an allocation method of scarce resources.[822] The regulator usually tries to solve it by introducing seemingly objective criteria, to provide a justifiable decision. The CAA in Great Britain considered aspects like legroom, pampering in the departure hall or the possibility that connecting flights are in the same terminal when selecting an airline. At this stage, the allocation process becomes absurd and one must raise the question, how far the regulatory bodies should go in the assessment of the quality of services. While some criteria like route integration, capacity or frequency are very objective, most other aspects require a subjective assessment. It is difficult to imagine, how regulators can assess the quality of in-flight services or the degree of promotion of innovation. It is hardly possible for an airline to offer non-stop flights with high quality in-flight services at a low price. Since this is not feasible, the regulator must decide, which aspect of a flight service is more valuable than other aspects. One may wonder, how the decision is taken, if one airline suggests a one-stop flight with a 30-minutes longer flight time compared to another applicant, who offers a non-stop flight that is maybe 15 per cent more expensive. And to top it all, some criteria applied in the selection process are even contradictory. In Germany, for example, the most suitable departure and arrival times are positively taken into account and, at the same time, it is advantageous, if slots at off-peak times are used. Unfortunately, one is not possible without the other. The best time for a flight from, for example, New York to Frankfurt is overnight, but then the arrival occurs in Frankfurt in the morning and thus at a high-peak time. The high score, an airline may receive for a beneficial departure and arrival time, is equalized when it comes to the evaluation of the slots an airline uses. To avoid such a contradictory result, it is firstly suggested

822 See above at 159f.

that criteria, which are incompatible with each other, should not be mixed and both applied in one single selection procedure. Secondly, any criterion that requires a subjective assessment, should be also avoided. As far as the assessment of the quality of services is concerned, there are only four criteria that fulfil this demand: capacity, frequency, fare level and route integration.

When it comes to competition, different approaches have been taken. In the past, the USA intended to promote competition by supporting weaker and smaller airlines. The downside of this strategy is that weaker airlines might have no incentive to increase efficiently and a certain sluggishness may evolve, if airlines have the security that governments will bail them out, if they get into financial troubles.[823] It is also difficult to define, when an airline is to be considered as weak. Initially, the CAB considered a carrier as weak, if it was small and operated on intrinsically unprofitable routes.[824] Subsequently, regard was also paid to a carrier's short-term financial conditions.[825] This resulted in that also bigger airlines, like for example TWA, could be considered as weak carriers.[826] Factors like profit, number of aircrafts or amount of international air traffic rights a carrier possesses can be all referred to, to determine whether a carrier is weak. The CAB was not very consistent in its evaluation, when a carrier was to be regarded as weak. This allowed the Board to manipulate the decision of which airline should be granted a route for competitive reasons. In some cases it was even argued that, to strengthen competition, a route should be given to a carrier simply because, when compared to its competitors, the respective airline had received fewer air traffic rights in previous allocations.[827] Other examples, where a distribution of air traffic rights to enhance competition was not successful can be found in Australia and Canada. The regulatory bodies of both countries tried to enhance competition by establishing a duopoly and sometimes granted routes with regard to the attainment of this objective. In the end, their strategies failed. Taking these developments into account, it has to be concluded that a government's deliberate interference with market forces in the air transport sector is not very efficient. Furthermore, when competition is taken into account, two aspects are usually brought forward: competition among national carriers and competition of national air carriers with foreign airlines. Having regard to the special situation in the European Community, it is not appropriate any more to take the competitive situation within a single Member State into account. Since the EC has established a single European market for air transportation services, it is necessary to take the competitive situation in the whole European Community into account. Concentrating on a single country bears the danger that Member States continue to favour their national carrier(s) arguing that a particular carrier is weak and needs to be supported in order to maintain a healthy competition. Thus, the aspect of competition could be abused for indirect discriminatory decisions. Besides, it can be doubted, whether it is necessary to take into account the competitive situation

---

823 Rollo, *supra* note 552 at 230.
824 *Ibid.* at 214.
825 *Ibid.* at 215.
826 Caves, *supra* note 611 at 212.
827 Rollo, *supra* note 552 at 214.

among Community air carriers at all. If the EC is perceived as what it is, a single European market, then a sufficient number of strong airlines exists that can compete with each other and it is not necessary that it is interfered, to guarantee future competition. It is very unlikely that, in a situation where several quite strong airlines exist, for example Air France-KLM, British Airways, Lufthansa and Iberia, one or two will be able to drive all other competitors out of the market and establish a competition threatening monopoly or duopoly. Apart from that, such an endeavour would be subject to competition laws and should be dealt with in this context and not when it comes to the distribution of air traffic rights. Hence, it is my opinion that the competitive situation among Community air carriers does not need to be taken into account, when air traffic rights are distributed. A different result might be, however, achieved, when it comes to competition between Community air carriers and non-EC airlines. Due to the significance of the airline industry for the economic development of the EC, Member States should have an interest in promoting Community air carriers and enhance their competitive ability in relation to non-EC carriers. The impact on the development of the EC could be tremendous, if a reliable air transportation system ceased to exist. Not only the economy could suffer from that, but also the European integration. Air transportation is needed to move goods and people, thereby increasing economic benefits and promoting understanding and tolerance. Thus, the preservation of an independent European airline industry should be taken into account, when international routes are distributed.

Finally, under a last set of criteria, overspill effects into other industries are considered. Australia and Germany are examples where this is practised quite openly to promote either tourism or trade. Despite the positive effects air transportation certainly has on other sectors, still, it is not coherent to deregulate an industry and privatize airlines, but continue using their operations to subsidize or support other industries. If deregulation and privatization mean that airlines are free from any governmental influence, then this must be applied consistently. Therefore, any economic aspects, unrelated to air transportation, shall be left out of the distribution process. Additional overspill effects, benefiting other industries or underdeveloped regions, are of course welcomed, but an approach, where the air transportation industry is used to actively promote other sectors or regions is rejected. Therefore, any criteria listed above under 'Externalities' should not be included in the selection process.

THE IMPACT OF THE GATS ON THE DISTRIBUTION OF AIR TRAFFIC RIGHTS

A first set of possible criteria was already set out in the previous chapter, when the distribution methods of various countries were examined. In addition to the criteria already found above, it shall be also scrutinized in the following, whether other principles exist that can influence the allocation process. Such principles could be derived from international law. Two regimes might have to be taken into consideration in that respect: the General Agreement of Trade in Services and the EU and the EC Treaty. Since it was concluded above that the distribution of air traffic rights will be undertaken either by the Member States alone, by the EC or by both institutions through a cooperation, both regimes can have an impact on the distributions procedure. The Member States are parties to GATS as well as to the EU and the EC Treaty. The European Community is party to GATS and also bound by the provisions of the EU and the EC Treaty. In this context it shall be noticed that any national laws of the Member States, that can have an impact on the distribution of air traffic rights, will not be further discussed here. In the following it will be assessed, whether the General Agreement on Trade in Services can have any impact on the distribution of air traffic rights.

GATS was concluded and signed within the framework of the revision of the General Agreement on Tariffs and Trade (GATT) and the creation of the World Trade Organization on 15 April 1994. It came into force, for the states that ratified it, on 1 January 1995.[828] While GATS does not contain a definition of services, it determines the scope of the term 'trade in services'. According to Article I, trade in services encompasses:[829]

> (1) across-the-border trade, whereby the service is offered from the territory of one state to the territory of another state and neither the consumer nor the service provider leave their countries;
>
> (2) domestic-established trade, whereby the consumer located in one state moves to the territory of another state, where the producer is established;
>
> (3) foreign-earnings trade, whereby the producer establishes a presence in the territory of the state of the consumer;
>
> (4) third-country trade, whereby consumer and service provider are both located in a foreign country.

Due to the fact that the term 'service' is not defined in GATS, all kinds of services, including air transportation services, are, at first, covered by this agreement.[830] International air transportation services would usually fall into the categories 2, 3 or 4. The regulation of trade in services under GATS is based on 'three pillars'.[831] The first pillar is a regulation directly provided in the GATS. It can be split into general obligations and specific commitments. The general obligations apply to all services in accordance with the sectoral annexes.[832] The sectoral annexes form the second pillar of GATS. The annexes are special regulations, which amend the application of GATS for certain services. Annexes exist for the movement of natural

828 Yang, *supra* note 30 at 50f.; Marrakech Agreement establishing the World Trade Institutionization, 15 April 1994, 1867 U.N.T.S. 31874 [GATS].

829 Alexandre Mencik von Zebinsky,' The General Agreement of Trade in Services: Its Implications for Air Transport', (1993) *Ann. Air & Sp. L.*, 359 at 361 [Zebinsky].

830 Yang, *supra* note 30 at 86.

831 See Zebinsky, *supra* note 829 at 373f.

832 Marie-Angélique, K. (1995), *The Economic Regulation of Air Transport: From the Chicago Convention to GATS* (LL.M. Thesis, Institut of Air and Space Law, McGill University [unpublished]) at 54 [Kolivakis].

persons providing services under GATS, financial services, telecommunication and air transportation services.[833] The most important provision in this set of rules is the most-favoured-nation clause (MFN) contained in Article II(1). This principle was already included in GATT (1947) and prohibits any discrimination among nations.[834] States are obliged to apply the most favourable regulation, they have agreed upon in a specific area with one country, to all other member states that are party to GATS. Thus, equal treatment is granted to all states without regard to reciprocity.[835] The specific-commitment rules are additionally applicable, if states undertake a special commitment to liberalize a particular area of services. Among this set of rules, provisions regarding market access and national treatment (NT) are the most important ones.[836] The regulation regarding market access prevents states from creating significant entry barriers and ensures access, pursuant to the terms, limitations and conditions specified by the states in their national schedules. National schedules are the third pillar upon which the GATS system is based. They contain, in detail, the specific commitments the states have agreed to undertake. Although the absence of a specific commitment does not mean that foreigners cannot provide services in a certain area, commitments provide for more security as they are legally binding.[837] The NT principle forbids discrimination between foreign and domestic services and is subject only to the conditions and qualifications set out in the national schedules.[838] Any national measure that restricts or distorts trade, thereby granting protection to domestic service providers or giving them a more favourable treatment compared to foreigners, shall be avoided by the NT principle.[839] However, governments are still allowed to differentiate between service providers, as long as the conditions of competition are not modified to favour domestic service providers.[840] Specific commitments are undertaken in two steps: Firstly, the respective state chooses a sector that it wants to liberalize and includes it in its national schedule. Secondly, it fixes the restrictions it wants to impose on market access and NT within these sectors.[841]

As far as aviation is concerned, the Annex on air transportation services is of utmost importance, as it specifies the degree, to which GATS is applicable in the air transportation sector. Generally, the Annex is restrictive and limits the application of GATS.[842] Any measures affecting air traffic rights or the exercise thereof, are explicitly excluded from the application of GATS. In the aviation sector GATS is

833 Zebinsky, *supra* note 829 at 378, n. 85.

834 *Ibid.* at 375.

835 Kolivakis, *supra* note 832 at 40.

836 Marek Zylicz (1992), *International Air Transportation Law*, (Dordrecht, The Netherlands: Kluwer Academia Publishers), 178 [Zylicz]; Kolivakis *ibid.* at 57.

837 Compare Zylicz, *ibid.* at 181f.

838 Zebinsky, *supra* note 829 at 377.

839 Kolivakis, *supra* note 832 at 38.

840 *Ibid.* at 57.

841 Dorothea Kreymborg, 'Developments Relevant to the International Air Transport in the World Trade Institutionization (WTO) – Analysis of the First Review of the GATS Annex', (2003) 28 *Ann. Air & Sp. L.*, 471 at 476 [Kreymborg].

842 *Ibid.* at 477.

limited to aircraft repair and maintenance services, the selling and marketing of air transport services and computer reservation system services.[843] Only here, the MFN principle is applicable.[844] In addition, the market access and NT regulations can be also made applicable, if a state undertakes a corresponding special commitment in these areas.[845] But, as far as the distribution of air traffic rights is concerned, GATS is not relevant. The biggest problem for extending GATS to air traffic rights is that an unconditioned application of the MFN clause would benefit restrictive and protectionist countries, while very liberal states and their airlines would have to bear the disadvantages.[846] Free-rider nations could make use of concessions given to other countries, without offering any benefits in return.[847] Under these circumstances it is easy to comprehend that states are less willing to liberalize their air transportation markets under GATS.[848] Unlike bilateral air service agreements, which tend to establish a sectoral balance and usually require reciprocity, GATS neglects any sectoral reciprocity and aims at establishing an overall balance.[849] Even the US, one of the leading aviation nations and usually a strong proponent of competition, had reservations in including any matters concerning air transportation into GATS.[850] If GATS was applicable, Member States of the European Union, due to the common market in air transportation, would be forced to grant cabotage rights to non-EC carriers. The far reaching consequences and the free-rider problem urged the states to limit the application of GATS only to soft-rights and exclude any hard-rights, that is air traffic rights, from its scope.[851]

Although presently GATS is not applicable for the assignment of international routes, one might ask, whether, *de lege ferenda*, such an inclusion is desirable. While some argue that this could enhance liberalization and bring benefits to consumers, others assume that the contrary could happen and that the free-rider problem resulting from an application of the MFN principle might retard any further air transport liberalization.[852] Proponents of an extension of GATS to the air transportation sector argue that Article II(2) GATS provides for a possibility to include exemptions to the MFN principle, to limit activities of free-rider states.[853] Others suggest a combination of the MFN principle and the reciprocity principle for the air transportation sector, that is the MFN principle would only apply to those countries that offer similar conditions in their air transportation relations.[854] The lack of reciprocity in the GATS system is one major obstacle against its further extension to the air transportation sector. In an industry that has been successfully regulated for over 50 years by an equal exchange

843 See GATS, *supra* note 828, Annex on Air Transportation Services, at 25 para 3.
844 Zebinsky, *supra* note 829 at 379.
845 Kolivakis, *supra* note 832 at 60; Yang, *supra* note 30 at 54.
846 Zylicz, *supra* note 836 at 176; Kolivakis, *ibid.* at 42; Yang, *ibid.* at 92.
847 Kolivakis, *ibid.* at 41.
848 *Ibid.* at 42, 62; Yang, *supra* note 30 at 91.
849 Kolivakis, *ibid.* at 61.
850 Zylicz, *supra* note 836 at 175.
851 Kolivakis, *supra* note 832 at 44.
852 Zebinsky, *supra* note 829 at 396; Zylicz, *supra* note 836 at 177.
853 Yang, *supra* note 30 at 56f., 89.
854 Serrano, *supra* note 22 at 213.

of rights and business opportunities, a new and different approach naturally causes scepticism. The idea of opening up the air transportation market of a state to all members of GATS, but not to obtain equal rights and business possibilities, might not be well perceived by all states. Indeed, it can be asked, whether the principles contained in GATS are appropriate for the air transportation sector. Of course, it can be argued that, if other industries could be subjected to the provisions of GATS, then there is no reason to treat aviation differently.[855] However, it is also necessary to have regard to the special significance that is attributed to air transportation and the high national interests in this industry. While the prestige factor and representative function of airlines might not have the same importance as in the past, due to the fact that states are privatizing their airlines and cross-border mergers are undertaken, there is still the infrastructural element of air transportation that might be of high relevance. Certainly, it is possible to draw an analogy to the telecommunication industry that is also part of the infrastructure of a country and that nevertheless was subjected to the provisions of GATS. The difference is that the means of production in the air transportation sector (aircrafts) are highly mobile and can be withdrawn from a market very easily, while facilities in the telecommunication sector are rather immobile and, once installed, cannot be transferred that easily. The danger is that, if a country decides to open up its aviation market to foreign competitors, its own national airlines will be fully exposed to competition and, in the worst case, could be driven out of the market. If, subsequently, the foreign airlines leave the respective market again, maybe due to low profitability, a country could be left without any air transportation service providers, which could have a detrimental impact on the state's economy. In this situation, a state could, of course, try to re-establish its own airline or to attract foreign carriers to offer services on subsidized routes. But, firstly, it is not guaranteed that an airline will be found for the requested purposes, and, secondly, foreign carriers, especially from developed countries, due to high labour costs, might be in need of subsidies that could be far higher than the upkeep of a state-owned carrier. In evaluating a further extension of GATS in the air transportation sector, the situation in different states must be taken into account and particularly the differences between developed and developing countries. Remarkably, during the negotiations on GATS the developed nations were very much in favour of including the air transportation sector into GATS, while the developing countries were less willing to do so.[856] Meanwhile, this situation has completely turned. While the only two countries having entirely opened up their passenger air transportation markets, Sierra Leone and the Gambia, can be classified as rather moderate powers with respect to air transportation, the big players with numerous and strong flag carriers, the US and the EC, have become very protective of their national markets.[857] It is uncertain, how an unconditioned application of GATS in the air transportation sector might affect different countries. On the one hand, it could turn out to be beneficial for developing states, as their airlines would be given access to the more lucrative markets, from which they may be presently foreclosed, due to the fact that

855 Yang, *supra* note 30 at 117f.; Kreymborg, *supra* note 841 at 496.

856 Zylicz, *supra* note 836 at 175.

857 Kreymborg, *supra* note 841 at 481.

developing countries can not offer similarly interesting markets for the air carriers from developed countries.[858] In addition, airlines from developing countries could have some competitive advantages, due to their lower labour costs. On the other hand, doubts remain, whether air carriers from developing countries are strong and efficient enough to compete with the airlines from the developed countries.[859] Many developing countries are slowly establishing their service industries and, like in the beginning of commercial aviation in the developed countries 50 years ago, need to protect and subsidize these infant industries before they can become competitive on the international level.[860] Thereby it should be also considered that international civil aviation is a commercial activity of great importance and, together with tourism, is the largest single world trade sector after petroleum. Introducing global competition in this sector, may deprive developing countries from the chance to participate in this segment by possibly driving their airlines out of the market, even before they can establish themselves.[861] Although Article IV of the GATS contains an obligation to increase participation of developing countries in world trade, it can be doubted, whether this is sufficient to make airlines from developing countries survive competition against air carriers from the developed nations. At this stage, it is difficult to evaluate the chances and risks for developed and developing countries, if the air transportation sector was fully included in the GATS system. In this context, it is interesting to have a look into the past. The notion of an exchange of air traffic rights in a multilateral treaty is not new and was brought forward for the first time almost 60 years ago in the negotiations on the Chicago Convention. While the US favoured such a liberal approach, other countries, especially Great Britain, dismissed this idea. The main concern was that the US airlines, with their large number of aircrafts, would simply flood the market and take control of international air transportation. Since that time much has changed, but this concern can be still brought forward today. The difference is that additional carriers, mostly from the European Union, Canada, Australia and some Asian states, have become strong enough to compete on a world-wide basis. The concerns, once raised by Great Britain, are now partially expressed by developing countries. They are afraid that airlines from developed countries will dominate the air transportation sector and will not leave sufficient business opportunities for their airlines. Indeed, it is hypocritical, if some developed countries now request the inclusion of air traffic rights under the GATS system, thereby denying a fair chance to the developing states to establish their own air transportation system, while the some of these countries once claimed this possibility for themselves a long time ago.

In the end, the question, whether air traffic rights should be subjected to GATS, is a political and not a legal one. Every country has to decide, whether it wants to take the risks and opportunities, such a re-organization of the air transportation sector may bring. One can only assume that a decision to extend the scope of GATS will not occur in the near future. When this matter was addressed at the fourth World-

858 Kolivakis, *supra* note 832 at 47.

859 *Ibid.*

860 Yang, *supra* note 30 at 85f.

861 *Ibid.* at 86.

Wide Air Transport Conference initiated by ICAO in 1994, only slight progress was achieved due to the opposition by African, Arab and Latin American countries.[862] The following fifth World-Wide ICAO Air Transport Conference in 2000 did not produce any significantly different results either and only a non-binding declaration was adopted, which urged the states to liberalize their air transportation markets, but left it to each state to decide upon the pace and the way it wanted to be achieve this objective. It was further suggested to pursue this goal by concluding bilateral, sub-regional, regional and multilateral treaties.[863] In the course of the revision of the Annex on air transportation in 2000, states were asked to present their opinions and proposals. Some countries such as Australia, Norway and New Zealand favoured, to different degrees, an expansion of the GATS system in the air transportation sector and an abolishment of the bilateral and regional approaches.[864] They argued that bilateral agreements discouraged the improvement of services and hampered competition and flexibility.[865] Regional agreements, on the other hand, had the disadvantage that non-founding members had little influence on the terms of the treaty. Furthermore, those agreements did not bring any additional benefits over bilateral relations, as reaching a consensus became more and more difficult, once a critical mass of members was reached. Therefore, the only solution to promote liberalization was to extend the applicability of GATS to the aviation sector.[866] It must be, however, added that the scope of such an extended GATS or Annex would be limited by and dependent on the special commitments undertaken by each state. Especially, if the most dominant nations in air transportation decide not to undertake any commitments and file in exemptions to the MFN principle, such a multilateral approach within the framework of GATS could share the fate of the Transport Agreement of 1944.[867] The only perceived advantage in favour of an extension of GATS is the ease with which a country could pursue the liberalization of its air transportation sector in the future. Instead of signing several liberal bilateral air service agreements with other states, it would be sufficient to enter into a specific commitment under the GATS system. Thus, every country could control the pace and degree of its liberalization process.[868] Another group of states, Japan, Columbia and the EC, on behalf of the Member States, proposed to extend the scope of GATS in aviation matters, but wanted to maintain the bilateral system for the exchange of air traffic rights. It was argued that the bilateral approach has proven to be efficient in the development of international air transportation and that there was no reason to abolish it. However, it could be appropriate to extend the applicability of GATS to ancillary services like ground handling.[869] Finally, a third group of states, in particular Brazil and the USA, rejected any further extension of GATS in the field of air transportation. While Brazil argued

---

862 Serrano, *supra* note 22 at 201.
863 Kreymborg, *supra* note 841 at 494.
864 *Ibid.* at 484–87.
865 *Ibid.* at 486.
866 *Ibid.* at 484.
867 See above at 7.
868 Kreymborg, *supra* note 841 at 485.
869 *Ibid.* at 487–90.

that the special interests and needs of developing countries with respect to air traffic rights could be best dealt with under the traditional bilateral system, the USA held the opinion that liberalization would be slowed down, if the GATS system was applied to air traffic rights. The USA argued that the degree of liberalization achieved so far, was due to the use of 'open-skies' agreements.[870] The US also announced that any further extension of GATS in the aviation sector would not be applicable in the US and even the inclusion of ancillary services like ground handling, as suggested by other states, was not negotiable.[871] It is remarkable that representatives from both sides of the spectrum, Brazil for the developing countries and the USA for the advanced states in the area of air transportation, considered the GATS system as inappropriate for further liberalization. Furthermore, the US position seems to confirm the fears mentioned above that an extension of GATS in the aviation sector might not be successful, since one of the most important countries in international air transportation would not participate in such an extension. At first, one may wonder about the protectionist position of the USA. The reason for this may be due to the fact that the US has much to lose, if the whole air transportation sector is liberalized at once. As was already indicated above, the consequence of the conclusion of 'open-skies' agreements with selected countries is that the US has already achieved a limited liberalized international air transportation sector for itself. The combination of all 'open-skies' agreements signed by the USA creates a partially liberalized air transportation area for US air carriers. If all air transportation markets are suddenly opened up, the US would lose this advantage it has secured for its airlines. In particular with respect to Community air carriers and Asian airlines, which have become stronger over the last years, the US airlines would lose their competitive advantage. Finally, when evaluating the possibility of an extended GATS one should, firstly, recall the very cumbersome and lengthy process of deregulation within the EC and, secondly, the negotiations between the EC and the US on the establishment of a TCAA, which lasted over a year and finally failed.[872] If the liberalization of the air transportation sector between like-minded and equally developed states is so troublesome, one can have only limited expectations for a world-wide liberalization that would require the consensus of states, which have different interests and where the industry development levels are not equal. The differences between developing and developed countries in the air transportation sector are still too big and air transportation is considered too important by many countries that they give up control over this sector. Under these circumstances it should be considered, whether the pursuit of a multilateral treaty should not be postponed for the time being and meanwhile other methods selected. A regional solution, where more equal states come together and agree on more liberalization, has proven to be very efficient as one can witness with respect to the European Community. Other states might also create their own regional pacts and these regional aviation areas could then try to reach a global consensus. In developing international aviation, the next step would be another attempt to establish the TCAA. This time maybe also other countries,

870 *Ibid.* at 490f.
871 *Ibid.* at 491f.
872 See above at 46ff.

apart from the US and the EC, like for example. Canada and Mexico, should be included in the negotiations. The establishment of such a TCAA would, in the end, bring together the EC and NAFTA. Hence, to achieve globalization, it might be useful to include a regionalism as an intermediate step. Such an approach is certainly a very long-term perspective, but the direct establishment of a multilateral global treaty could not have been achieved during the last 60 years and there is no sign that this will change soon.

THE IMPACT OF EC LAW ON THE DISTRIBUTION OF AIR TRAFFIC RIGHTS

It was concluded above that the distribution of air traffic rights will have to be accomplished either by the Member States alone, by the European Community or by both institutions together. *De lege lata* Article 5 of Regulation 847/2004 prescribes that the distribution of air traffic rights must be conducted in a transparent and non-discriminatory manner. In the following it shall be examined, whether in addition to these two criteria, further requirements may be applicable with respect to the allocation of air traffic rights.

SUBJECT MATTER LINK

As already mentioned above, air traffic rights are commercial public rights. In addition to air traffic rights, states possess many other kinds of commercial public rights that they distribute among citizens and entities. Hence, the allocation of air traffic rights is only another procedure in which states distribute a valuable resource. Under these circumstances it is necessary to have regard to the regulations existing with respect to other distribution procedures in the EC. For this purpose an examination of the Community's public procurement laws will be undertaken.

The EC has adopted two directives on the award of public contracts. Directive 2004/17/EC provides for the procurement procedures of entities operating in the water, energy, transport and postal service[873] and Directive 2004/18/EC includes provisions on the coordination of procedures for the award of public work contracts, public supply contracts and public service contracts.[874] These two directives were adopted only recently and consolidate and substitute all former acts that existed in the area of public procurement. Although these former acts are not in force any more, they must be still taken into regard as they will be relevant in the following.

873 EC, Directive 2004/17 of the European Parliament and of the Council of 31 March 2004 coordinating the procurement procedures of entities operating in the water, energy, transport and postal service, [2004] O.J. L 134/1 [Directive 2004/17].

874 EC Directive 2004/18 of the European Parliament and of the Council of 31 March 2004 on the coordination of procedures for the award of public work contracts, public supply contracts and public service contracts, [2004] O.J. L 134/114 [Directive 2004/18].

93/37/EEC[875] was applicable to contracts granted in the field of constructions, Directive 93/36/EEC[876] covered contracts on the supply of goods, Directive 93/38/EC[877] referred to contracts for the supply of water, energy as well as transport and telecommunication services and Directive 92/50/EEC[878] was applicable to contracts related to the supply of services. The respective provisions on the applicable criteria for the awarding of public contracts were almost identical. They prescribed that public contracts should have been awarded to the tenderer offering either the economically most advantageous tender or to the one with the lowest price. In deciding, what was to be considered as the economically most-advantageous tender various criteria could have been taken into considerations depending on the contract in question, like, for example: quality; technical merit; aesthetic and functional characteristics; technical assistance and after-sales service; delivery date; delivery period; period of completion; running costs; cost-effectiveness; commitment with regard to spare parts; security of supplies or price. The current directives mostly adopted the respective regulations. Several disputes, which were finally decided by the ECJ, arose on whether additional criteria under the economically most advantageous tender test, besides those contained in the respective directives, could be applied. Mostly, the question was whether social criteria could be introduced in the allocation process. In the ECJ's first judgement on public contracts, the Beentjes judgement, the judges ruled that the criteria listed in the respective directive to be used under the economically most advantageous tender test were not exhaustive, but only exemplary.[879] The ECJ further concluded that only those criteria were permissible that aimed at 'identifying the offer which is economically the most advantageous'.[880] Based on this statement one may have assumed that these criteria must be of an economic nature. This understanding was, however, precluded by the ECJ itself, when the judges further elaborated that it was permissible to take into account the ability of a company to employ long-term unemployed persons. Unfortunately, at this point, the judges caused additional confusion, when they referred to this aspect as a 'condition' and added that it had no relation to the criteria listed in respect of the economically most advantageous tender test.[881] In particular this last statement caused more disputes in the aftermath, which were clarified in another judgement of

875 EEC, Council Directive 93/37 of 14 June 1993 concerning the coordination of procedures for the award of public works contracts, [1993] O.J. L 199/54 [Directive 93/37].

876 EEC, Council Directive 93/36 of 14 June 1993 coordinating procedures for the award of public supply contracts, [1993] O.J. L 199/1 [Directive 93/36].

877 EEC, Council Directive 93/38 of 14 June 1993 coordinating the procurement procedures of entities operating in the water, energy, transport and telecommunications sectors, [1993] O.J. L 199/84 [Directive 93/38].

878 EEC, Council Directive 92/50 of 18 June 1992 relating to the coordination of procedures for the award of public service contracts, [1992] O.J. L 209/1 [Directive 92/50].

879 ECJ, *Gebroeders Beentjes BV v The Netherlands*. 31/87, [1988] E.C.R. 4635 at para 27 [Beentjes].

880 *Ibid.* at para 19.

881 *Ibid.* at para 28.

the ECJ 12 years later.[882] Even the Commission tried to ignore the Beentjes judgement and intended to apply it according to its own understanding of how public contracts should be awarded. While some authors held the inclusion of social criteria in the distribution process as valid according to the Beentjes judgement, the Commission wanted to introduce these aspects as conditions of contract, but not as criteria in the allocation procedure.[883] The winning bidder could be thus only obliged to conclude a contract with specific conditions referring to social aspects. The Commission further suggested that any bidder, who had infringed or did not comply with national regulations in the areas of environmental protection or social policy in the past, could be precluded from a public procurement.[884] Some authors also simply considered the Beentjes judgement as a lapse of the ECJ that should not be given any further attention. The confusion caused was finally clarified, when the ECJ had to decide another case dealing with public procurement This case also included the question of the legality of social criteria in the allocation process. In the Nord-Pas-de-Calais judgement, the Court set out explicitly that the directives on public procurement did not prevent the public entity awarding a contract from taking social criteria into account, if they were in accordance with the principles of EC law.[885] For this purpose it was necessary that the applicable criteria did not lead to discrimination and that they were made public before the decision on the granting of a public contract was taken, so the bidders knew about the conditions of the tender and could adapt their proposals accordingly.[886] The ECJ explicitly disagreed with the Commission, who believed that the Beentjes judgement allowed to introduce social criteria only as conditions of contract, but not as criteria in the distribution process.[887] This second landmark judgement on public procurement caused again vivid discussions. Some authors welcomed the ruling as clarifying once and for all that social criteria were permissible in tenders on the award of public contracts. Especially women applauded the judges for their ruling, since this would give a blessing to various existing national laws that foresaw advantages in the award of public

882 See overview in Christoph Benedict, 'Anmerkung zu EuGH Rs. C-225/98' (Remarks on ECJ decision C-225/98) (2000) *Europäisches Wirtschafts – und Steuerrecht* (*Europ. J. of Commercial and Tax L.*) at 514 [Benedict].

883 *Ibid.*; Meinrad Dreher, 'Anmerkung zu EuGH, Rs. C-225/98' (Remarks on the ECJ Decision C-225/98) (2000) *Juristenzeitung* (*Lawyer's Journal*) 140, 141 [Dreher]; EC, Commission, *Public procurement in the European Union* of 11 March 1998 [1998] Bulletin of the European Union 3 1.2.42, at 27 [Commission, public procurement]; EC, Commission, *Community law applicable to public procurement and the possibilities for integrating social considerations into public procurement (COM (2001) 566 final)* of 15 October 2001 [2001] O.J. 333 C 27 at 3, 10, 18f. [Commission, Interpretative Communication].

884 Commission, Interpretative Communication, *ibid.* at 11, 12.

885 ECJ, *Commission v France,* C-225/98, [2000] E.C.R. I-7445 at para 50 [Nord-Pas-de-Calais].

886 *Ibid.* at paras 51, 73

887 *Ibid.* at paras 52, 54; Ursula Rust, 'Sozialpolitische Kriterien in landesgesetzlichen Vergabevorschriften' (Sociopolitical Criteria in National Public Procurement Regulations) (2001) *Neue Justiz* (*New Justice*) 113–116 [Rust].

contracts for companies that promoted women in their working life.[888] Others criticized the ECJ and described its verdict as 'tremendously wrong' [889] It was argued that the provisions regarding the applicable criteria with respect to the award of public contracts explicitly did not contain any criteria that were possibly alien to the subject matter of the contract and thus only those criteria should be taken into account that had a link with the subject matter of the contract.[890] The Commission argued that the criteria exemplary listed in the respective directives did not provided for any social criteria and thus concluded that the latter kind of criteria was not permitted in a public procurement.[891] According to the Commission it might be permissible for social criteria to be considered in some cases, but only if they were linked to the subject-matter of the respective contract in question.[892] This might be, for example, the case with criteria referring to environmental protection.[893] Furthermore, it was added that the inclusion of criteria into the allocation process alien to the subject-matter would contravene the objective of the directives on public procurement. The main goal of these acts was to award public contracts based on the best price/quality ratio. Additional socio-political criteria had the contrary effect and rendered the offers more expensive and thus less beneficial.[894] Only shortly after the Pas-Nord-de-Calais judgement, the ECJ delivered another verdict in the area of public procurement. In the Concordia judgement a bus company complained about the tender procedure conducted in Finland for the award of a public contract to provide urban bus transport services.[895] It mainly criticized the inclusion of ecological criteria into the selection process. The ECJ firstly underlined again that the list of criteria with respect to the economically most advantageous tender set out in the respective directives on public procurement was not exhaustive.[896] Secondly, the judges explicitly allowed for ecological aspects to be taken into account under the economically most advantageous tender test. The ECJ justified this criterion by referring to Article 6 EC arguing that it required the integration of environmental protection measures in the other policies of the Community.[897] It shall be stressed that the Commission and some Member States concurred with the ECJ as far as this reasoning was concerned. The Finish Government argued in this respect that ecological aspects needed to be included under the economically most advantageous tender test, because increased pollution and emissions caused by urban traffic incurred damages to the health of the inhabitants. Any impairment of the health of the citizens increased the expenses of the state for the services it subsequently had to provide in the medical sector.[898]

888 Rust, *ibid.* at 113ff.

889 Dreher, *supra* note 883 at 140.

890 *Ibid.*

891 Commission, Interpretative Communication, *supra* note 883 of at 15.

892 *Ibid.* at 16.

893 Dreher, *supra* note 883 at 140.

894 *Ibid.*

895 ECJ, *Formerly Stagecoach Finland Oy Ab v Helsingin kaupunki and HKL-Bussiliikenne Concordia Bus Finland Oy Ab*, C-513/99, [2002] E.C.R. I-7213 [Concordia].

896 Benedict, *supra* note 882 at 514; Concordia, *ibid.* paras 45, 54.

897 Concordia, *ibid.* at paras 45, 57.

898 *Ibid.* at para 46.

Taking ecological aspects into account that help reducing pollution and emissions is not only beneficial for the health of the citizens and the situation of the environment, but also has direct economical benefits for a state. The Commission had already provided in its communication on public procurement in the European Union of 1998 for the possibility to take into account environmental aspects, if they lead to economic advantages.[899] Irrespective of any economic benefits, the ECJ however followed the proposal of its Advocate General and allowed for the inclusion of ecological criteria in the allocation procedure, even if they did not provide for any economic advantages.[900] The third aspect that is of high interest with respect to the judgement, is the conviction of the Austrian, the British and the Swedish Government as well as of the Commission that the criteria applied in a public procurement must be directly linked to the subject matter of the respective contract.[901] The Commission tried to introduce a subject matter link since 1998.[902] Contrary to the opinion of its Advocate General[903] the ECJ concurred with this approach. The Court argued that 'since a tender necessarily relates to the subject matter of the contract, it follows that the award criteria which may be applied in accordance with that provision must themselves also be linked to the subject-matter of the contract'.[904] Thereby, the judges explicitly adopted the wording of the Commission's observation delivered in the proceedings.

The subject matter link introduced by the Court of Justice in the Concordia judgement is of particular interest. The question is, whether this constitutes an additional principle, according to which criteria applied in public procurements must be linked to the subject matter of the respective public contract or whether it is a condition that was only specifically applicable in the cases decided by the ECJ. To answer this question, the applicable provisions in the directives in force and dealing with public procurements at the time when the ECJ delivered its judgement, shall be analysed. As far as the regulations with respect to the economically most advantageous tender test were concerned, the respective provisions of the different directives were very similar. Differences existed only insofar, as the criteria specifically applicable were not totally the same, which is not necessarily surprising as the corresponding directives referred to different kinds of subject matters for which public contracts were awarded. Furthermore, with regard to a possible subject matter link, the wordings of the respective provisions were not the same. Article 36(1)(a) of Directive 92/50 said that 'various criteria relating to the contract' could be taken into account. Article 34(1)(a) of Directive 93/38 suggested that 'various criteria depending on the contract in question' should be set up. Article 26(1)(b) of Directive 93/36 and Article 30(1)(b) of Directive 93/37 prescribed that 'various criteria according to the

899 Commission, Public Procurement, *supra* note 883 at 27; Commission, Interpretative Communication, *supra* note 883 at 14, 16.

900 ECJ, *Opinion of Advocate General Jean Mischo of 13 December 2001*, [2002] E.C.R. I-7215 at I-7238f. [Opinion of AG Mischo]; Concordia, *supra* note 895 at paras 55f.

901 Concordia, *ibid.* at paras 49, 52.

902 Commission, Public Procurement, *supra* note 883 at 27; Commission, Interpretative Communication, *supra* note 883 at 14f.

903 Opinion of AG Mischo, *supra* note 900 at I-7240.

904 Concordia, *supra* note 895 at paras 59, 64.

contract in question' were to be chosen. However, none of the provisions referred to the subject matter of the contract, but only to the contract itself. One could interpret the references to 'contract' as a subject matter link. However, such an interpretation is not convincing for two reasons. First, the reference to 'contract' was included to express that the criteria applicable depend on the nature of the contract in question. This is further substantiated by the fact that the subsequently mentioned criteria were only of an exemplary nature, as can be seen from the inclusion of the term 'such as' before the listed criteria. The second reason, why the references formerly used did not provide for a subject-matter link, is the fact that in the new directives on public procurements the corresponding provisions have a different wording. Article 53(1)(a) of Directive 2004/18 and Article 55(1)(a) of Directive 2004/17 explicitly set out that 'various criteria linked to the subject-matter of the public contract in question' shall be applied. When the Commission presented its draft for the above-mentioned directives, with respect to the allocation procedure it mostly took over the provisions from the directives in force at that time.[905] As far as the criteria for the economically most advantageous tender were concerned, the Commission's proposal did not provide for any additional socio-political aspects apart from the possibility to take into account environmental characteristics of the tender. The only significant amendment was the inclusion of an explicit subject matter link into the catalogue of the criteria for the economically most advantageous tender.[906] The Commission's proposal was highly disputed during the legislative proceedings. In respect of the applicable criteria, the Economic and Social Committee criticized that the draft did not provide for any socio-political criteria.[907] The European Parliament suggested to explicitly include the equal treatment policy of an entity as a criterion, under the economically most advantageous tender test.[908] While the European Parliament's Committee on Legal Affairs and Internal Market suggested the deletion of the subject-

---

905 See e.g. EC, Proposal for a Directive of the European Parliament and of the Council on the coordination of procedures for the award of public supply contracts, public service contracts and public works contracts [2001] O.J. C 29 E/11 [Proposal Directive on public contracts]; the following reference to the legislative history will concentrate on the proceedings in respect of the adoption of Directive 2004/17.

906 *Ibid.* at 32.

907 EC, Economic and Social Committee, 'Proposal for a Directive of the European Parliament and of the Council on the coordination of procedures for the award of public supply contracts, public service contracts and public works contracts' of 26 April 2001 [2001] O.J. 193, online: European Parliament <http://www2.europarl.eu.int/omk/sipade2?SAME_LEVEL=1&LEVEL=0&NAV=X&DETAIL=&PUBREF=-//EP//TEXT+REPORT+A5-2001-0378+0+DOC+XML+V0//EN> (date accessed: 4 March 2005) 7, 8.

908 EC, European Parliament, 'Position of the European Parliament adopted at first reading on 17 January 2002 with a view to the adoption of European Parliament and Council Directive .../.../ EC on the coordination of procedures for the award of public supply contracts, public service contracts and public works contracts' of 17 January 2002 [2002] O.J. 271 E 176 at 37 [EP Position]; see also EC, European Parliament's Committee on Legal Affairs and the Internal Market, 'Report on the proposal for a directive of the European Parliament and of the Council on the coordination of procedures for the award of public supply contracts, public service contracts and public works contracts (COM (2000) 275 – C5-0367/2000 – 2000/0115 (COD))' of 29 October 2001 [2001] at 68f. [EP Committee Report].

matter link[909], the European Parliament did not adopt this proposal.[910] In its opinion on the amendments suggested by the European Parliament, the Commission refused to accept the Parliament's notion of a criterion referring to the equal treatment policy of an entity. Such a criterion would favour companies for reasons that had nothing to do with their proposals and thus would distort the allocation procedure and the state or public entity awarding a contract would be maybe obliged to accept a contract that did not provide for the best price/quality ratio, which originally it the main objective of a public procurement.[911] The Council's opinion on the draft proposal contained a very interesting grammatical deviation from the Commission's proposal. Although, as far as the content of the respective provision was concerned, the Council agreed with the Commission in respect of the listed criteria, it changed the wording of the subject-matter link from 'various criteria directly linked to the subject of the public contract in question,' as was suggested by the Commission, to 'various criteria justified by the subject of the public contract in question'.[912] However, this suggested amendment was not adopted and the original wording of the subject matter link as foreseen by the Commission in it proposal was maintained. The same is true for any socio-political criteria. The new directives did not provide for any such criteria with the only exception being the reference to the environmental characteristics of the subject of the tender.[913] Apart from that, the lists of suggested criteria are exactly the same as they were in the previous public procurement directives. The inclusion of environmental aspects in the list of permissible criteria was justified, among others, by the obligation of the Community laid down in Article 6 EC to integrate environmental protection requirements into the implementation of the other policies listed in Article 3 EC.[914] *De lege lata*, when awarding public contracts, no socially motivated criteria can be used in the allocation procedure and the criteria applied must be linked to the subject matter of the contract. Although under current EC law an explicit reference to the subject matter of a contract is laid down, this was not the case when the ECJ delivered its Concordia judgement. A subject matter link was not incorporated in any of the regulations on public procurement being in force at that time. The acknowledging of a subject matter link by the ECJ in the Concordia judgement can be thus only justified, when it is regarded as a generally applicable principle under public procurement law. Under these circumstances, the incorporation of the subject-matter link in the new Directives 17/2004 and 18/2004 would be only of declaratory nature. This understanding is confirmed by the fact that under the

---

909 EP Committee Report, *ibid.* at 68f.

910 See EP Position, *supra* note 908 at 212.

911 EC, Amended Proposal for a European Parliament and Council Directive Concerning the Coordination of Procedures for the Award of Public Contracts, Public Service Contracts and Public Works Contracts, [2002] O.J. C 203 E/210 at 238f.

912 EC, Council, 'Common Position (EC) No 33/2003 adopted with a view to adopting Directive 2003 / ... / EC of the European Parliament and of the Council on the coordination of procedures for the award of public works contracts, public supply contracts and public service contracts' of, 20 March 2003 [2003] O.J. C 147 E 1 at 33.

913 See Directive 2004/18, *supra* note 874, Article 53 (1) (a); Directive 2004/17, *supra* note 873, Article 55 (1) (a).

914 Directive 2004/18, *ibid.*, recital 5 of the preamble.

first recital in the preambles of the current directives on public procurements it is declared that the directives are also based on case law of the European Court of Justice and in particular on case law on award criteria.

The existence of an unwritten principle, according to which criteria used in a procedure to award public contracts must be linked to its subject matter, could have an impact for the procedure of the distribution of air traffic rights, if this procedure is also considered as a public procurement. To answer this question it is first necessary to have regard to Directive 17/2004, which refers to the procurement procedures of entities operating in the water, energy, transport and postal service sectors. According to Article 31 of this Directive it is applicable to those service contracts that are listed in Annex XVIIA. Annex XVIIA Category 3 mentions 'air transport services of passengers and freight except transport of mail' and Annex XVIIA Category 4 includes the 'transport of mail by land and by air'. Insofar the distribution of air traffic rights could fall under the scope of Directive 17/2004. This, however, is not the case. Recital four of the preamble of Directive 17/2004 sets out that through the adoption of different EC regulations it has been undertaken to introduce a higher level of competition in the air transport sector and to ensure the attainment of this objective, it is held as inappropriate to include air carriers in the scope of this directive. Article 5(1) of Directive 17/2004 further limits the scope of the directive to the 'transport by railway, automated systems, tramway, trolley bus, bus or cable'. Air transport services are not included in this enumeration. Thus, Directive 17/2004 is not applicable on the distribution of air traffic rights. The more general Directive 18/2004 provides for a similar scheme as Directive 17/2004. Article 1(2)(d) of Directive 18/2004 defines public service contracts as 'public contracts other than public works or supply contracts having as their object the provision of services referred to in Annex II'. Annex IIA Categories 3 and 4 contain the same reference to air transportation services as Directive 17/2004 and one could ask again, whether the distribution of air traffic rights could be regarded as the award of a public contract to provide air transportation services under Directive 18/2004. Unlike with respect to Directive 17/2004, this hypothesis cannot be rejected that easily. Recital 20 of the preamble of Directive 18/2004 provides that contracts in the area of transport generally fall into the scope of Directive 17/2004. Article 12 of Directive 18/2004 further stipulates that only contracts awarded by contracting authorities exercising one of the activities described in Articles 3 to 7 of Directive 17/2004 are excluded from the application of Directive 18/2004. The activities referred to in Directive 17/2004 include the provision or supply of gas, electricity, heat, water, transport services, postal services, services on airports and maritime or inland ports as well as the exploitation of geographical areas for the exploration and extraction of mineral resources. Since the provision of air transportation services is explicitly not included in Articles 3 to 7 of Directive 17/2004, it could mean that this kind of services is not exempted from the scope of Directive 18/2004. Although such a result would be in accordance with the regulation included in Article 12 of Directive 18/2004, it is not convincing that the distribution of air traffic rights for the provision of air transportation services was intended to fall under the scope of Directive 18/2004. Firstly, recital 20 of the preamble of Directive 18/2004 extends the scope of this directive in the transport sector explicitly only to maritime, coastal and river transport. Taking into account that regulations on maritime transportation and on air transportation are adopted under the same provision in the EC Treaty, that is Article 80(2) EC, both means of transportation are, from a legal perspective, comparable. The explicit inclusion of maritime transportation under the scope of Directive 18/2004 without mentioning air transportation at the same time, is an indication that the latter means of transportation should not be included in the applicability of this directive. Furthermore, pursuant to its Article 17, Directive 18/2004 is not applicable, when service concessions are granted. According to the definition in Article 1(4) of Directive 18/2004 a service concession 'is a contract of the same type as a public service contract except for the fact that the consideration for the provision of services consists either solely in the right to exploit the service or in this right together with payment.' Air traffic rights allocated to air carriers usually provide for the right of the air carriers to receive compensation for the services they provide directly from the customers that make use of the air transport services. Hence, air traffic rights granted by a state to an airline are to be classified as service concessions. Under these circumstances, the distribution of air traffic does not fall under Directive 18/2004. The only EC act applicable on the allocation of air traffic rights remains Regulation 847/2004 and in particular its Article 5.

It remains now to determine, how the subject matter link established by the ECJ can influence the distribution of air traffic rights. In the Concordia judgement, the ECJ referred to the former directives regulating public procurement and explained that 'those directives taken as a whole constitute the core of Community law on public contracts and are intended to attain similar objectives in their respective fields'.[915] Thus, according to the ECJ the directives on public procurement establish a coherent EC law on the award of public contracts. As a result, the principle of a subject matter link must be applied equally to all awards of public contracts, no matter under which Directive they fall. Under these circumstances it shall be scrutinized, whether Regulation 847/2004, which contains a regulation in its Article 5 on the allocation of air traffic rights, could be also part of this specific Community law on public contracts. In this respect it can be firstly observed that Directives 17/2004 and 18/2004 as well as Regulation 847/2004 require that the award of public contracts or the distribution of air traffic rights respectively is based on the principles of non-discrimination and transparency. These two principles lie, according to the ECJ, at the very heart of the public procurement directives.[916] Secondly, public contracts for the provision of urban transportation services can be awarded through tenders based on Directive 17/2,004 or 18/2004. In principle there is no difference between air transportation and bus transportation services, apart from the means used and the distances that are covered by the two types of services. Both types of services are part

915 Concordia, *supra* note 895 at paras 90, 92.

916 Juliane Hilf, in: Hans von der Groeben and Jürgen Schwarze, eds. (2003), *Vertrag über die Europäische Union und Vertrag zur Gründung der Europäischen Gemeinschaft* (*Treaty on the European Union and Treaty establishing the European Community*), vol. 3, 6th edn. (Baden-Baden, Germany: Nomos Verlagsgesellschaft) Article 163 EC at para 26; Concordia, *ibid.* at paras 81, 92; Directive 2004/18, *supra* note 874, Article 2; Directive 17/2004, *supra* note 873, Article 10.

of the transportation system of a state and the right to undertake these activities and offer the corresponding services are conferred to the respective market players by the states. Bus companies need concessions to offer public transportation services. The same is true for airlines. With respect to international air transportation these concessions are called air traffic rights. Furthermore, states used to, and partially still do, provide air and bus transportation services by themselves. Transportation services have been partially regarded as part of the infrastructure of a country in the first place and only on a secondary level also as an economical branch of their own. Gradually, states have privatized their formerly state-owned airlines first and are now continuing this process by outsourcing bus transportation services to private entities. These similarities between bus transportation and air transportation services could suggest to treat both industries equal, when it comes to market access. Indeed, there is no immediate reason why criteria for the allocation of rights to provide urban transportation services shall be necessarily linked to the subject matter, while in the allocation of air traffic rights the possibility shall be provided to include criteria that are alien to the provision of air transportation services. Therefore, Directive 847/2004 shall be also be also considered as part of Community law on public procurement limited to international air transportation services. Under these circumstances the unwritten principle of a subject-matter link is also applicable on the allocation of air traffic rights. Hence, when allocating air traffic rights pursuant to Article 5 of Regulation 847/2004, the Member States shall not apply criteria that are alien to the provision of air transportation services.

ENVIRONMENTAL PROTECTION

As was shown above, in a judgement dealing with the procedure for the award of public service contracts, the European Court of Justice ruled that, when deciding about the economically most advantageous tender, the scope of environmental protection could be one of the criteria, upon which the decision was based.[917] In another judgement with respect to the common agriculture policy, the Court concluded that, when implementing an EC regulation, Member States could favour farmers that applied ecological methods.[918] Finally the ECJ also ruled that in the common transport policy economical, social and ecological problems must be taken into account.[919] In

917 Concordia, *supra* note 895 at para 55ff.

918 ECJ, Karlsson et al., C-292/97, [2000] E.C.R. I-2737 at para 51.

919 ECJ, *Commission v Federal Republik of Germany*, and C-195/90, [1992] E.C.R. I-3141 at para 29 [*Commission v Germany*]; see also ECJ, *Pinaut Wieger Spedition GmbH v Bundesanstalt für den Güterfernverkehr*, C-17/90, [1991] E.C.R. I-5253 at para 11 [Pinaut Wieger]; see also Pieper, in: Albert Bleckmann, ed. (1997), *Europarecht* (*European Law*), 6th edn. (Cologne, Berlin, Bonn, Munich: Carl Heymanns Verlag) at para 2838 [Pieper, European Law]; Christian Calliess, 'Die neue Querschnittsklausel des Article 6 ex 3c EGV als Instrument zur Umsetzung des Grundsatzes der nachhaltigen Entwicklung' (The New Integration Clause of Article 6 ex 3c EC as an Instrument to Implement the Principle of Sustainable Development) (1998) *Deutsches Verwaltungsblatt* (*German J. of Admin. L.*) 559 at 567 [Calliess, Integration Clause].

general the European Court of Justice had a very environmental friendly position and declared environmental protection as a fundamental objective of the EC to be pursued within the other policies already in 1985.[920] Hence, environmental protection has been an objective of EC law for a long time and since the adoption of the Single European Act and the Treaty of Maastricht it is explicitly laid down in the EC Treaty. Under current EC law Article 6 EC contains an obligation to include environmental protection requirements in the other Community policies listed in Article 3 EC. All policies shall be not only developed according to their specific objectives, but shall also contribute to environmental protection.[921] Any environmental policy is doomed to fail, if in other areas like the transport, agriculture or energy policy, no regard is given to environmental issues.[922] This is especially true in regard of policies that serve the establishment of the internal market. The increasing use of the market freedoms, in particular the right to free movement of goods and the right to free movement of services, leads to an increase in traffic and transportation activities, which finally results in more pollution and renders the achievement of environmental protection and sustainable development, another objective of the EC pursuant to Article 2 EC, more and more difficult.[923] Article 6 EC intends to stop this vicious circle by prescribing that all policies shall contribute to environmental protection.[924] Insofar Article 6 EC, which is commonly referred to as the integration clause, is maybe the

---

920 Jutta Jahns-Böhm and Siegfried Breier, 'Die umweltrechtliche Querschnittsklausel des Article 130r II 2 EWGV', (The environmental integration clause of Article 130r II 2 EEC treaty), (1992) *Europäische Zeitschrift für Wirtschaftsrecht* (*Europ. J. of Commercial L.*) at 49 [Jahns-Böhm and Breier]; see also Commission v. Germany, *ibid.* at para 29; Pinaut Wieger, *ibid.* at para 11.

921 Ludwig Krämer, in: Hans von der Groeben and Jürgen Schwarze, eds. (2003), *Vertrag über die Europäische Union und Vertrag zur Gründung der Europäischen Gemeinschaft* (*Treaty on the European Union and Treaty establishing the European Community*), vol. 1, 6th edn. (Baden-Baden, Germany: Nomos Verlagsgesellschaft) Article 6 EC at para 12 [Krämer, in: Groeben/Schwarze, Article 6]; Christian Calliess, in: Christian Calliess and Matthias Ruffert, eds. (2002), *Kommentar zum EU-Vertrag und EG-Vertrag* (*Commentary to the EU Treaty and EC Treaty*), 2nd edn. (Neuwied, Kriftel, Germany: Luchterhand Verlag,) Article 6 EC at para 6 [Calliess, in: Calliess/Ruffert, Article 6]; Pieper, European Law, *supra* note 919 at para 2837; Martin Wasmeier, 'Die Integration des Umweltschutzes als allgemeine Auslegungsregel des Gemeinschaftsrechts' (The Integration of Environmental Protection as a General Interpretation Rule of Community Law) (2000) *Europäisches Wirtschafts – und Steuerrecht* (*Europ. J. of Commercial and Tax L.*) at 47f. [Wasmeier]; Astrid Epiney, 'Umweltrechtliche Querschnittsklausel und freier Warenverkehr: die Einbeziehung umweltpolitischer Belange über die Beschränkung der Grundfreiheit' (Environmental Integration Clause and Right to Free Movement of Goods: the Inclusion of Environmental Matters by Limiting a Market Freedom) (1995) *Nature und Recht* (*Environment and L.*) 497 at 501, 503 [Epiney]; Calliess, Integration Clause, *supra* note 919 at 565.

922 Epiney, *ibid.* at 497.

923 *Ibid.* at 498.

924 See in particular with reference to the transport policy Epiney, *ibid.* at 504; Wasmeier, *supra* note 919 at 47, 48; Dieter H. Scheuing, 'Umweltschutz auf der Grundlage der Einheitlichen Europäischen Akte' (Environmental Protection based on the Single European Act) (1989) *Europarecht* (*Europ. L. Rev.*) 152–176, 179.

most important provision regarding the regulation of environmental issues, although it itself does not provide for any legislative or administrative powers. Article 6 EC is binding for the European Community as well as for the Member States. The obligation of the Member States can be derived from the fact that Article 6 EC refers to the 'implementation of the Community policies,' which can be understood as 'execution of EC law'. Considering that there are only few areas, where the European Community directly executes EC law, the scope of Article 6 EC with respect to the implementation alternative would be very limited. The *effet-utile* principle requires to extend the scope of Article 6 EC also to the Member States when they implement EC law.[925] Additionally, since Article 6 EC is included in the title on the principles of the European Community, it must be taken into account when EC law is applied.[926] Thus, primary and secondary law shall be construed in a way that environmental protection objectives are sufficiently regarded. This means that, when air traffic rights are distributed pursuant to Article 5 of Regulation 847/2004, environmental protection requirements shall be considered.

To determine, what is understood by 'environmental protection requirements' one has to refer to the objectives of the environmental policy set out in Articles 2, 3 and 174(1), (2) EC.[927] Firstly, the EC shall preserve, protect and improve the quality of the environment. This means that present pollutions shall be reduced and future burdens of the environment avoided, as well as existing damages inflicted on the environment shall be repaired and any lost quality of the environment has to be restored.[928] Article 174(1) EC further sets out that the protection of human health is a particular aspect of the environmental policy and protection of the environment always contributes to the protection of human health. Another objective of an EC environmental policy is to promote a prudent and rational use of natural resources. The term 'resources' covers all kinds of natural goods available, the perishable as well as the inexhaustible ones, thus including for example air or oil.[929] When

925 Calliess, in: Calliess and Ruffert, Article 6, *supra* note 921 at para 12; Siegfried Breier and Hendrik Vygen, in: Carl-Otto Lenz and Klaus-Dieter Borchardt, eds. (2003), *EU – und EG-Vertrag* (*EU and EC Treaty*), 3rd edn. (Cologne: Bundesanzeiger) Article 6 EC at para 7 [Breier and Vygen, in: Lenz/Borchard, Article 6]; Calliess, *Integration Clause*, *supra* note 919 at 566.

926 Wasmeier, *supra* note 919 at 47f.; Epiney, *supra* note 921 at 501.

927 Krämer, in: Groeben/Schwarze, Article 6, *supra* note 921 at para 6; Calliess, in: Calliess/Ruffert, Article 6, *supra* note 921 at para 5; Breier and Vygen, in: Lenz and Borchardt, Article 6, *supra* note 925 at para 5; Epiney, *ibid.* at 499; Jahns-Böhm and Breier, *supra* note 920 at 50; Calliess, *Integration Clause*, *supra* note 919 at 565.

928 Calliess, in: Calliess/Ruffert, Article 6, *supra* note 921 at para 8; Siegfried Breier and Hendrik Vygen, in: Carl-Otto Lenz and Klaus-Dieter Borchardt, eds. (2003), *EU – und EG-Vertrag* (*EU and EC Treaty*), 3rd edn. (Cologne: Bundesanzeiger) Article 174 EC at para 5 [Breier/Vygen, in: Lenz/Borchard, Article 174]; Calliess, Integration Clause, *ibid.* at 562.

929 Christian Calliess, in: Christian Calliess and Matthias Ruffert, eds. (2002), *Kommentar zum EU-Vertrag und EG-Vertrag* (*Commentary to the EU Treaty and EC Treaty*), 2nd edn. (Neuwied, Kriftel, Germany: Luchterhand Verlag) Article 174 EC at para 10 [Calliess, in: Calliess/Ruffert, Article 174]; Ludwig Krämer, in: Hans von der Groeben and Jürgen Schwarze, eds. (2003), *Vertrag über die Europäische Union und Vertrag zur Gründung*

pursuing the goals of the environmental policy, Article 174(2) EC provides for some guiding principles. These principles can be also referred to, to determine the meaning of 'environmental protection requirements'. The precautionary principle obliges the EC to undertake measures to reduce existing risks for the environment and to ensure a gentle use of natural resources. The principle of preventive action requires that, where an imminent danger for the environment exists, the Community shall take immediate action to prevent any damage to the environment.[930] Finally under the polluter-pays principle, costs for the pollution of the environment shall be burdened on the person or entity that is responsible for causing the respective pollution. The intention of this principle is to relieve the general public from the costs necessary for cleaning the environment and to increase the awareness for the use of the environment. Through this mechanism the polluter shall have a financial incentive to reduce its pollutions and to develop technologies or procedures that are more environmental friendly.[931]

As far as air transportation in general and aircraft operation in particular are concerned, several burdens for the environment are caused. Firstly, aircraft engines emit gases that have a negative impact on the atmosphere, contributing to the greenhouse effect and the depletion of the ozone layer.[932] Secondly, the operations of aircrafts require kerosene, which is a limited and depletable resource. Thirdly, approaching and departing aircrafts cause noise, which can have an impact on the health of persons living in the vicinity of airports. Having regard to the objectives and principles under the environmental policy, the Member States and the Community are obliged to fight these negative side effects. For this purpose it is suggested that, when air traffic rights are allocated, those airlines shall be preferred, which use more modern aircrafts with lower noise and gas emission levels and a high fuel efficiency.[933] Alternatively, the size of the aircraft the airlines intend to operate on a route and the frequency of proposed flights might be also taken into regard. Although bigger aircrafts usually have a higher nominal noise and gas emission as well as fuel consumption levels, their use may result in fewer flights and thus a lower overall noise and gas emission, and fuel consumption levels calculated on a per passenger basis for a certain period.[934] Such an approach would be also in accordance with the polluter-pays-principle. To fulfil the three criteria, an airline will have to invest in new and modern aircraft models, which usually are more expensive than older types. It is only consequent to benefit and allocate air traffic rights to an

---

der Europäischen Gemeinschaft (*Treaty on the European Union and Treaty establishing the European Community*), vol. 3, 6th edn. (Baden-Baden, Germany: Nomos Verlagsgesellschaft) Article 174 EC at para 3 [Krämer, in: Groeben/Schwarze, Article 174]; Calliess, Integration Clause, *ibid.*; Breier and Vygen, in: Lenz/Borchard, Article 174, *ibid.* para 9.

930 Calliess, in: Calliess/Ruffert, Article 174; *ibid.* para 28f; Breier and Vygen, in: Lenz/Borchardt, Article 174; *ibid.* at para 14; Calliess, Integration Clause, *ibid.* at 563.

931 Breier and Vygen, in: Lenz/Borchardt, Article 174; *ibid.* at para 16; Krämer, in: Groeben/Schwarze, Article 174, *supra* note 929 at para 51.

932 Commission, Sustainable Development, *supra* note 576 at 5, 8, 41f.

933 See also *ibid.* at 9, 11.

934 See also similarly with respect to the allocation of slots Commission, Sustainable Development, *ibid.* at 23.

airline, which has incurred higher expenses to lower the impact on the environment. It would be contradictory to favour air carriers, which use older technologies, thus have lower expenses and, at the same time, cause more exhaust gases and noise as well as require more kerosene. To comply with the polluter-pays principle, the air carrier causing more pollution to the environment must incur higher costs. This can be accomplished by withholding business opportunities from such an airline, which finally will result in less profit. The Commission describes this approach as 'reward the best – punish the worst'.[935]

FOSTERING COMPETITION

As was shown in the previous chapter on the distribution of air traffic rights in other countries, it was sometimes chosen to allocate these rights to foster competition. In particular in the US airlines were granted international routes in order to grow and to be able to compete effectively with other US airlines. The US Government intended to establish a sound air transportation industry by increasing competitive pressure on the airlines. The question is, whether such an approach would be also possible and feasible in the European Community. One principle that could preclude such an approach is the principle of open market and free competition set out in Articles 4, 98 EC. Article 4 EC imposes a legal obligation on the Member States and the EC to develop their economic policies in accordance with the principle of open market and free competition.[936] This principle that commercial activities are undertaken in a competitive surrounding. It precludes the Member States and the EC from introducing a planned economy. The reason for the introduction of the principle of open market and free competition is given in Article 98 s. 2 EC: it favours the efficient allocation of resources.[937] Furthermore, pursuant to Article 4 EC an open market is expected to help attaining the objectives laid down in Article 2 EC. The distribution of air traffic rights based on grounds of competition could not only be incompatible with the principle of open market and competition, but also contravene several objectives of the EC. Firstly, if air traffic rights are considered

935 *Ibid.* at 11.

936 Helmut Wittelsberger, in: Hans von der Groeben and Jürgen Schwarze, eds. (2003), *Vertrag über die Europäische Union und Vertrag zur Gründung der Europäischen Gemeinschaft* (*Treaty on the European Union and Treaty establishing the European Community*), vol. 1, 6th edn. (Baden-Baden, Germany: Nomos Verlagsgesellschaft) Article 4 EC at paras 1, 5 [Wittelsberger, in: Groeben/Schwarze, Article 98]; Ulrich Häde, in: Christian Calliess and Matthias Ruffert, eds. (2002), *Kommentar zum EU-Vertrag und EG-Vertrag* (*Commentary to the EU Treaty and EC Treaty*), 2nd edn. (Neuwied, Kriftel, Germany: Luchterhand Verlag) Article 98 EC at para 4 [Häde, in: Calliess/Ruffert, Article 98]; Rüdiger Bandilla, in: Eberhard Grabitz and Meinhard Hilf, eds. (January 2000), *Das Recht der Europäischen Union* (*The Law of the European Union*), looseleaf (Munich: Verlag C.H. Beck, 15th supplement), Article 4 EC at para 3 [Bandilla, in: Grabitz/Hilf, Article 4].

937 Wittelsberger, in: Groeben and Schwarze, Article 98, *ibid.* para 11; Häde, in: Calliess/Ruffert, Article 98; *ibid.* para 4; Bandilla, in: Grabitz/Hilf, Article 4, *ibid.* at para 5.

as a resource[938], contrary to Article 98 s. 2 EC their most efficient allocation may be precluded. Granting these rights for reasons of competitive opportunity and not by considering which airline could use them most efficiently, may result in a waste of this resource. Secondly, an allocation of air traffic rights not to the most efficient air carriers, could impair the achievement of a high degree of competitiveness and convergence of economic performance, which is an objective of the EC pursuant to Article 2 EC. Distributing air traffic rights to less competitive air carriers would not only keep them alive artificially, but also impair the more efficient airlines from a stronger and faster development.[939] Both results hamper the development of a more competitive economy in the EC. On the other hand, it must be also taken into account that according to Article 2 EC a harmonious, balanced and sustainable development of economic activities throughout the EC shall be promoted. This could require to prevent the disappearance of too many airlines and the possible emergence of a monopoly or oligopoly. Especially the importance of aviation as part of the infrastructure may oblige the EC and the Member States to ensure that a healthy air transportation structure is maintained.[940] On the other hand, taking into regard that the main means of production in the air transportation sector, aircrafts, are extremely mobile, a danger for the economy of a Member State or even the whole European Community as result of the disappearance of some airlines resulting from competition is not considered as very likely. Due to the high number of air carriers in the EC, it can be expected that the loss of transportation capacities in one Member State would be compensated quite fast by other airlines. Thus, the objective of ensuring a harmonious, balanced and sustainable development of economic activities does not require the EC to allocate air traffic rights to specific air carriers to keep them alive. Additionally, the difficulty would exist under which circumstances an airline would be considered as being in need of air traffic rights. In this respect the principle of open market and free competition does not only require the EC and the Member States to omit any unnecessary interference in the market, but also demands equal treatment. Undistorted competition requires that all market players are treated equally.[941] This principle would be infringed, if some airlines would be favoured by the EC through the preferred allocation of air traffic rights. Thus, it is concluded that an allocation of air traffic rights with the objective to strengthen competition is not possible and precluded by the principle of open market and free competition.

---

938 See above at 164.

939 Jörg Ukrow, in: Christian Calliess and Matthias Ruffert, eds. (2002), *Kommentar zum EU-Vertrag und EG-Vertrag* (*Commentary to the EU Treaty and EC Treaty*), 2nd edn. (Neuwied, Kriftel, Germany: Luchterhand Verlag) Article 2 EC at para 15 [Ukrow, in: Calliess/Ruffert, Article 2].

940 Manfred Zuleeg, in: Hans von der Groeben and Jürgen Schwarze, eds. (2003), *Vertrag über die Europäische Union und Vertrag zur Gründung der Europäischen Gemeinschaft* (*Treaty on the European Union and Treaty establishing the European Community*), vol. 1, 6th edn. (Baden-Baden, Germany: Nomos Verlagsgesellschaft) Article 2 EC at para 9 [Zuleeg, in: Groeben/Schwarze, Article 2].

941 *Ibid.* at para 13; Ukrow, in: Calliess/Ruffert, Article 2, *supra* note 939 at para 11.

CONCLUSION

In the previous chapter it was examined, whether there are any provisions and principles in EC law that can influence the procedures for the allocation of air traffic rights. It was concluded that the allocation of air traffic rights is comparable to the award of public contracts and that Regulation 847/2004 is partially part of the Community law on public procurement. As a consequence, an unwritten principle is applicable, according to which the criteria chosen in the allocation procedure, must be linked to the subject-matter of the allocation, which is here the provision of international air transport services. Furthermore, when deciding on the distribution of air traffic rights, for reasons of environmental protection, attention shall be given to the noise and exhaust gas levels as well as the fuel efficiency of the aircrafts the applying airlines intend to use. Since aircraft operations have an impact on the environment, this criterion is also in accordance with the subject-matter link. Finally, when distributing air traffic rights, it is not possible under EC law to have regard to the competitive situation in the air transportation sector in the European Community.

Chapter 6

Proposal for a Procedure to Allocate Air Traffic Rights in the European Community

In this section, it will be undertaken to develop a comprehensive method for the allocation of air traffic rights. Although it is intended to adapt this method to the particular situation of the EC and the Member States, the following solution may be also applicable in third countries that have to allocate air traffic rights. As was already the case in the previous analysis, it will be concentrated in the following on the allocation of air traffic rights for passenger or passenger/cargo operations. Unlike passenger and passenger/cargo services, states grant each other seventh freedom traffic rights for all-cargo flight operations, which results in different strategies on how airlines structure their all-cargo flights.

BEAUTY CONTEST

GENERAL CONSIDERATIONS

As a result of a sectoral and state-based comparison, it was concluded in previous chapters that international air traffic rights shall be distributed through beauty contests. This procedure has the advantage that it usually results in the selection of the most appropriate applicant. On the downside it bears the danger that the selection can be influenced by subjective preferences of the persons in charge of the distribution on the side of the administration or that the applicants appear to be all equally well qualified. Both effects could be found in the allocation of telecommunication frequency licences as well as in the allocations of air traffic rights in various countries. As far as the allocation of air traffic rights in the EC and the Member States *de lege lata* is concerned, Regulation 847/2004 stipulates that it must be transparent and non-discriminatory. For the transparency condition to be fulfilled, it is therefore necessary that the criteria, upon which the distribution of international air traffic rights is based, are disclosed. Under these circumstances the danger exists that route allocations in the Member States, undertaken through a beauty contest, will produce the same disadvantages. Apart from any subjective preferences of the respective officials that may influence the distribution, due to the transparency criterion, the applying airlines may also adapt to the conditions of every distribution procedure and thus appear to be all equally well qualified to obtain the respective rights. The solution presented in the following will try to prevent the emergence of a situation, where the selection process would be burdened with these two obstacles. At the same time it is undertaken to develop a procedure, which would allow to be conducted in a short time, to enable the airlines to make use of the available air traffic rights as soon as possible. The most difficult part of the beauty contest is to determine the criteria that will govern the selection. According to Breyer, every regulator, facing this problem, has, at first, the desire to find criteria, which could enable an objective selection process. At the same time, it is also believed that a subjective influence will be always necessary, since a purely objective approach cannot take all particularities into regard that may arise in the different situations.[942] The result of this dilemma was shown in the state-based comparison: a multitude of criteria is introduced, to originally allow for an objective selection and to cover all kinds of different situations that can occur, as the conditions of each allocation depend very much on the underlying air service agreements. Before it is undertaken to develop the applicable criteria in the following, two general conditions to be applicable in all distribution procedures are proposed. The first measure suggested to minimize subjective influence, is to attribute a fixed weight to the applicable criteria. Secondly, it is believed that a regulator is neither adequately equipped to undertake extensive studies of markets, future expectations, traffic growth and so on, nor should such an analytical activity be imposed on a governmental agency or Ministry.[943] Therefore, the applicable criteria should be easy to compare and not require additional studies by the regulator. Not only would this render the selection procedure faster and less expensive on the administrative level, but has also the advantage that, if a losing applicant decides to contest the regulator's decision in court, the less complicated the distribution procedure on the administrative level is, the easier it is to prove, why a decision was finally taken in a certain manner. Any external factors, unrelated to the selection, could be then easier detected in a judicial review.[944]

CRITERIA

Classification of different air service agreements

The dilemma of the responsible civil aviation administrations is fully acknowledged that, due to the high number of different types of air service agreements, it is very difficult to establish one coherent procedure that covers all possible scenarios. In order to establish one single procedure, it might be more appropriate to have different procedures, depending on the features of the respective air service agreements. For this purpose, a classification of the different kinds of air service agreements with respect to the allocation of air traffic rights shall be undertaken. The allocation

942 Breyer, *supra* note 522 at 567; see also Kahn and Roach, *supra* note 580 (according to Kahn and Roach it is not at all possible to work out a certain scheme for the allocation of air traffic rights, at 106).

943 See also Kahn and Roach, *ibid.* 105f.

944 CAB Report, *supra* note 597 (the subcommittee reviewing the performance of the CAB advocated already in 1975 the development of clear, simple and coherent route award standards, at 618).

of air traffic rights mainly depends on three aspects contained in the air service agreements:

- designation possibilities;
- capacity; and
- frequency.

Regulations with respect to these three aspects can be either liberal or restrictive and the combination of these aspects, with the possibility of being either liberal or restrictive, allows for nine different types of air services agreements. However, usually the contracting states of an air service agreement take a coherent approach and regulate all of these aspects in the same manner, thereby concluding in total either a liberal or a restrictive agreement.[945] Under these circumstances, four different kinds of air service agreements can be finally established:

1. Single-restrictive

The air service agreement provides for a single designation possibility only. One or both of the two other aspects, capacity and frequency, are predetermined and restricted.

2. Single-liberal

The air service agreement provides for a single designation possibility. None of the other two aspects, capacity and frequency, are predetermined or restricted.

3. Multiple-restrictive

The air service agreement provides for a multiple designation possibility, but the number of airlines that can be designated is limited. One or both of the other two aspects, capacity and frequency, are predetermined and restricted.

4. Multiple-liberal

The air service agreement provides for a multiple designation possibility and the number of airlines that can be designated is unlimited. None of the other two aspects, capacity and frequency, are predetermined or restricted.

Interests of different user groups of air transportation services

In addition to a classification of the different air service agreements, the interests of different user groups of air transportation services shall be also taken into account. One category of users can be composed of business travellers. Usually they

945 See e.g. CAA Licensing, *supra* note 771 at 42; *contra* Werner Guldimann, 'A flexible multilateral in Traffic Rights?' (1993) *Ann. Air and Sp. L.* 125 (according to Guldimann, there are three different positions a government can take: a restrictive, an intermediate or a liberal one, at 125).

coordinate their flights with the operational needs and activities of their businesses. Decisions to undertake a flight are often taken on a short-term basis. This requires a lot of flexibility. Therefore, business travellers are interested in having a high choice among available flights. High frequency and high capacity (seat-availability) are their major concern. Another important aspect is route integration so that travels to more remote destinations are also possible. Fares, on the other hand, are less significant, since a company usually pays the travels and these expenses can be deducted from the taxable income.[946] Another group of travellers are tourists. Members of this group put more emphasis on the price of their tickets, since they cannot deduct these expenses from their taxable income. Furthermore, their flights are usually planned well in advance. Thus, capacity and frequency that provide for flexibility are less significant.[947] Route integration with respect to European destinations can be beneficial as well, but is less important than for business travellers. Since tourists do not travel as often as business travellers, they can be expected to undertake longer journeys to the respective European departure airports. Thereby it shall be taken into account that, firstly, unlike for example in the US or in Canada, most Member States in the EC have a very dense railway and highway network and, secondly, international airports are widespread and thus only few places will be found that are completely remote and isolated in Europe. Therefore, leisure travellers, who are more flexible and who usually plan their journeys in advance, may benefit from route integration, but this aspect has a lesser significance than with respect to business travellers. The predominant factor, as far as tourists are concerned, is the fare of a flight. Finally, the last group of users having an interest in air transportation services are companies, involved in the trade of goods. They are mostly interested in cargo services. As indicated above and since cargo services are increasingly dealt with in separate air service agreements, the interests of this group of users will not be taken particularly into account here. However, since airlines can operate combined passenger/cargo services, considerations with respect to cargo services will be also made, where this is possible.

The allocation of air traffic rights

In the following, the allocation of air traffic rights shall be discussed. Thereby, a distinction will be made between the four different possible types of air service agreements. Furthermore, the special interests of the two main user groups, business travellers and tourists, will be also taken into account.

Multiple-liberal agreement

The least complicated situation exists under a multiple-liberal air service agreement. Since no restrictions are laid down in the air service agreement with respect to designation, capacity and frequency, there is no reason to introduce any restrictions in the allocation process. It would be contradictory, if, on the one hand, a country concluded a liberal agreement that is intended to liberalize air transportation between

946 Giemulla'Schmid/Mölls, *supra* note 42 at para 6.

947 *Ibid.*; Matte, *supra* note 10 at 164.

the contracting states and, on the other hand, imposed limitations with respect to the entrance into this respective market, which it originally intended to liberalize. Hence, in this situation, every eligible air carrier applying for air traffic rights should be designated and granted all air traffic rights requested.

Multiple-restrictive agreement

The situation becomes more complicated, if an air service agreement is more restrictive. A first difficulty arises with respect to the number of air carriers to be designated, if capacity or frequency is limited. One solution would be, to designate the maximum number of airlines allowed under the agreement. This would allow for maximum competition and a high degree of choice for the passengers. Competition would furthermore lead to lower fares, which is in the interests of tourists. The downside is that the existence of limited capacity or frequency might results in a situation, where every airline can maybe operate only once a week and/or would need to use smaller aircrafts so the overall capacity limits are not exceeded. Not only that passenger convenience would be lowered as a result of the use of smaller aircrafts, frequency and capacity (seat availability) per airline could be also very low. This is clearly not in the interest of the business traveller group. Therefore, it is suggested that the available frequencies and/or capacities are divided between two categories of air carriers and that in each category only one airline is designated. One air carrier should mainly concentrate on offering services for business travellers. The other airline would specialize on services for tourists. The distribution of air traffic rights within each category would occur independently from the other category and be based on different criteria according to the needs of the relevant travellers. Air carriers, however, would be able to apply for air traffic rights in both categories. However, in order not to impede competition, air traffic rights of both categories shall not be granted to one single airline. Hence, if an air carrier is successful in both categories, it would have to chose for one category and shall give up the air traffic rights allocated to it in the other category. Before an allocation of air traffic rights to the applying air carriers could be undertaken, it would have to be first decided upon the contingent of air traffic rights available in each category. To ensure a more objective distribution of air traffic rights among the air carriers, it is recommended to include an intermediate step and to distribute the available air traffic rights between the two categories. One possibility would be, to divide the available air traffic rights in proportion to the amount of business and leisure travellers undertaken in the past flight periods. Statistics on the ratio of both kinds of travel could provide the necessary basis for such a decision. However, such an isolated approach does not reflect the importance that is attributed to and goes along with the different flights. While tourists are mostly beneficial for the economy of the host country, business travellers and their activities have positive effects for the economies of both states. Taking into account that air transportation services offered for business travellers require higher frequencies and capacities than services for the leisure travellers, the first-mentioned category shall be given a higher number of air traffic rights. How much higher this is supposed to be, cannot be pre-determined or fixed here. Much will depend on the respective situation as well as regulations contained in the corresponding air service agreements, since limitations can exist with respect to capacities, frequencies or both combined. Therefore, the decision must be taken on a case-by-case basis. This should even allow, to attribute all available air traffic rights to one category, and thus finally to one airline only, if this is deemed necessary. A fixed and pre-determined number of air traffic rights available in each category will also give the air carriers a reliable basis for their planning and will enable them to file in realistic proposals for the suggested services. Although, at first, such an approach might sound anti-competitive, this is not the case. Nothing precludes the two airlines finally selected to compete against each other and to try to divert traffic from the other. The 'leisure airline' can introduce a business class and offer high-quality services to business travellers, while the 'business airline' can offer particular low fare tariffs to attract tourists. In the end, it will be left to the airlines, to make the best use of their air traffic rights through innovation and effective managements.

In the category for business travellers, the main criteria for the allocation of air traffic rights should be capacity, frequency, route integration and characteristics of the proposed aircraft to be used with respect to environmental protection. The inclusion of the last criterion in the allocation procedure is, as set out above, due to Article 6 EC. The airline using the most modern, advanced and environmental-friendly aircrafts shall be benefited, when it comes to the allocation of air traffic rights.[948] The airline scoring best in these fields should be given all air traffic rights available in this category. If the applicants are all equally qualified, the respective fares should be also taken into account and the air carrier offering the lowest fares shall be favoured.[949] As far as environmental protection requirements are concerned, the aircrafts proposed by the air carriers to be used can be assessed in two manners. On the one hand it could be possible to directly compare the types of aircrafts with respect to their exhaust fumes, noise and fuel consumption levels. Such a comparison would be however distorted, if the respective air carriers propose to offer different frequency and capacity levels. While one air carrier may decide to offer more flights, but less capacity per operation, another airline may opt for offering fewer flights, but greater capacity for each flight. In such a case it is very likely that the aircrafts both air carriers would propose to use would be of a different size. Smaller aircrafts will have an advantage over bigger aircrafts with respect to exhaust fumes, noise and fuel consumption levels on a single-flight basis. On the other hand, on a per-passenger basis bigger aircrafts usually should score higher. To be able to compare both situations with each other, it is suggested that the aircrafts are compared with each other on an overall, per-passenger basis. It shall be evaluated how high the exhaust fume, noise level and fuel consumption levels of the respective aircrafts are on a per-passenger/per-seat basis during a specific period, for example one week, one month, one flight period. Such an approach would ensure that different strategies of the airlines with respect to frequencies and capacities are sufficiently taken into account. When deciding which airline offers the greatest benefits, a point system as it is applied in Germany, could be helpful. But unlike in Germany, where the applicant's proposals are graded independently for each other with points between

948 See above at 253ff.

949 See below at 268 for the calculation of the lowest fare level.

one and 10, it is suggested that all applications are graded in relation to each other. For this purpose the available points shall depend on the number of applicants. If for example five air carriers apply for air traffic rights, the air carrier offering the greatest benefits with respect to frequencies shall be given five points, the second best four points, and so on. Such an evaluation shall be undertaken with respect to all above-mentioned criteria and, in the end, all points shall be summed up. This approach has the advantage over the German solution that no subjective influence can occur, when the proposals are independently evaluated and when no reference exists on how high a proposal shall be graded.[950] No disputes can arise, whether a proposal foreseeing for example a departure time of 6.00 a.m. is worth six or eight points. If, at this stage, an airline is still not selected, it might be possible to take certain objective aspects of the proposed services into account, like for example category of flight (non-stop, one-stop, code-sharing), flight time and arrival and departure times. It is beneficial for business travellers, if flights are scheduled in a way that they can conduct their businesses in the most efficient way. Insofar an overnight flight would be for example an ideal flight. It shall be ensured that these supplementary criteria referring to the proposed services still enable an objective decision. If, after all these assessments, it is still not possible to allocate the respective air traffic rights, the interests of the industry in cargo services shall be considered, whereby available cargo space and fares would be the decisive factors. After this assessment no further criteria should be introduced. If all air carriers still are equally qualified, it is suggested to conduct a lottery. This ensures the most objective outcome. Any other additional criteria would very likely contain a subjective assessment by the allocating regulatory body and shall be thus avoided. Also, the above-mentioned disadvantages of a lottery would not occur at this stage any more. The previous selection process, based on the facts submitted by the airlines, ensures that the applicants are all equally qualified and any resulting randomness of the selection process at this stage is not harmful any more. Participation for speculative purposes, another disadvantage of lotteries, is, under these circumstances, also unlikely. Still, another condition shall be introduced at this stage. Since lotteries are based on luck, it would be unfair, if air carriers might be continuously allocated their air traffic rights as a result of the outcome of a lottery. Therefore, if a lottery is to be conducted, those airlines having previously received air traffic rights should be excluded from any subsequent lotteries, unless all other airlines that are still in the competition for air traffic rights have also previously been granted international routes through a lottery.

In the category of air transportation services for leisure travellers, the decisive factor should be the suggested fares and, hence that airline designated, which submits the lowest fare proposal. Thereby, not only the proposed fares, but also the respectively foreseen frequency and capacity made available under this fare will have to be taken into account. When assessing the advantages of a low fare, it must be also taken into account, how many tourists will be able to benefit from. The airline, suggesting the highest capacity and frequency at the lowest price, offers the highest benefits to the leisure travellers. To determine, which proposal is most

950 See above at 227.

beneficial, it is necessary to include these three variables, that is fares, capacity and frequency, in the following calculation:

$$\frac{\text{fare}}{\text{capacity x frequency}}$$

capacity: available seats for the respective fare
frequency: available flights in a certain period, for example a week

The lower the coefficient of this calculation is, the higher are the benefits for tourists. Apart from the fare level, environmental protection requirements must be also taken into account. Secondary consideration should be given to route integration. In addition, the already above-mentioned aspects can be taken into account, like category of service, flight times, available cargo space and suggested cargo fares. For the assessment of the greatest benefit point system, as described above, can be applied, whereby originally only the fare level and the characteristics of the proposed aircrafts with respect to environmental protection shall be taken into account. Only after this evaluation when no air carrier could be determined shall the other aspects be regarded. As with the business-traveller-category, a lottery should be conducted, if, after all other assessments, several applicants are equally well qualified.

Unlike in the previous case, when a multiple-liberal air service agreement is concluded, the same level of competition cannot be maintained due to the limited number of operating air carriers. Although, for the reasons given above,[951] competitive aspects should not be made part of the allocation decision, competition in the air transportation sector can be still indirectly influenced and a high level of competition can be maintained by granting route licences only on a temporary basis. While Germany explicitly rejects such a possibility, the USA and Australia have opted for such an approach. From the perspective of competition and having regard to the Theory of Contestable Markets, a time limitation of the granted air traffic rights is more convincing. According to the Theory of Contestable Markets, it is assumed that a player in a certain market will behave, as if it finds itself in real competition, although no competitor might be actually present. The reason for such behaviour is the theoretical possibility that new competitors may enter the market at any time.[952] The possible entry of a competitor creates an incentive with any existing market player, to act as competitive as possible, to deter any new entrants from the start. The only condition, for the Theory of Contestable Markets to function, is that entry and exit to and from the respective market must be free and without any sunk costs, which could impede other entities from entering the market.[953] The experience, gained in the deregulation of the US domestic market, showed that, for several reasons, the airline industry did not provide for contestable markets. Lack of contestability was due to the existence of economies of scale, predatory behaviour by the incumbents, frequent-flyer programs, and shortage of slots and gates at the airports, caused by

951 See above at 257ff.
952 Button and Swann, *supra* note 35 at 102.
953 *Ibid.*; Petsikas, *supra* note 725 at 125ff.

warehousing by the incumbents.[954] These circumstances put the new entrants at a disadvantage and prevented any effective competition against the existing air carriers. Therefore, free entrance might have been *de jure* free, but *de facto* this was not the case. However, a similar situation cannot occur in the international air service sector. The difference between the US domestic sector, where the Theory of Contestable Markets could not prevail, and international aviation is a limitation of access possibilities *de jure*. Although the Theory of Contestable Markets requires that market access must be free, a total freedom is not always beneficial. It might be even more successful, if a market remains foreclosed to other entrants for a while and where it is examined on a regular basis, whether the operations of the current market player are satisfying. If this is denied, the incumbent shall be substituted with a new entrant. Such an approach could be selected in international air transportation, if route permissions are granted only for limited period. Unlike in an open market, the incumbents could no defend 'their' markets with predatory behaviour, as they would be obliged to give up their air traffic rights and would have to leave the respective market completely. The incumbent would have no possibility to react to a new entrant and to disturb it in the development of its services. The only chance to defend a market against other competitors would be to maintain a highly competitive approach and to operate as efficient, successful and satisfying as possible, so that no reason arise for a route licence to be withdrawn or not to be extended. Under these circumstances the Theory of Contestable Markets could be made applicable in international aviation.[955] The necessity to grant air traffic rights on a temporal basis, exists only, where air service agreements foresee restrictions. Thus, under the previously described multiple-liberal air service agreement, air traffic rights can be issued without any time limitations.

As far as the duration of international route licenses is concerned, a time limit of five years, as is applied in the US and partially in Australia, seems reasonable and recommendable. The problem that follows then and that also needs to be answered is, how the renewal procedure is to be organized and how the incumbent's prior operations shall be taken into account.[956] One possibility would be that the administration considers the incumbent's operations as a positive decisional factor, if its past performance was indeed satisfying, but undertakes a new distribution procedure, which can lead to the result that the bonus that is attributed to the incumbent is finally set-off by other aspects. Another solution, which has also been adopted by the US Department of Transport and the Australian International Air Service Commission, is a rebuttable presumption in favour of the incumbent, unless the air carrier has deviated from its original fare and service proposal without any justified and comprehensible reasons. The choice for the rebuttable presumption is based on the conviction that only this approach gives adequate credit to the

---

954 See e.g. Forsyth, *supra* note 697 at 81; Petsikas, *ibid.* at 129–31; Dempsey, Thoms and Clapp, *supra* note 613 at 122.

955 See also the policy statements of the USA, 14 C.F.R. § 399.120 (1986), and of Great Britain, CAA Licensing, *supra* note 771 at 42, where explicit reference to the Theory of Contestable Markets is made).

956 14 C.F.R. § 399.120 (1986).

incumbent's commitment. The incumbent might have had to develop a whole new market, thereby mastering different challenges that go along with such an endeavour. Its performance during this time is the best indicator, whether it acted according to the market's needs and its track record should therefore be given adequate weight when evaluating the proposals of other air carriers.[957] Under these circumstances, competitors, who want to challenge the incumbent's performance, must show that they will provide substantially improved services in the future. Due to the incumbent's commitment on a route, a substitution should only take place, if the challenger's proposal is clearly superior. Unlike in the initial allocation procedure, there should be no lottery, if the challenger and the incumbent are equally qualified. In such a case, the incumbent should enjoy the merits of having developed a route and be granted the requested air traffic rights.

Single-liberal agreement[958]

It is doubtful, whether the combination of a liberal air service agreement as far as frequency and capacity are concerned and a single designation clause is very likely, as this would be quite a contradictory approach. Still, the situation is possible. In such a case, one air carrier would have to fulfil the needs of both travelling groups, business travellers and tourists. Unfortunately, as seen above, their interests are quite opposite to each other. Therefore, the chosen air carrier must offer a compromise, which is satisfying for both groups. The criteria, which would have to be taken into account, are: capacity, frequency, route integration, environmental protection and fares. The above-mentioned grade system would be applicable again. If, in this evaluation, two or more air carriers have scored equally, the other 'soft criteria' can be taken into regard: category of flight, flight time, arrival and departure times, available cargo space and suggested cargo fares. If, at this stage, no air carrier could be determined, a lottery should be conducted. Furthermore, for the reasons given above, the granted air traffic rights should be limited for a period of five years.

Single-restrictive agreement

A restrictive air service agreement with a single designation provision provides the biggest danger for a situation, in which all applicants will look alike. The more aspects of air transportation are regulated and pre-determined, the less possibilities remain for airlines to offer different products and develop new strategies. The criteria that are chosen for the selection procedure to distinguish between the applying airlines will therefore depend very much on the underlying air service agreement and the foreseen restrictions. In the worst case it could happen that the only aspect directly related to the distribution of air traffic rights, which may remain unregulated, is route integration. And, although not very likely, even this criterion could be foreclosed, if sixth freedom traffic was prohibited or restricted in the respective air service agreement. In addition to route integration, the above-mentioned 'soft criteria' could be applied to an increased extent. Again, if in such an evaluation no airline emerges

957 *Ibid.*

958 See also DOT Order 90-8-57, Dkt. 46701 at 8, 20.

ahead of the others, the most objective solution would be to undertake a lottery among the (remaining) applicants. More than ever, it would be artificial to try to introduce criteria for choosing an air carrier in a situation, where all relevant conditions of the envisaged air transportation service are pre-determined. Any of such established criteria that could enable a distinction among the applicants, would be only remotely related to the respective air transportation service. A lottery, on the other hand, has at least the advantage that it is fair and that, due to few controversial issues, no time consuming legal disputes should emerge, thus allowing for a fast implementation of the new services for the benefit of the consumers. A time limitation of the granted air traffic rights is also recommended here.

This previous classification into four categories is considered to cover the majority of all air service agreements. A distribution of air traffic rights granted in air service agreements that do not fall into this scheme should thus remain an exception. In these exceptional situations, the administration shall distribute air traffic rights at its discretion. However, since the above-mentioned classification is held to cover most of the air service agreements, the administration would need to be obliged to give reasons, why it assumes the existence of an exceptional situation that requires an allocation at its discretion.

PROCEDURE

Particular attention must be given to the procedure. In the US the procedure used to be exceedingly tedious and it took the CAB very long to evaluate the different public interests and finally to decide, which air carrier should obtain a route.[959] This was mainly due to the fact that a hearing was usually conducted before an administrative law judge. As already pointed out above, such a procedure is not mandatory. The uncomplicated matter of air transportation allows to allocate air traffic rights mainly based on written applications without any personal appearance of the airlines' representatives. The fact that the US used administrative hearings was based on their national legislation and was not because the subject matter was air transportation.[960] Another advantage is that no airline will be able to see the proposal of another applicant, before the final decision is taken. This will preclude any attempts to discredit another air carrier's application. The airlines shall provide the facts, upon which their proposals are based, but should not be permitted to point out, why their proposal is more appropriate than the one of a competitor. Such an approach will guarantee that each application is based only on the data and assumptions of the respective applicant and the airline having undertaken its research most thoroughly and having developed the best business model, will be able usually to obtain the air traffic rights in question. To ensure that the procedure is conducted legally and, in particular, complies with the conditions set out in Article 5 of Regulation 847/2004, namely that it is transparent and non-discriminatory, the participating airlines shall be granted access to the other applications at the end of the process, in order to

959 Thomas, *supra* note 598 at 70.

960 Bernard Schwartz, 'Administrative law' (1965) 55 *Ann. Surv. Am. L.* at 59.

appeal the decision, if they do not agree with the outcome. As far as the applications are concerned, it shall be precluded that airlines submit too optimistic proposals and then simply amend their suggested services once they are granted the requested air traffic rights. For this purpose they must be obliged to comply with their proposals for the timeframe the air traffic rights are granted. Any negative amendments compared with the initial proposal shall be only permitted upon approval by the administration in exceptional cases. The danger to incur losses shall deter airlines from submitting proposals that do not correspond to the available traffic on a specific route. Under these circumstances airlines will have to undertake extensive studies of the prospective markets and will only assign such aircrafts to the routes, whose use can be justified with respect to the available demand and the existing business opportunities. Thus, a more efficient utilization of the available equipment will be guaranteed as well.

When it comes to the question, which kind of institution shall be responsible for the distribution of air traffic rights, a subordinated department in the Ministry of Transport or an independent commission or agency, this matter is of secondary importance. Since, in the here-presented solution, subjective preferences are avoided to a great extent, it is generally not important, which institution is in charge of the allocation decision.

START-UP SERVICES

Airlines may have the incentive not only to serve a given route, but also to develop it further and to expand business opportunities. Since this can beneficial for other branches of the economy, such activities shall be supported. For this purpose and to grant the airlines' the necessary flexibility, timely limited start-up licences should be granted that allow the airlines to withdraw from a route again, if a market does not produce sufficient traffic and thus is not profitable enough. Under such a start-up licence, air carrier should be still obliged to continue offering the services they have proposed, but the period, for which such an obligation exists, would be shorter. Thus, any possible losses are limited. An airline shall indicate in its application that it intends to use a certain number of air traffic rights for the development of a new market and that it requests a start-up licence. The duration of the requested start-up licence should be determined by the respective air carrier, since it will know best, how much time is needed to develop the relevant market. Of course, this period cannot be longer than the duration of a normal licence, that is five years. If an air carrier enters a wholly new market, where no comparable air transportation services are offered and no indication is available on the demand of such services, it might be necessary to foreclose any other air carrier from this market for a certain period, to limit the risk of the air carrier engaged on such a route. If, on the other hand, air transportation services are already available in a market, the risk of failing is limited and there is no need to shield the new entrant from competition that may arise in the future.

AIRPORTS, SLOTS AND INTER-GATEWAY COMPETITION

To ensure that applicants use the granted air traffic rights as soon as possible and the consumers can benefit from the implemented services, every applicant shall have the necessary slots for the services it proposes to offer in its application. This is also to avoid that air carriers file in proposals with respect to service to and from highly demanded airports, for example London–Heathrow, which might put the respective air carrier at an advantage regarding the route-integration-criterion, but then fail to offer these services due to a lack of the necessary slots. Unless a route is predetermined by an air service agreement, airlines should be also free to choose the airports they want to operate to and from. Air carriers will favour those airports, where they can expect maximum number of users without having to rely too much on additional feeder traffic, where costs are low and where the turnaround time is short. Under these circumstances, air traffic rights are used in the most efficient manner. Airlines may also decide to further develop a smaller airport by establishing a hub, thereby relieving other congested airports. Hence, inter-gateway competition would be also increased and an incentive would be created for airports to improve their services, to attract more airlines. New procedures and technologies may be developed to reduce traffic delays, increase efficiency and lower charges, imposed on air carriers. All these measures would help reducing the costs of travelling. However, the possibility for the air carriers to freely select the airports they intend to use, must be limited to airlines offering flights for tourists only. Since time and easy accessibility to an airport is important to business travellers, airlines that intends to offer flights for business travellers shall be obliged to use an airport closely located to a business centre as a gateway. For this purpose it is possible that administrations give out lists of all eligible airports and that the airlines are allowed to select the airport(s) they prefer and want to operate to and from.

RESULT

The previously presented proposal mainly refers to the particular situation in the European Community. Still, it is not precluded that the suggested approach can be also applied in other countries. I am aware of the difficulties that exist with respect to the distribution of air traffic rights. The great variety of different air service agreements makes it almost impossible to develop a coherent and comprehensive method that can cover all possible situation. Nevertheless, it is believed that a standardized approach can produce reliability and is beneficial for all parties involved. In the countries examined, air traffic rights are distributed on a case-by-case basis, without that any reference is made to previous cases. This approach leads to uncertainty and unpredictability. The objective of this proposal is to overcome these disadvantages by introducing a more objective method that should lead to more coherent decisions.

SUMMARY

The here-presented work deals with two problems that currently exist or will arise in the air transportation sector in the European Community in the short-term future. The first one concerns the conclusion of air service agreements. Although in the recent 'open-skies' judgement the European Court of Justice denied a comprehensive treaty-making power of the European Community to conclude air service agreements, the judges extended the EC's treaty-making powers in this area to three subject matters, namely regulations regarding slots, fares applicable on intra-Community routes and computerized reservation systems. Opinions expressed in the aftermath of the ruling were quite diverse. Some scholars blamed the ECJ for intending to destroy the system of bilateral air service agreements that has been successfully applied in the area of air transportation for the last 50 years, while others saw the possibility for the beginning of a new era in regulating international air transportation. With respect to the conclusion of air service agreements, most scholars held the opinion that those agreements would have to be concluded as mixed agreements in the future. It is shown here that a more discriminated approach is necessary. It is first shown that international agreements of the European Community do not necessarily need to be concluded as mixed agreements, only because the Community does not have the treaty-making power for all subject matters included in such an agreement. Subordinated clauses, which have only an auxiliary function cannot trigger an additional participation of the Member States. In this respect it is further concluded that the main aspect, regulated in air service agreements, is the granting of air traffic rights. Thus, with respect to the partition of treaty-making powers between the Community and the Member States relating to the conclusion of air service agreement, the scope of air traffic rights granted in an air service agreement is decisive. If the first five freedoms are granted, the European Community can conclude the air service agreement alone, whereby the fifth freedom must be included in the agreement. If, on the other hand, the fifth freedom is not granted, the Member States alone remain responsible for the conclusion of the agreement.

The second problem resulting from the 'open-skies' judgement relates to the distribution of air traffic rights granted in air service agreements. The ECJ was of the opinion that the nationality clause, which is usually contained in air service agreements and which precludes an allocation of air traffic rights to air carriers not majority owned by the allocating Member States and/or its citizens, infringed the freedom of establishment and needed to be amended, to allow for an equal treatment of all Community air carriers. As a result, and once the respective clause is changed in the Member States' air service agreements, it could be possible in the future that all Community air carriers are eligible for designation in every Member State. Theoretically, the Member States could have to deal with 130 Community air carriers applying for air traffic rights. Since only few air service agreements provide for unlimited designation possibilities, a procedure will be necessary for the distribution of air traffic rights. Since, as concluded in Part Two, the European Community could be partially responsible for concluding air service agreements with third countries, it is first discussed, whether the EC or the Member States will be responsible for the distribution of air traffic rights. The following examination leads to a very complex result. It is concluded that the administrative power to allocate air traffic rights depends on the content of the concluded air service agreements. In one case, if a liberal air service agreement is concluded that provides for unlimited air traffic rights and a multiple designation clause or if an air service agreement is concluded between the EC and solely one Member States on one side and a non-EC state on the other side, the distribution of air traffic rights can be sufficiently accomplished by the respective Member State(s) alone. If a multilateral air service agreement is concluded, in which air traffic rights are limited, the distribution of these rights would require a two-step approach and a cooperation between the Member States and the EC. Finally, if air traffic rights are granted in an air service agreement, which foresees a cap on the number of possible designations, an allocation by the EC alone is suggested. As far as the allocation procedure itself is concerned, Article 5 of Regulation 847/2004, which was adopted in the aftermath of the 'open-skies' judgement, prescribes that the distribution of air traffic rights must be undertaken in a transparent and non-discriminatory manner. Having these conditions in mind, different allocation methods are examined. Attention is given to the first-come/first-served approach, lotteries, user fees, beauty contests and auctions. In this respect, a comparison between the telecommunication and the airline industry is undertaken. Since in the telecommunication industry auctions have proven to be the most suitable method for the distribution of frequency licenses, it is assumed that, considering also the similarities found between both industries, an auction may be also the most appropriate procedure to allocate air traffic rights. But, due to the international nature of aviation and for specific legal reasons, it is finally concluded that auctions can not be applied for the distribution of international air traffic rights. Instead beauty contests are considered to be the most suitable approach. This result is further confirmed in the subsequent chapter, where an analysis of allocation methods for air traffic rights in various countries is undertaken. In all states examined, not only beauty contests are used to distribute air traffic rights, but also the criteria applied by the respective regulators are very similar. Unfortunately, it is also found that an allocation through beauty contest bears two disadvantages: the possibility of subjective preferences by the persons in charge of the allocation and the danger that, due to a publication of the applicable criteria, all applicants may end up to be equally qualified. These two issues are later taken into regard, when an own proposal is presented. Since, as concluded before, the European Community and/or the Member States might distribute air traffic rights, it is also scrutinized, whether specific rules and principles exist in EC law that may have to be observed in the allocation procedure. As a result two aspects are found that need to be considered. Firstly, the applicable allocation criteria must be linked to the subject matter of the allocation, that is air transportation services. Secondly, due to Article 6 EC, environmental protection requirements must be taken into regard.

Based on the results found in the previous examinations, an own proposal for the allocation of air traffic rights is finally presented. It is suggested to distribute air traffic rights by way of a beauty contest, whereby special attention is given to avoid the inherent disadvantages of this procedure. For this purpose, four different categories of air service agreements are established, depending on the regulations included therein with respect to designation possibilities, capacity and frequency.

For each category of air service agreements, a specific approach to distribute air traffic rights is suggested. Thereby the specific interests of the two different user groups of air transportation services, business travellers and tourists, are particularly taken into account. It is my intention that the thus developed approach results in a more objective, more reasonable and less time consuming distribution of air traffic rights benefiting the air carriers and the consumers to the same extent.

Bibliography

EUROPEAN COMMUNITY LEGISLATION, PROCEEDINGS AND COMMUNICATIONS

Treaty Establishing the European Economic Community, 25 March 1957, 294 U.N.T.S. 4300, as amended by the Treaty Establishing the European Community, 7 February 1992 [1992] 1 C.M.L.R. 573, as amended by the Treaty of Amsterdam, 2 October 1997 [1997] O.J. C 340, as amended by the Treaty of Nice, 24 December 2002 [2002] O.J. C 325/33.

Treaty on the European Union, 7 February 1992 [1992] 1 C.M.L.R. 719, as amended by the Treaty of Amsterdam, 2 October 1997 [1997] O.J. C 340, as amended by the Treaty of Nice, 24 December 2002 [2002] O.J. C 325/33.

Agreement between the European Community and the Swiss Confederation on Air Transport [2002] O.J. L 114/73.

Amended proposal for a European Parliament and Council Directive concerning the coordination of procedures for the award of public contracts, public service contracts and public works contracts [2002] O.J. C 203 E/210.

Commission, Communication 'Public procurement in the European Union' of 11 March 1998 [1998] Bulletin of the European Union 31.2.42.

Commission, Communication 'Air Transport and the Environment – Towards meeting the Challenge of Sustainable Development', COM (1999) 640 final of 1 December 1999, online: Eur-Lex <http://europa.eu.int/eur-lex/lex/LexUriServ/site/en/com/1999/com1999_0640en01.pdf> (date accessed: 26 October 2005).

Commission, Regulation No 756/99/EC of 12 April 1999 on the implementation of Council Regulation 404/93 regarding the common market for bananas in the EC [1999] O.J. L 98/10.

Commission, Interpretative Communication 'Community law applicable to public procurement and the possibilities for integrating social considerations into public procurement (COM (2001) 566 final)' of 15 October 2001 [2001] O.J. C 333/27.

Commission, Communication 'The consequences of the Court judgements of 5 November 2002 for European air transport policy,' COM (2002) 649, online: Eur-Lex <http://europa.eu.int/eur-lex/lex/LexUriServ/site/en/com/2002/com2002_0649en01.pdf> (date accessed: 2 June 2005).

Commission, Communication 'The relation between the Community and third countries in the field of air transport', COM (2003) 94, online: Eur-Lex <http://europa.eu.int/eur-lex/en/com/pdf/2003/com2003_0094en01.pdf> (date accessed: 27 July 2003).

Commission, Communication 'To the European Parliament', COM (2004) 08, online: European Commission <http://europa.eu.int/eur-lex/en/com/pdf/2004/com2004_0008en01.pdf> (date accessed: 27 July 2004).

Commission, Communication, 'To the European Parliament and the Council on a framework for developing relations with the Russian Federation in the field of air transport', COM (2005) 77 final of 14 March 2005, online: Eur-Lex <http://europa.eu.int/eur-lex/lex/LexUriServ/site/en/com/2005/com2005_0077en01.pdf> (date accessed: 8 August 2005).

Commission, Communication 'On a civil aviation policy towards the People's Republic of China – strengthening co-operation and opening markets', COM (2005) 78 final of 11 March 2005, online: Eur-Lex <http://europa.eu.int/eur-lex/lex/LexUriServ/site/en/com/2005/com2005_0078en01.pdf> (date accessed: 8 August 2005).

Commission, 'Developing the agenda for the Community's external aviation policy', COM (2005) 79 final of 11 March 2005, online: Eur-Lex <http://europa.eu.int/eur-lex/lex/LexUriServ/site/en/com/2005/com2005_0079en01.pdf>.

Commission, Staff working document: Annex to the Communication from the Commission 'Developing the agenda for the Community's external aviation policy' (COM (2005) 79 final), of 11 March 2005, online: European Commission <http://www.europa.eu.int/comm/transport/air/international/doc/sec_2005_336_en.pdf> (date accessed: 8 August 2005).

EC, Commission, Press Release IP/05/783, 'EU aviation agreements with Australia and New Zealand', of 23 June 2005, online: European Commission, Press Releases <http://www.europa.eu.int/rapid/pressReleasesAction.do?reference=IP/05/783andformat=HTMLandaged=1andlanguage=ENandguiLanguage=en> (date accessed: 26 October 2005).

EC, Commission, Press Release IP/05/1190, 'The European Commission proposes to open aviation negotiations with Ukraine', of 27 September 2005, online: European Commission, <http://www.europa.eu.int/comm/transport/air/international/doc/2005-09-28/ip_1190_en.pdf> (date accessed: 4 November 2005).

EC, Commission, Communication 'Developing a Common Aviation Area with Ukraine (COM (2005) 451 final)' of 27 September 2005, online: European Commission <http://www.europa.eu.int/comm/transport/air/international/doc/2005-09-28/com_2005_0451_en.pdf> at 6 (date accessed: 5 November 2005)

Commission, Directorate-General for Energy and Transport, 'Towards a Coherent External Policy for Aviation: Recent Developments – Bringing Air Service Agreements in Line with Community Law' online: European Commission <http://europa.eu.int/comm/transport/air/international/dev_en.htm#conformity> (date accessed: 27 July 2004).

Council Directive 83/416/EEC of 25 July 1983 concerning the approval of inter-regional scheduled air service for the transport of passengers, mail and cargo between EEC countries [1983] O.J. L 237/19.

Council Regulation 3975/87/EEC of 14 December 1987 laying down the procedure for the application of the rules on competition to undertakings in the air transport sector [1987] O.J. L 374/01.

Council Regulation 3976/87/EEC of 14 December 1987 on the application of Article 85(3) of the Treaty to certain categories of agreements and concerted practices in the air transport sector [1987] O.J. L 374/09.

Council Directive 87/601/EEC of 14 December 1987 on fares for scheduled air

services between Member States [1987] O.J. L 374/12.
Council Decision 87/602/EEC of 14 December 1987 on the sharing of passenger capacity between air carriers on scheduled air services between Member States and on access for air carriers to scheduled air-service routes between Member States [1987] O.J. L 374/19.
Council Regulation 2299/89/EEC of 24 July 1989 on a Code of Conduct for Computerized Reservation Systems [1989] O.J. L 220/01.
Council Regulation 2342/90/EEC of 24 July 1990 on fares for scheduled air services [1990] O.J. L 217/01.
Council Regulation 2343/90/EEC of 24 July 1990 on access for air carriers to scheduled intra-Community air service routes and on the sharing of passenger capacity between air carriers on scheduled air services between Member States [1990] O.J. L 217/08.
Council Regulation 2344/90/EEC of 24 July 1990 amending Regulation (EEC) No. 3976/87 on the application of Article 85(3) of the Treaty to certain categories of agreements and concerted practices in the air transport sector [1990] O.J. L 217/15.
Council Directive 92/50/EEC of 18 June 1992 relating to the coordination of procedures for the award of public service contracts [1992] O.J. L 209/1.
Council Regulation 2407/92/EC of 23 July 1992 on Licensing of Air Carriers [1992] O.J. L 240/01.
Council Regulation 2408/92/EC of 23 July 1992 on Access for Community Air Carriers to Intra-Community Air Routes [1992] O.J. L 240/08.
Council Regulation 2409/92/EC of 23 July 1992 on Fares and Rates for Air Services [1992] O.J. L 240/15.
Council Decision 92/384/EEC of 22 June 1992 concerning the conclusion of an Agreement between the European Economic Community, the Kingdom of Norway and the Kingdom of Sweden on civil aviation [1992] O.J. L 200/20.
Council Directive 93/36/EEC of 14 June 1993 coordinating procedures for the award of public supply contracts [1993] O.J. L 199/1.
Council Directive 93/37/EEC of 14 June 1993 concerning the coordination of procedures for the award of public works contracts [1993] O.J. L 199/54.
Council Directive 93/38/EEC of 14 June 1993 coordinating the procurement procedures of entities operating in the water, energy, transport and telecommunications sectors [1993] O.J. L 199/84.
Council Regulation 95/93/EEC of 18 January 1993 on Common Rules for the Allocation of Slots at Community Airports [1993] O.J. L 014/01.
Council, Resolution on problems caused by congestion and crisis situations in air traffic in Europe of 17 November 1995 [1995] O.J. C 317/1.
Council Regulation No 2587/2001/EC of 19 December 2001 on the common market for bananas [2001] O.J. L 345/13.
Council Regulation No 2371/2002/EC of 20 December 2002 on the conservation and sustainable exploitation of fisheries resources under the Common Fisheries Policy [2002] O.J. L 358/59.
Council, 'Common Position (EC) No 33/2003 adopted with a view to adopting Directive 2003/ .../EC of the European Parliament and of the Council on the

coordination of procedures for the award of public works contracts, public supply contracts and public service contracts' of 20 March 2003 [2003] O.J. C 147 E 1.

Council, Conclusions on developing the agenda for the Community's external aviation policy, online: European Commission <http://www.europa.eu.int/comm/transport/air/international/doc/2005_06_council_conclusions_en.pdf> (date accessed: 26 October 2005).

Economic and Social Committee, Opinion 'Proposal for a Directive of the European Parliament and of the Council on the coordination of procedures for the award of public supply contracts, public service contracts and public works contracts' of 26 April 2001 [2001] O.J. C 193/, online: European Parliament <http://www2.europarl.eu.int/omk/sipade2?SAME_LEVEL=1andLEVEL=0andNAV=XandDETAIL=andPUBREF=-//EP//TEXT+REPORT+A5-2001-0378+0+DOC+XML+V0//EN> (date accessed 4 March 2005).

Directive 2004/17/EC of the European Parliament and of the Council of 31 March 2004 coordinating the procurement procedures of entities operating in the water, energy, transport and postal service [2004] O.J. L 134/1.

Directive 2004/18/EC of the European Parliament and of the Council of 31 March 2004 on the coordination of procedures for the award of public work contracts, public supply contracts and public service contracts [2004] O.J. L 134/114.

European Parliament, 'Position of the European Parliament adopted at first reading on 17 January 2002 with a view to the adoption of European Parliament and Council Directive …/ …/EC on the coordination of procedures for the award of public supply contracts, public service contracts and public works contracts' of 17 January 2002 [2002] O.J. C 271 E/176.

European Parliament and Council Regulation 847/2004/EC of 29 April 2004 on the Negotiation and Implementation of Air Service Agreements between Member States and Third Countries [2004] O.J. L 157/7.

European Parliament Committee on Legal Affairs and the Internal Market, 'Report on the proposal for a directive of the European Parliament and of the Council on the coordination of procedures for the award of public supply contracts, public service contracts and public works contracts (COM (2000) 275 –C5-0367/2000 - 2000/0115 (COD))' of 29 October 2001.

Proposal for a Directive of the European Parliament and of the Council on the coordination of procedures for the award of public supply contracts, public service contracts and public works contracts [2001] O.J. C 29 *E*/11.

OTHER LEGISLATION, PROCEEDINGS AND COMMUNICATIONS

Air Transport Agreement Between Germany and the United States, 7 July 1955 (as amended on 25 April 1989, on 23 May 1996 and on 10 October 2000), online: <http://luftrecht-online.de/regelwerke/d-us-e.htm> (date accessed: 2 November 2005).

Civil Aviation Authority (1984), *Airline Competition Policy* (Cheltenham, UK: Civil Aviation Authority).

Civil Aviation Authority, Official Record Series 1, *Licensing Airlines in the UK* (12 August 2002, revised 31 March 2004), online: Civil Aviation Authority <http://

www.caa.co.uk/docs/213/Series per cent201- per cent2031 per cent20March per cent202004.pdf> (date accessed: 6 August 2004).

Convention on International Civil Aviation, 7 December 1944, 15 U.N.T.S. 102

U.S. Federal Aviation Act, 49 USC § 41102 (2003).

German Federal Ministry of Transportation and Construction, *Guidelines for the Allocation of Air Traffic Rights and Frequencies to German Air Carriers of 13.2.2001*, LS 12/20.00.50.

International Air Transport Agreement, 7 December 1944, 171 U.N.T.S. 502.

International Air Service Transit Agreement, 7 December 1944, ICAO Doc. 7500.

International Air Services Commission Act 1992, Act No. 103 of 1992 as amended, online: International Air Services Commission <http://www.iasc.gov.au/iascleg.aspx#IASCAct> (date accessed: 3 August 2004).

International Air Services Commission Act 1992 Policy Statement, online: International Air Services Commission <http://www.iasc.gov.au/pubs/PSV520040519.pdf> (date accessed: 4 August 2004).

Explanatory Statement to the International Air Services Commission Policy Statement No. 5, online: International Air Services Commission <http://www.iasc.gov.au/pubs/PSV5EM20040519.pdf> (date accessed: 4 August 2004).

International Air Service Commission, *Procedures of Handling Applications for Capacity in the Start-Up Phase*, online: International Air Services Commission <http://www.iasc.gov.au/pubs/instart.pdf> (date accessed: 4 August 2004).

German model air service agreement, online: Luftrecht-Online <http://www.luftrecht-online.de/regelwerke/pdf/e-muster.pdf> (date accessed: 2 November 2005).

Luftverkehrszulassungsgesetz (German Aeronautics Act) of 27 March 1999, Bundesgesetzblatt I page 550, as amended on 21 August 2002, Bundesgesetzblatt I page 3355.

Luftverkehrszulassungsordnung (German Air Traffic Licensing Regulations) of 27 March 1999, Bundesgesetzblatt I, page 610, as amended on 21 August 2002, Bundesgesetzblatt I, page 3355.

Marrakech Agreement establishing the World Trade Organization, 15 April 1994, 1867 U.N.T.S. 31874.

Transport Minister announces new Policy for International Scheduled Air Services, online: Transport Canada <http://www.tc.gc.ca/mediaroom/releases/nat/2002/02_h051e.htm> (date accessed: 6 August 2004).

U.S. Senate Subcommittee on Administrative Practices and Procedures, Report: 'Airline Regulation by the Civil Aeronautics Board – Summary', reprinted in (1975) 41 *J. Air L. and Com.* at 607.

U.S. Secretary of Transport Norman Mineta, 'Bush seeks more foreign ownership of U.S. airline stock' of 2 June 2003, online U.S. Mission to the EU <http://www.useu.be/Categories/Transportation/June0203AirlineOwnership.html> (date accessed: 2 June 2005).

U.S. Secretary of Transportation Norman Mineta, 'Mineta urges successful conclusion of U.S.-EU open skies talks' of 6 April 2004, online: U.S. Mission to the EU <http://www.useu.be/Article.asp?ID=06364054-1B62-43D3-9F6A-44AB36AEEC63> (date accessed: 2 June 2005).

U.S. Department of Transportation, Secretary Norman Y. Mineta, 'U.S., France sign open skies aviation agreement' of 22 January 2002, online: U.S. Mission to the EU <http://www.useu.be/Categories/Transportation/Jan2202USFranceOpenSkies.html> (date accessed: 2 June 2005).

U.S. Department of Transportation, Assistant Deputy Secretary Jeffrey Shane, 'U.S. Official comments on EU "open skies" ruling' of 8 November 2002, online: U.S. Mission to the EU <http://www.useu.be/Categories/Transportation/Nov0602USEUOpenSkies.html> (date accessed: 2 June 2005).

U.S. State Department, Communication, 'U.S. calls for discussion of open skies agreement in Europe' of 20 November 2002, online: U.S. Mission to the EU <http://www.useu.be/Categories/Transportation/Nov2002USOpenSkies Commission.html> (date accessed: 2 June 2005).

U.S. Department of State, Chief negotiator for transportation at the U.S. State Department Byerly, 'U.S. aims for comprehensive accord in air service talks with EU' of 29 September 2003, online: U.S. Mission to the EU <http://www.useu.be/Categories/Transportation/Sept2903ByerlyOpenSkies.html> (date accessed: 2 June 2005).

U.S. Under Secretary of Commerce Grant Aldona, Unofficial transcript of a press conference 'Interview with Secretary Aldona on trade talks with EU' of 17 December 2003, online: U.S. Mission to the EU <http://www.useu.be/Categories/Biotech/Dec1702AldonasPress.html> (date accessed: 2 June 2005).

U.S. Department of State, Deputy Assistant Secretary for Transportation Affairs John Byerly, 'U.S. Determining How to Proceed on Open Skies Talks with EU' of 13 July 2004, online: U.S. Mission to the EU <http://www.useu.be/Article.asp?ID=A1024B73-08FC-4485-B500-ACC710FDA142> (date accessed: 2 June 2005).

U.S. Department of Transport, Under Secretary for Policy Jeffrey Shane, 'U.S. urges more liberalization in air services with China, EU' of 3 December 2004, online: U.S. Mission to the EU <http://www.useu.be/ Articles.Asp?ID = 2C502D09-9A64-4C2B-8,109-E9020A29CA5F > (date accessed: 15 June 2005).

U.S. Department of State, Deputy Assistant Secretary for Transportation Affairs John Byerly, 'U.S.-EU Air Service Negotiations' of 8 February 2005, online: U.S. Mission to the EU <http://www.useu.be/Categories/Transportation/Feb0805ByerlyAirTalksSpeech.html> (date accessed: 3 June 2005).

EUROPEAN COURT OF JUSTICE RULINGS

Commission v Italy, 10/61 [1962] E.C.R. 5.

Commission v Council, 22/70 [1971] E.C.R. 263.

Opinion of Advocate General Lamothe of 10 March 1971 in 22/70 [1971] E.C.R. 284.

Commission v French Republic, C-167/73 [1974] E.C.R. 359.

Opinion 1/76 [1977] E.C.R. 741.

Opinion 1/78 [1979] E.C.R. 2871.

Ruling 1/78 [1978] E.C.R. 2151.

Attorney General v Juan C. Burgoa, 812/79 [1980] E.C.R. 2787.

Ministère Public v Asjes, C-209-213/84 [1986] E.C.R. 1425.

Ahmed Saeed v Office for combating unfair competition, C-66/86 [1989] E.C.R. 229.

Gebroeders Beentjes *BV v The Netherlands*, 31/87 [1988] E.C.R. 4535.

Pinaut Wieger Spedition GmbH v Bundesanstalt für den Güterfernverkehr, C-17/90 [1991] E.C.R. I-5253.

Commission v Federal Republic of Germany, C-195/90 [1992] E.C.R. I-3141.

Ministère public and Direction du travail and de l'emploi v Levy, C-158/91 [1993] E.C.R. I-4287.

Opinion 2/91 [1992] E.C.R. I-1061.

Opinion 2/92 [1995] E.C.R. I-525.

Office National de l'Emploi v Madleine Minne, C-13/93 [1993] E.C.R. I-371.

Ex parte Evans Medical Ltd. and MacFarlan Smith Ltd. The Queen v Secretary of State for the Home Department, C-324/93 [1995] E.C.R. I-563.

Opinion 1/94 [1994] E.C.R. I-5267.

Ex parte Centro-Com Srl v H.M. Treasury and Bank of England The Queen, C-124/95 [1997] E.C.R. I-81.

T. Port GmbH and Co. KG v Hauptzollamt Hamburg-Jonas, C-365/95 [1998] E.C.R. I-1023.

Karlsson et al., C-292/97 [2000] E.C.R. I-2737.

Commission v France, C-225/98 [2000] E.C.R. I-7445.

Commission v United Kingdom et al., C-466-69, 471, 472, 475 and 476/98 [2002] E.C.R. I-9427.

Opinion of Advocate General Tizzano of 31 January 2002 in C-476/98 [2002] E.C.R. I-9431.

Formerly Stagecoach Finland Oy Ab v Helsingin kaupunki and HKL-Bussiliikenne Concordia Bus Finland Oy Ab., C-513/99 [2002] E.C.R. I-7213.

Opinion of Advocate General Jean Mischo of 13 December 2001 in C-513/99 [2002] E.C.R. I-7215.

ADMINISTRATIVE DECISIONS

14 C.F.R. § 399.120 (1986)

DOT Order 96-4-23, Docket OST-95-422, online: Department of Transport <http://dms.dot.gov/search/searchResultsSimple.cfm> (date accessed: 1 August 2004).

DOT Order 90-8-57, Dkt. 46701.

DOT Order 93-12-27, Dkt. 49052.

Civil Aviation Authority, 'Decision on Air Transport Licences and Route Licences 2/98' (25 August 1998), online: Civil Aviation Authority <http://www.caa.co.uk/docs/213/ald9802.pdf> (date accessed: 6 August 2004).

Civil Aviation Authority, 'Decision on Air Transport Licences and Route Licences 3/98' (25 August 1998), online: Civil Aviation Authority <http://www.caa.co.uk/docs/213/ald9803.pdf> (date accessed: 6 August 2004).

Civil Aviation Authority, 'Decision on Air Transport Licences and Route Licences 4/98' (26 November 1998), online: Civil Aviation Authority <http://www.caa.co.uk/docs/213/ald9804.pdf> (date accessed: 6 August 2004).

Civil Aviation Authority, 'Decision on Air Transport Licences and Route Licences 5/98' (10 December 1998), online: Civil Aviation Authority <http://www.caa.co.uk/docs/213/ald9805.pdf> (date accessed: 6 August 2004).

Civil Aviation Authority, 'Decision on Air Transport Licences and Route Licences 1/99' (20 September 1999), online: Civil Aviation Authority <http://www.caa.co.uk/docs/213/ald9901.pdf> (date accessed: 6 August 2004).

Civil Aviation Authority, 'Decision on Air Transport Licences and Route Licences 3/99' (25 October 1999), online: Civil Aviation Authority <http://www.caa.co.uk/docs/213/ald9903.pdf> (date accessed: 6 August 2004).

Civil Aviation Authority, 'Decision on Air Transport Licences and Route Licences 4/99' (14 December 1999), online: Civil Aviation Authority <http://www.caa.co.uk/docs/213/ald9904.pdf> (date accessed: 6 August 2004).

ESSAYS, MONOGRAPHS, DISSERTATIONS

Achterberg, N., 'Die Annex-Kompetenz' (The Implied Power) (1966) *Die Öffentliche Verwaltung* (*J. of Public Admin.*), 695.

Akehurst, M.A. (1992), *Modern Introduction to International Law*, 6th edn., (London: Routledge).

Arnold, R., 'Der Abschluss gemischter Verträge durch die Europäischen Gemeinschaften' (The Conclusion of Mixed Agreements by the European Community) (1980/81) *Archiv des Völkerrechts* (*Archives of Intl. L.*), 419.

Aust, A. (2000), *Modern Treaty Law and Practice*, (Cambridge: Cambridge University Press).

Bailey, E.O., 'Article 15 of the Chicago Convention and the Duty of States to Avoid Discriminatory User Charges: The US–UK London Heathrow Airport User Charges Arbitration', (1994) 19(2) *Ann. Air and Sp. L.*, 81.

Baldwin, R. (1985), *Regulating the Airlines – Administrative Justice and Agency Discretion*, (Oxford : Clarendon Press).

Becher, G., 'Luftverkehrspolitische Entwicklungstendenzen in Asien im Spannungsfeld zwischen Liberalismus und Dirigismus' (Air Traffic Policy Development Tendencies in Asia between Liberalization and State Control) (1979) *Zeitschrift für Luft- und Weltraumrecht* (*Air and Space L. Rev.*), 317.

Becker, J. (1976), *Die Anwendbarkeit der Theorie von den implied powers im Recht der Europäischen Gemeinschaft* (*The Application of the Implied-Powers Theory in EC Law*), (Münster)

Benedict, C., 'Anmerkung zu EuGH Rs C-225/98' (Remarks on ECJ Decision C-225/98) (2000) *Europäisches Wirtschafts- und Steuerrecht* (*Europ. J. of Commercial and Tax L.*), 514.

Bernhardt, R., 'Die Europäische Gemeinschaft als neuer Rechtsträger im Geflecht der traditionellen zwischenstaatlichen Rechtsbeziehungen' (The European

Community as a New Legal Entity in the Context of Traditional Bilateral Relations) (1983) *Europarecht* (*Europ. L. Rev.)*, 199.

Beutler, B., Epiney, A., Bieber, R. and Haag, M., eds. (2001), *Die Europäische Union*, (*The European Union*), 5th edn., (Baden-Baden, Germany: Nomos Verlagsgesellschaft).

Bin Cheng (1962), *The Law of International Air Transport*, (London: Stevens and Sons).

Blake, J., 'FCC Licensing: from Comparative Hearings to Auctions', (1994) 47 *Fed. Comm. L.J.*, 179.

Bleckmann, A., 'Der gemischte Vertrag im Europarecht' (The Mixed Agreement in European Law) (1976) *Europarecht* (*Europ. L. Rev.*), 301.

Bleckmann, A., 'Die Kompetenz der Europäischen Gemeinschaft zum Abschluß völkerrechtlicher Verträge' (The Power of the European Community to Conclude International Agreements) (1977) *Europarecht* (*Europ. L. Rev.)*, 109.

Bleckmann, A., 'Teleologie und dynamische Auslegung des Europäischen Gemeinschaftsrechts' (Teleological and Dynamic Interpretation of European Community Law) (1979) *Europarecht* (*Europ. L. Rev.*), 239.

Bleckmann, A. (1997), *Europarecht* (*European Law*), 6th edn., (Cologne: Carl Heymanns Verlag).

Bleckmann, A. 'Zur Auflage im europäischen Beihilfenrecht', (The condition in European anti-dumping law) (2004) *Neue Zeitschrift für Verwaltungsrecht* (*New J. of Admin. L.*), 11.

Bogosian, R.W., 'Aviation Negotiations and the U.S. Model Agreement' (1981) 46 *J. Air L. and Com.*, 1007.

Böhm, R. (1985), *Kompetenzauslegung und Kompetenzlücken im Gemeinschaftsrecht*, (*Interpretation of Competences and Omissions in Community Law*), (Frankfurt, Bern, New York).

Bonder, T., 'A "Better" Marketplace Approach to Broadcast Regulation' (1984) 36 *Fed. Comm. L. J.*, 27.

Breyer, S., 'Analyzing Regulatory Failure: Mismatches, Less Restrictive Alternatives, and Reform' (1978–79) 92 *Harv. L. Rev.*, 547.

Bullinger, M., 'Ungeschriebene Kompetenzen im Bundesstaat', (Implied powers in federal states), (1971) *Archiv des öffentlichen Rechts* (*Archives of Public L.*), 237.

Burkhardt, R. (1974), *The Civil Aeronautics Board*, (Virginia: Green Hills Company).

Button, K. and Swann, D., 'Aviation Policy in Europe', in: Kenneth Button, ed. (1991), *Airline Deregulation – International Experiences*, (Washington Square, NY: New York University Press), 85.

Calliess, C. and Ruffert, M., eds. (2002), *Kommentar zu EU-Vertrag und EG-Vertrag* (*Commentary to the EU Treaty and EC Treaty*), 2nd edn., (Neuwied: Luchterhand Verlag).

Calliess, C., 'Der Schlüsselbegriff der "ausschließlichen Zuständigkeit" im Subsidiaritätsprinzip des Article 3b II EGV', (The core element of the 'exclusive competence' in the principle of subsidiarity stipulated in Article 3b II EC Treaty), (1995) *Europäische Zeitschrift für Wirtschaftsrecht* (*Europ. J. of Commercial L.*), 693.

Calliess, C., 'Die neue Querschnittsklausel des Article 6 Ex 3c EGV als Instrument zur Umsetzung des Grundsatzes der nachhaltigen Entwicklung', (The new integration clause of Article 6 ex 3c EC-Treaty as an instrument to implement the principle of sustainable development), (1998) *Deutsches Verwaltungsblatt* (*German J. of Admin. L.*), 559.

Campbell, D., 'Competition Policy in Domestic Aviation or The Rise and Fall of the Two Airline Policy', in: S.G. Corones ed. (1992),*Competition Policy in Telecommunication and Aviation* (Sydney: Federation Press), 17.

Caves, R.E. (1968), *Air Transport and its Regulators – An Industry Study*, (Cambridge, MA: Harvard University Press).

Churchill, R.R. (1987), *EEC Fisheries Law*, (Dordrecht, The Netherlands: Martinus Nijhoff Publishers).

Dagtoglou, P., 'Der Luftverkehr und die Europäische Gemeinschaft – eine Bestandsaufnahme' (Air Transportation and the European Community) in: Bieber, Roland, Nickel and Dietmar, eds. (1981), *Gedächnisschrift für Christoph Sasse – Das Europa der zweiten Generation* (*Liber amicorum Christoph Sasse – The Europe of the Second Generation*), vol. 1, (Baden-Baden, Germany: Nomos Verlagsgesellschaft), 539.

Dauses, M.A., ed., *Handbuch des EU-Wirtschaftsrechts (Manual of EU Commercial Law)*, vol. 1, looseleaf, (Munich: Verlag C.H. Beck).

de Marolles, O. and Lenoir, A. (1993), *The Deregulation of Air Transport in Australia*, (Paris, France: Institute of Air Transport).

Degenhart, C. (2001), *Staatsrecht I – Staatsorganisationsrecht (Constitutional Law I – State Organization Law)*, 17th edn., (Heidelberg: C. F. Müller Verlag).

Dempsey, P.S. and Gesell, L.E. (2004), *Air Commerce and the Law*, (Chandler, AZ: Coast Aire Publications).

Dempsey, P.S., Thoms, W.E. and Clapp, S., 'Canadian Transport Liberalization: Planes, Trains, Trucks and Buses Rolling Across the Great White North', (1990–91) 19 *Transp. L. J.*, 113.

Diederiks-Verschoor, I.H. Philepina (1993), *An Introduction to Air Law*, 5th edn., (Deventer, The Netherlands: Kluwer).

Dörr, O., 'Die Entwicklung der ungeschriebenen Außenkompetenzen der EG', (The development of implicit external powers of the EC), (1996) *Europäische Zeitschrift für Wirtschaftsrecht* (*Europ. J. of Commercial L.*), 39.

Douglas, G.W. and Miller, J.C., III (1974), *Economic Regulation of Domestic Air Transport: Theory and Policy*, (Washington: Brookings Institution).

Dreher, M., Anmerkung zu EuGH, Rs, C-225/98 (Remarks on the ECJ Decision C-225/98), (2000) *Juristenzeitung* (*Lawyer's Journal*), 140.

Eisermann, K.S., 'Die Luftfahrtaußenkompetenz der Gemeinschaft', (The external power of the community in air transportation), (1995) *Europäische Zeitschrift für Wirtschaftsrecht* (*Europ. J. of Commerical L.*), 331.

Epiney, A., 'Umweltrechtliche Querschnittsklausel und freier Warenverkehr: die Einbeziehung umweltpolitischer Belange über die Beschränkung der Grundfreiheit' (Environmental Integration Clause and Right to Free Movement of Goods: the Inclusion of Environmental Matters by Limiting a Market Freedom), (1995) *Natur und Recht* (*Environment and Law*), 497.

Ezor, J.I., 'Costs Overhead: Tonga's Claiming of Sixteen Geostationary Orbital Sites and the Implications for U.S. Space Policy', (1993) 24 *Law and Pol'y Int'l Bus.*, 915.

Ferrar, T.A., 'Route Assignment and the C.A.B', (1973) 5 *Transp. L. J.*, 215.

Finnsson, G., 'Airports and Route Facilities: International Cost Recovery Policies and their Applicability in the Framework of New Forms of Infrastructure Provision', (1994) 19(2) *Ann. Air and Sp. L.*, 283.

Fiorita, D. (Dan) M. (1995), *Safety and Economic Regulation of Air Transportation in Canada* (LL.M. Thesis, Institute of Air and Space Law, McGill University) [unpublished].

Forsyth, P., 'The Regulation and Deregulation of Australia's Domestic Airline Industry', in: Kenneth Button, ed. (1991), *Airline Deregulation – International Experiences*, (Washington Square, NY: New York University Press), 48.

Foster Rollo, V.A., (2000), *Law – An Introduction*, 5th edn., (Lanham, MD: Maryland Historical Press).

Frederick, J.H. (1961), *Commercial Air Transportation*, 5th edn., (Homewood, IL: Richard D. Irwin).

Fuß, E.-W., 'Die Befugnis der Europäischen Wirtschaftsgemeinschaft zum Abschluß völkerrechtlicher Verträge mit Drittstaaten', (The competence of the European Economic Community to conclude international agreements with third countries) (1972) *Deutsches Verwaltungsblatt* (*German J. of Admin. L.*), 237.

Gassner, U. and Deichstetter, A., 'Luftverkehrsabkommen als Gemeinschaftsaufgabe', (Air service agreements as a task of the Community) (2003) *Europäisches Wirtschafts- und Steuerrecht (Europ. Commercial and Tax L.)*, 265.

Geiger, R., 'Außenbeziehungen der Europäischen Wirtschaftsgemeinschaft und auswärtige Gewalt der Mitgliedstaaten', (External relations of the European Economic Community and the treaty-making power of the Member States) (1977) *Zeitschrift für ausländisches öffentliches Recht und Völkerrecht* (*J. of Foreign Public L. and Inl. L.*), 657.

Geiger, R., 'Vertragsschlusskompetenzen der Europäischen Gemeinschaft und auswärtige Gewalt der Mitgliedstaaten', (Treaty-making powers of the European Community and external powers of the Member States) (1995) *Juristenzeitung* (*Lawyer's Journal*), 973.

Giemulla, E. and Schmid, R., *Frankfurter Kommentar zum Luftverkehrsrecht – Luftverkehrsgesetz (Commentary to Aeronautics Law – Aeronautics Act)* looseleaf, (Neuwied, Germany: Luchterhand Verlag).

Giemulla, E., Schmid, R. and Mölls, W., *Europäisches Luftverkehrsrecht (European Air Law)* looseleaf, Neuwied, Germany: Luchterhand Verlag, 32nd supplement).

Giemulla, E., Schmid, R. and Mölls, W. (1993), *European Air Law*, looseleaf, (Deventer, The Netherlands: Kluwer Law and Taxation Publishers).

Gilsdorf, P.W., 'Die Außenkompetenzen der EG im Wandel – Eine kritische Auseinandersetzung mit Praxis und Rechtsprechung' (The External Powers of the EC Changing – a Critical Discussion with Practice and Jurisprudence) (1996) *Europarecht* (*Europ. L. Rev.*), 145.

Goh, J. (2001), *The Single Aviation Market of Australia and New Zealand*, (London: Cavendish Publishing).

Goodman, E.P., 'Digital Television and the Allure of Auctions: The Birth and Stillbirth of DTV Legislation', (1997) 49 *Fed. Comm. L.J.*, 517.

Goppel, T., 'Die Bedeutung des Subsidiaritätsprinzips', (The meaning of the principle of subsidiarity) (1993) *Europäische Zeitschrift für Wirtschaftsrecht* (*Europ. J. of Commercial L.*), 367.

Grabitz, E. and Hilf, M., eds., *Das Recht der Europäischen Union (The Law of the European Union)*, looseleaf, (Munich: Verlag C.H. Beck).

Guldimann, W., 'Luftverkehrspolitik' (Air Traffic Policy) (1993) *Zeitschrift für Luft- und Weltraumrecht* (*Air and Space L. Rev.),* 369.

Guldimann, W., 'A Flexible Multilateral in Traffic Rights?', (1993) 18(1) *Ann. Air and Sp. L.*, 125.

Hailbronner, K., Klein, E., Magiera, S. and Müller-Graff, P.-C., eds. (April 1995), *EUV/EGV (EU-Treaty/EC-Treaty)*, (Cologne: Carl Heymanns Verlag).

Herdegen, M. (2003), *Europarecht* (*European Law*), 5th edn., (Munich: C.H. Beck Verlag).

Hilf, M., 'EG-Außenkompetenzen in Grenzen', (The limits of the treaty-making powers of the EC) (1995) *Europäische Zeitschrift für Wirtschaftsrecht* (*Europ. J. of Commercial L.*), 7.

Hirsch, G., 'Das Subsidiaritätsprinzip' (The Principle of Subsidiarity), in Bötcher, R., Hueck, G. and Jähnke, B. eds. (1996), *Festschrift für Walter Odersky zum 65. Geburtstag (Liber Amicorum Walter Oderksy to his 65th Birthday)*, (Berlin: Walter de Gruyter), 197.

Hofmann, M. and Grabherr, E. (2004), *Luftverkehrsgesetz (Aeronautics Act)*, looseleaf, (Munich: Verlag C.H. Beck).

Ipsen, H.P. (1972), *Europäisches Gemeinschaftsrecht*, (*European Community Law*), (Tübingen: J.C.B. Mohr (Paul Siebeck)).

Ipsen, K. (2004), *Völkerrecht (International Law)*, 5th edn., (Munich: Verlag C.H. Beck).

Jahns-Böhm, J. and Breier, S., 'Die umweltrechtliche Querschnittsklausel des Article 130r II 2 EWGV', (The environmental integration clause of Article 130r II 2 EEC-Treaty) (1992) *Europäische Zeitschrift für Wirtschaftsrecht* (*Europ. J. of Commercial L.*), 49.

Jakhu, R.S. and Serrano, V.R. (2000), 'International Regulation of Radio Frequencies for Space Services' in: *Proceedings of the Project 2001 Workshop on Legal Framework for Commercial Satellite Telecommunications*, (Cologne, Germany: Institute of Air and Space Law of the University of Cologne/Deutsches Zentrum für Luft- und Weltraumrecht), 41.

Janda, R., 'Auctioning Airport Slots: Airline Oligopoly, Hubs and Spokes, and Traffic Congestion', (1993) 18(1) *Ann. Air and Sp. L.*, 153.

Jarass, H.D., 'Die Kompetenzverteilung zwischen der Europäischen Gemeinschaft und den Mitgliedstaaten', (The distribution of powers between the European Community and the Member States) (1996) 121 *Archiv des Öffentlichen Rechts* (*Archives of Public L.*), 173.

Jones, R., An International Perspective on Anti-Competitive Pricing Practices by Dominant Carriers and Regulatory Rules to Facilitate Competitive Entry – Australia, (2001–02) 14 *DePaul Bus. L. J.*, 243.

Jones, W.K. (1962), *Licensing of Domestic Air Transportation by the Civil Aeronautics Board* [unpublished, archived at Columbia University School of Law].

Jürgensen, T., 'Gemeinschaftrechtliche Grenzen der Bananenmarktordnung' (Limitation of the Common Market for Bananas Derived from Community Law) (1998) *Europäisches Wirtschafts- und Steuerrecht* (*Europ. J. of Commercial and Tax L.*), 357.

Kahl, W., 'Hat die EG die Kompetenz zur Regelung des Allgemeinen Verwaltungsrechts?', (Does the EC has the power to adopt a regulation on general administration law?) (1996) *Neue Zeitschrift für Verwaltungsrecht (New J. of Admin. L.)*, 865.

Kahn, A.E. and Roach, M., 'A Paean to Legal Creativity', (1979) 31 *Admin. L. Rev.*, 91.

Kane, R.M. and Vose, A.D. (1975), *Air Transportation*, 5th edn., (Dubuque: Kendall/ Hunt).

Kenntner, M., 'Das Subsidiaritätsprinzip des Amsterdamer Vertrages - Anmerkungen zum Begrenzungscharakter des gemeinschaftlichen Subsidiaritätsprinzips', (The Principle of Subsidiarity in the Treaty of Amsterdam – Remarks on the Ability to Restrict of the Principle of Subsidiarity) (1998) *Neue Juristische Wochenschrift* (*New Weekly L. Rev.*), 2871.

Klingman, R.A., 'Predatory Pricing and other Exclusionary Conduct in the Airline Industry: Is Antitrust Law the solution?', (1991–92) 4 *DePaul Bus. L.J.*, 281.

Klösters, A. (1994), *Kompetenzen der EG-Kommission im innerstaatlichen Vollzug von Gemeinschaftsrecht* (*The Powers of the EC-Commission in the National Implementation of Community Law*) (Cologne: Carl Heymanns Verlag).

Köck, H.F., 'Die "implied powers" der Europäischen Gemeinschaften als Anwendungsfall der "implied powers" interntaionaler Organisationen überhaupt' (The 'implied powers' of the European Community as an Application of 'implied powers' of International Organizations), in: Böckstiegel, KH.-A., Folz, H.-E., Mössner, J. M. and Zemanek, K. eds. (1988), *Völkerrecht, Recht der Internationalen Organisationen, Weltwirtschaftsrecht – Festschrift für Ignaz Seidl-Hohenveldern* (*Law of Nations, Law of International Organizations, World Economy Law – Liber amicorum Honouring Ignaz Seidl-Hohenveldern*) (Cologne: Heymann Verlag), 279.

Kolivakis, MA.-A. (1995), *The Economic Regulation of Air Transport: From the Chicago Convention to GATS* (LL.M. Thesis, Institut of Air and Space Law, McGill University) [unpublished].

Konow, G., 'Zum Subsidiaritätsprinzip des Vertrages von Maastricht' (The Principle of Subsidiarity in the Treaty of Maastricht) (1993) *Die Öffentliche Verwaltung* (*J. of Public Admin.*) at 405.

Kraus, B.R., 'Pyrrhic Victory in Spectrum Auction', (25 August 1997) *The National Law Journal*.

Krenzler, H.G. and da Fonseca-Wollheim, H., 'Die Reichweite der gemeinsamen Handelspolitik nach dem Vertrag von Amsterdam – Eine Debatte ohne Ende?' (The Scope of the Common Commercial Policy after the Treaty of Amsterdam – an Endless Debate?) (1998) *Europarecht* (*Europ. L. Rev.*), 223.

Kreymborg, D., 'Developments Relevant to the International Air Transport in the World Trade Organization (WTO) – Analysis of the First Review of the GATS Annex', (2003) 28 *Ann. Air and Sp. L.*, 471.

Krück, H. (1977), *Völkerrechtliche Veträge im Recht der Europäischen Gemeinschaften* (*International Agreements in the Law of the European Communities*), (Berlin: Springer-Verlag).

Kruse, H., 'Implied Powers und Implied Limitations' (Implied Powers and Implied Limitations) (1953/54) *Archiv des Völkerrechts* (*Archives of Intl. L.*), 169.

Küchenhoff, E., 'Ausdrückliches, stillschweigendes und ungeschriebenes Recht', (Explicit, implicit and unwritten law) (1957) *Archiv des öffentlichen Rechts* (*Archives of Public L.*), 413.

Kummel, W., 'Spectrum Bids, Bets, and Budgets: Seeking an Optimal Allocation and Assignment Process for Domestic Commercial Electromagnetic Spectrum Products, Services, and Technology', (1996) 48 *Fed. Comm. L.J.*, 511.

Kwerel, E. and Felker, A., 'Using Auctions to Select FCC Licensees', in Krattenmacher, T. G. ed. (1994), *Telecommunications Law and Policy* (Durham, NC: Carolina Academia Press), 121.

Kwerel, E. and Felker, A. , 'A Free Market solution – FCC Should Auction Licences for Radio', (1986) *Legal Times*, (22 September).

Lambers, H.-J., 'Subsidiarität in Europa – Allheilmittel oder juristsiche Leerformel?' (Subsidiarity in Europe – Omnipotent Instrument or Insignificant Legal Formula?) (1993) *Europarecht* (*Europ. L. Rev.*), 229.

Le Goueff, S., 'Licensing Global Mobile Personal Communications by Satellite: The Quest for the Holy Grail?', (1997) 22 *Ann. Air and Sp. L.*, 417.

Lecheler, H., 'Die Pflege der auswärtigen Beziehungen in der Europäischen Union', (The fostering of external relations in the European Union) (1994) *Archiv des öffentlichen Rechts* (*Archives of Public L.*), 1.

Lenz, C.-O. and Borchardt, K., eds. (2003), *EU- und EG-Vertrag (EU- and EC Treaty)*, 3rd edn., (Cologne: Bundesanzeiger).

Loewenstein, A. (1991), *European Air Law – Towards a New System of International Air Transportation Regulation* (Baden-Baden, Germany: Nomos).

Lyall, F. (2000), Expanding Global Communications Services, in: *Proceedings of the Workshop on Space Law in the Twenty-first Century*, (New York: Office of Outer Space Affairs United Nations Office at Vienna).

Lyall, F., 'On the Privatisation of INTELSAT', (2000) 28 *J. Space L.*, 101.

Margo, R.D. and MacCary, W.T., 'The Chicago Convention and the Resolution of Landing Fee Disputes', (1994) 19(2) *Ann. Air and Sp. L.*, 501.

Matte, N. and Mateesco (1981), *Treatise on Air-Aeronautical Law*, (Montréal: McGill University and Carswell).

McGoldrick, J.T., Jr, 'The Impact of the Civil Aeronautics Board's Regulations on Price Structure in the Airline Industry', (1975–76) 10 *Ga. L. Rev.*, 619.

Meessen, K.M., 'Das Abkommen von Lomé als gemischter Vertrag' (The Treaty of Lomé as a Mixed Agreement) (1980) *Europarecht* (*Europ. L. Rev.*), 36.

Michalopoulos, P., 'Money for Nothing?', (23 September 1996) *Legal Times*.

Möschel, W., 'Subsidiaritätsprinzip und europäisches Kartellrecht', (The principle of subsidiarity and european anti-trust law) (1995) *Neue Juristische Wochenschrift* (*New Weekly L. Rev.*), 281.

Möschel, W., 'Zum Subsidiaritätsprinzip im Vertrag von Maastricht', (The principle of subsidiarity in the Treaty of Masstricht) (1993) *Neue Juristische Wochenschrift* (*New Weekly L. Rev.*), 3025.

Munson, M.W., 'A legacy of lost opportunity: Designated entities and the Federal Communications Commission's broadband PCS spectrum auction', (2000/01) *Mich. Telecomm. and Tech 7 L. Rev.*, 217.

Nicolaysen, G., 'Zur Theorie von den implied powers in den Europäischen Gemeinschaften' (The Theory of Implied Powers in the European Communities) (1966) *Europarecht* (*Europ. L. Rev.*), 129.

Obuchowski, J., 'The Unfinished Task of Spectrum Policy Reforms', (1994) 47 *Fed. Comm. L.J.*, 325.

Ophüls, C.F., 'Quellen und Aufbau des Europäischen Wirtschaftsrechts', (Sources and structure of European commerical law) (1963) *Neue Juristische Wochenschrift* (*New Weekly L. Rev.*), 1697.

Oppermann, T. (1999), *Europarecht* (*European Law*), 2nd edn., (Munich: C. H. Beck Verlag).

Ostertun, D. (1996), *Gewohnheitsrecht in der Europäischen Union* (*Customary Law in the European Union*), (Göttingen, Germany: Peter Lang).

Oum, T., Stanbury, W. and Tretheway, M., 'Airline Deregulation in Canada', in Kenneth Button ed. (1991), *Airline Deregulation – International Experiences*, (Washington Square, NY: New York University Press), 124.

Paas, H.-W., 'Rechtsgrundlage des Charterluftverkehrs' (Legal Basis for Non-Scheduled Air Transportation) (1972) *Zeitschrift für Luft- und Weltraumrecht* (*Air and Space L. Rev.*), 55.

Petsikas, G. (1989), *Airline Deregulation and Competition in the Canadian Air Transport Industry Today, and Prospects for the Future* (LL.M. Thesis, Institute of Air and Space Law, McGill University) [unpublished].

Pieper, S., 'Subsidiaritätsprinzip – Strukturprinzip der Europäischen Union', (The principle of subsidiarity – structural principal of the European Union) (1993) *Deutsche Verwaltungsblätter* (*German J. of Admin. L.*), 705.

Pipkorn, J., 'Das Subsidiaritätsprinzip im Vertrag über die Europäische Union - rechtliche Bedeutung und gerichtliche Überprüfbarkeit', (The principle of subsidiarity in the EU-Treaty – legal meaning and judicial control) (1992) *Europäische Zeitschrift für Wirtschaftsrecht* (*Europ. J. of Commercial L.*), 697.

Pitschas, C., 'Anmerkungen zum "open-skies" Urteil des EuGH, (Remarks on the 'Opens-Skies' Judgement of the ECJ)', (2003) *Europäische Zeitschrift für Wirtschaftsrecht* (*Europ. J. of Commercial L.*), 92.

Pogue, L.W and Davison, C., 'User Charges in International Aviation', (1979) 73 *Am. J. Int'l L.*, 42.

Rabe, H.-J. (1963), *Das Verordnungsrecht der Europäischen Wirtschaftsgemeinschaft* (*The Legislation Law of the European Economic Community*), (Hamburg: Kommissionsverlag Ludwig Appel).

Rapp, L. and Vellas, F. (1992), *Airline Privatisation in Europe*, 2nd edn., vol. 28, (Paris, France: Institute of Air Transport).

Rijsdijk, O., 'EC Aviation Scene', (2000) 15(4–5) *Air and Space L.*, 184.

Rijsdijk, O., 'EC Aviation Scene', (2001) 16(6) *Air and Space L.*, 333.

Roberts, L.D., 'A Lost Connection: Geostationary Satellite Networks and the International Telecommunication Union', (2000) 15 *Berkeley Tech. L.J.*, 1095.

Rudolf, A., Die sogenannte Pauschalreise-(IT)Charter im Spannungsfeld zwischen Fluglinien – und Gelegenheitsverkehr (The so Called Holiday Package Charter Flight in the Tension between Scheduled and Non-Scheduled Air Transportation) (1970) *Zeitschrift für Luft- und Weltraumrecht* (*Air and Space L. Rev.*), 110.

Rust, U., Sozialpolitische Kriterien in landesgesetzlichen Vergabevorschriften (Sociopolitical Criteria in National Public Procurement Regualtions) (2001) *Neue Justiz* (*New Justice*), 113.

Sandell, H. (1976), *Government Regulation of Air Carriers, in Canada* (LL.M. Thesis, Institute of Air and Space Law, McGill University) [unpublished].

Sasse, C., 'Zur auswärtigen Gewalt der Europäischen Gemeinschaft' (The Treaty-Making Power of the European Community) (1971) *Europarecht* (*Europ. L. Rev.*), 208.

Scheuing, D.H., 'Umweltschutz auf der Grundlage der Einheitlichen Europäischen Akte' (Environmental Protection Based on the Single European Act) (1989) *Europarecht* (*Europ. L. Rev.),* 152.

Schmidhuber, P.M. and Hitzler, G., 'Binnenmarkt und Subsidiaritätsprinzip', (Internal market and the principle of subsidiarity) (1993) *Europäische Zeitschrift für Wirtschaftsrecht* (*Europ. J. of Commercial L.*), 8.

Schmidhuber, P.M. and Hitzler, G., 'Die Verankerung des Subsidiaritätsprinzips im EWG-Vertrag – ein wichtiger Schritt auf dem Weg zu einer föderalen Verfassung der Europäischen Gemeinschaft', (The inclusion of the principle of subsidiarity in the EC Treaty – an important step towards a federal constitution of the European Community) (1992) *Neue Zeitschrift für Verwaltungsrecht* (*New J. of Admin. L.*), 720.

Schmidhuber, P.M., 'Das Subsidiaritätsprinzip im Vertrag von Maastricht', (The principle of subsidiarity in the treaty of Maastricht), (1993) *Deutsche Verwaltungsblätter* (*German J. of Admin. L.),* 417.

Schmidt-Aßmann, E. (2004), *Das allgemeine Verwaltungsrecht als Ordnungsidee* (*General Administration Law as Means of Structuring*), 2nd edn., (Berlin: Springer).

Schnabel, A., 'Das Subsidiaritätsprinzip im Vertrag über die Europäsiche Union', (The principle of subsidiarity in the EU Treaty) (1992) *Bayerische Verwaltungsblätter* (*Bavarian J. of Admin. L.),* 393.

Schockweiler, F., 'Zur Kontrolle der Zuständigkeitsgrenzen der Gemeinschaft' (The Control of the Limits of the Competences of the Community) (1996) *Europarecht* (*Europ. L. Rev.*), 123.

Schreiber, S. (1997), *Verwaltungskompetenzen der Europäischen Gemeinschaft* (*Administrative Powers of the European Community*), (Baden-Baden, Germany: Nomos Verlagsgesellschaft).

Schuster, G. and Stoll, P.-T., 'Gemeinschaftskompetenz und Altverträge mit Drittstaaten', (Competences of the Community and international agreements with third countries) (1996) *Recht der internationalen Wirtschaft* (*L. of Intl. Commerce*), 89.

Schwartz, B., 'Administrative Law', (1965) *Ann. Surv. Am. L.*, 55.

Schwarze, J. (1988), *Europäisches Verwaltungsrecht* (*European Administration Law*), Vol. 1, (Baden-Baden, Germany: Nomos Verlagsgesellschaft).

Schwarze, J., ed. (2000), *EU-Kommentar*, (EU-Commentary), (Baden-Baden, Germany: Nomos Verlagsgesellschaft).

Schweitzer, M. and Hummer, W. (1996), *Europarecht – Das Recht der Europäischen Union – Das Recht der Europäischen Gemeinschaften (EGKS, EG, EAG)* (*European Law – The Law of the European Union – The Law of the European Communities*), 5th edn., (Neuwied, Germany: Luchterhand Verlag).

Schwenk, W. and Giemulla, E. (2005), *Handbuch des Luftverkehrsrechts* (*Manual of Aeronautics Law*), 3rd edn., (Cologne: Carl Heymanns Verlag).

Sedding, R.G. (1971), *Regulatory Police Making in the Civil Aeronautics Board*, (Ph.D. Thesis, political science, Princeton University) [unpublished].

Serrano, V.R., 'Trade in Air Transport Services: Liberalizing Hard Rights', (1999) 24 *Ann. Air and Sp. L.*, 199.

Settanni, A.M., 'Competitive Bidding for the Airwaves: Meeting the Budget and Maintaining Policy Goals in a Wireless World', (1994) 2 *CommLaw Conspectus*, 117.

Shenefield, J.H., 'Regulation and Deregulation – Where Do We Stand?', (1976) 45 *Antitrust L.J.*, 244.

Sinclair, P.B., 'Case Notes – Administrative Law', (1978–79) 44 *J. Air L. and Com.*, 849.

Sochor, E.,'From the DC-3 to Hypersonic Flight: ICAO in a Changing Environment', (1989–90) 55 *J. Air L. and Com.*, 407.

Solon, D., 'What Next for Europe – Wide Multilateral Air Service Agreements?', (2002) 7(2) *The Avmark Aviation Economist*, 2.

Sorensen, F., van Weert, W. and Cheng-Jui Lu, A., 'ECJ Ruling on Open Skies Agreements v. Future International Air Transport', (2003) 28(2) *Air and Space L.*, 3.

Staniland, M., (2003), *Government of Birds – Air Transport and the State in Western Europe*, (Lanham, MD: Rowman and Littlefield).

Stein, K.D. (1986), *Der gemischte Vertrag im Recht der Außenbeziehungen der Europäischen Gemeinschaft* (*The Mixed Agreement in External Relations of the European Community*), (Berlin: Duncker and Humblot).

Stein, L.R., 'The Airline Deregulation Act of 1978', (1979) 20 *Harv. Int.*, 385.

Stern, M.L. and Rice, D., 'Fallout from the C-Block Debacle Threatens to Cost Small Businesses their Best Chance to Participate in Spectrum-Based Telecom Services', (13 October 1997) *Legal Times*, .

Stevens, T., 'Regulation and Licensing of Low-Earth-Orbit Satellites', (1994) 10 *Santa Clara Computer and High Tech. L.J.*, 401.

Stewing, C., 'Das Subsidiaritätsprinzip als Kompetenzverteilungsregel im Europäischen Recht', (The application of the principle of subsidiarity to distribute powers in European law) (1992) *Deutsche Verwaltungsblätter* (*German J. of Admin. L.*), 1516.

Streinz, R. (2001), *Europarecht* (*European Law*), 5th edn., (Heidelberg, Germany: C. F. Müller Verlag).

Suerbaum, J. (1998), *Die Kompetenzverteilung beim Verwaltungsvollzug des Europäischen Gemeinschaftsrechts in Deutschland* (*The Distribution of Powers in the Execution of EC Law*), (Berlin: Duncker and Humblot).

Sydow, V., Schmitt, H. and Institutionen, D. (1980), *Erweiterten Europäischen Gemeinschaften – die Kommission* (*Institutions of the Extended European Communities – The Commission*), (Baden-Baden, Germany: Nomos Verlagsgesellschaft).

Thomas, A.J., Jr (1951), *Economic Regulation of Scheduled Air Transport*, (Buffalo: Dennis).

Thym, D., 'Der Binnenmarkt und die "Freiheit der Lüfte"' (The Internal Market and the Freedoms of the Air) (2003) *Europarecht* (*Europ. L. Rev.*), 277.

Tomuschat, C., 'Anmerkungen zum Gutachten 1/76' (Remarks on Opinion 1/76) (1977) *Europarecht* (*Europ. L. Rev.*), 157.

Tomuschat, C. (1972), *Verfassungsgewohnheitsrecht?*, (*Constitutional customary law?*), (Heidelberg, Germany: Carl Winter Universitätsverlag).

Tomuschat, C., 'Völkerrechtliche Grundlagen der Drittlandbeziehungen der EG' (Legal Basis in International Law for the External Relations of the EC), in Hilf, M. and Tomuschat, C. eds. (1991), *EG und Drittstaatsbeziehungen nach 1992* (*EC and External Relations after 1992*), (Baden-Baden, Germany: Nomos-Verlagsgesellschaft), 139.

Tourtellot, C.T., 'Competitive Policy in Airline Deregulation', (1978–79) 28 *Am. U. L. Rev.*, 537.

Trevett, J., 'EC Dark Cloud across Open Skies', (2001) 10(18) *The Avmark Aviation Economist*, 18.

Usher, J.A. (2001), *EC Agricultural Law*, (Oxford: Oxford University Press).

Vedder, C.W. (1980), *Die auswärtige Gewalt des Europa der Neun* (*The Treaty-Making Power of the Europe of the nine*), (Göttingen, Germany: Schwartz).

Verdross, A. and Simma, B. (1984), *Universelles Völkerrecht* (*Universal International Law*), 3rd edn., (Berlin: Duncker and Humblot).

von Arnim, C., Die Vertragsschließungskompetenz der Europäischen Wirtschaftsgemeinschaft in der Rechtsprechung des Europäischen Gerichtshofes, (The treaty-making power of the European Economic Community in rulings of the European Court of justice) (1972) *Außenwirtschaftsdienst des Betriebs-Beraters* (*J. of Foreign Trade and Investment*), 215.

von Borries, R., 'Verwaltungskompetenzen der Europäischen Gemeinschaft' (Administrative Powers of the European Community), in: Due, O., Lutter, M. and Schwarze, J. eds. (1995), *Festschrift für Everling (Liber amicorum Everling)* (Baden-Baden, Germany: Nomos Verlagsgesellschaft), 127.

von Der Groeben, H. and Schwarze, J., eds. (2003), *Vertrag über die Europäische Union und Vertrag zur Gründung der Europäischen Gemeinschaft* (*Treaty on the European Union and Treaty Establishing the European Community*), 6th edn., (Baden-Baden, Germany: Nomos Verlagsgesellschaft).

von Münch, I. and Kunig, P. (2003), *Grundgesetz-Kommentar* (*Basic Law Commentary*), 5th edn., (Munich: Verlag C.H. Beck).

Ward, J.J., 'Transportation Law – Agency Interpretation of Regulations for Awarding International Air Carrier Routes', (1995–96) 64 *Geo. Wash. L. Rev.* at 1450.

Wasmeier, M., 'Die Integration Des Umweltschutzes als allgemeine Auslegungsregel des Gemeinschaftsrechts' (The Integration of Environmental Protection as a General Interpretation Rule of Community Law) (2000) *Europäisches Wirtschafts- und Steuerrecht* (*Europ. J. of Commercial and Tax L.*) at 47.

Wassenbergh, H., 'A Mandate to the European Commission to Negotiate Air Agreements with Non-EU States: International Law versus EU Law', (2003) 28(3) *Air and Space L.*, 139.

Wassenbergh, H., 'Common Market, Open Skies and Politics: A Bald Eagle's-Eye View of Today's Air Transport Regulation', (2000) 25(4–5) *Air and Space L.*, 174.

Wassenbergh, H., 'The Decision of the ECJ of 5 November 2002 in the "Open Skies" Agreements Cases', (2003) 28(1) *Air and Space L.* 19.

Werbke, A., 'Staatliche und gemeinschaftliche Kompetenz zum völkerrechtlichen Handeln', (Treaty-making powers of States and of the community), (1971) *Neue Juristische Wochenschrift* (*New Weekly L. Rev.*), 2103.

Williams, G. (1993), *The Airline Industry and the Impact of Deregulation*, (Brookfield, VA: Ashgate).

Wimmer, K.A. and Tiedrich, L.J., 'Competitive Bidding and Personal Communications Services: A New Paradigm for FCC Licensing', (1994) 3 *CommLaw Conspectus*, 17.

Wohlfahrt, E., Everling, U., Glaesner, H.J. and Sprung, R. (1960), *Die Europäische Wirtschaftsgemeinschaft* (The European Economic Community), (Berlin: Verlag Franz Vahlen GmbH).

Yang, J. (1995), *Air Transport Regulation in a New Era*, (LL.M. Thesis, Institute of Air and Space Law, McGill University).

Zebinsky, V. and Mencik, A., 'The General Agreement of Trade in Services: Its Implications for Air Transport', (1993) 18(1) *Ann. Air and Sp. L.*, 359.

Zippelius, R., Zum Problem der Rechtsfortbildung, (The problem of customary law), (1964) *Neue Juristische Wochenschrift* (*New Weekly L. Rev.),* 1981.

Zuleeg, M. (1969), *Das Recht der Europäischen Gemeinschaften im innerstaatlichen Bereich* (*The Law of the European Communities in the National Sphere*), (Cologne: Carl Heymanns Verlag).

Zuleeg, M., Die Auslegung des Europäischen Gemeinschaftsrechts (The Interpretation of European Community Law) (1969) *Europarecht* (*Europ. L. Rev.*), 97.

Zuleeg, M. (1994), 'Deutsches und europäisches Verwaltungsrecht – wechselseitige Einwirkungen' (German and European Administration Law – Interrelated Impacts) in: *Veröffentlichungen der Vereinigung der deutschen Staatsrechtslehrer* (*Publications of the Association of German Scholars for Constitutional L.*), vol. 53, (Berlin: Walter de Gruyter), 154.

Zuleeg, M., 'Das Subsidiaritätsprinzip im Europarecht – Inhalt, Justiziabilität, Entwicklung', (The Principle of Subsidiarity in European Law – Content, Judicial Control, Development) in: Rodríguez Iglesias, G. C., Due, O., Schintgen, R. and

Elsen, C. eds. (1999), *Mélanges en hommage à Fernand Schockweiler* (*In Memory of Fernand Schockweiler*), (Baden-Baden: Nomos Verlagsgesellschaft), 635.

Zuleeg, M. (2004), *Der rechtliche Zusammenhalt in der Europäischen Union* (*The Legal Unity in the European Union*), (Baden-Baden, Germany: Nomos Verlagsgesellschaft).

Zuleeg, M., 'Die Kompetenzen der Europäischen Gemeinschaften gegenüber den Mitgliedstaaten', (The Powers of the European Communities towards the Member States) (1971) *Jahrbuch der öffentlichen Rechts* (*Yearbook of Public L.*), 1.

Zuleeg, M. 'Verkehrspolitische Außenbeziehungen der Europäischen Gemeinschaft', (External relations in the transport policy of the European Community), in: *Europäischen Verkehrspolitik* (*European Transport Policy*) J. Basedow ed. (1987), (Tübingen: J.C.B. Mohr (Paul Siebeck)), 59.

Zylicz, M. (1992), *International Air Transportation Law*, (Dordrecht, The Netherlands: Kluwer Academia Publishers).

NEWSPAPER ARTICLES

Eine Frage der Ehre, (A question of honour), *Handelsblatt* (*Trade Gazette*) of 11 March 2005, 10.

Lufthansa übernimmt Swiss (Lufthansa takes over Swiss), *Financial Times Deutschland* (*Financial Times Germany*) of 23 March 2005, 1.

Lufthansa übernimmt Swiss (Lufthansa takes over Swiss), *Handelsblatt* (*Trade Gazette*) of 23 March 2005 , 1.

Index